The Molecular Biology of the Bacilli

MOLECULAR BIOLOGY

An International Series of Monographs and Textbooks

Editors: BERNARD HORECKER, NATHAN O. KAPLAN, JULIUS MARMUR, AND HAROLD A. SCHERAGA

A complete list of titles in this series appears at the end of this volume.

The Molecular Biology of the Bacilli

Volume I: *Bacillus subtilis*

Edited by

DAVID A. DUBNAU

Department of Microbiology
The Public Health Research
Institute of the City of New York, Inc.
New York, New York

1982

ACADEMIC PRESS

A Subsidiary of Harcourt Brace Jovanovich, Publishers

New York London

Paris San Diego San Francisco São Paulo Sydney Tokyo Toronto

ACADEMIC PRESS, INC.
111 Fifth Avenue, New York, New York 10003

United Kingdom Edition published by
ACADEMIC PRESS, INC. (LONDON) LTD.
24/28 Oval Road, London NW1 7DX

Library of Congress Cataloging in Publication Data
Main entry under title:

The Molecular biology of the Bacilli.

(Molecular biology series)
Includes bibliographies and index.
Contents: v. 1. Bacillus subtilis / edited by David Dubnau.
1. Bacillaceae. 2. Molecular biology. I. Dubnau, David. II. Series. [DNLM: 1. Bacillus. QW 127.5.B2 M718]
QR82.B3M64 589.9'5 81-22815
ISBN 0-12-222701-8 (v. 1) AACR2

PRINTED IN THE UNITED STATES OF AMERICA

82 83 84 85 9 8 7 6 5 4 3 2 1

Contents

Contributors

Numbers in parentheses indicate the pages on which the authors' contributions begin.

V. G. Debabov (331), Institute of Genetics and Selection of Industrial Microorganisms, Moscow 113543, USSR

Roy H. Doi (71), Department of Biochemistry and Biophysics, University of California at Davis, Davis, California 95616

David Dubnau (147), Department of Microbiology, The Public Health Research Institute of The City of New York, Inc., New York, New York 10016

Peter E. Geiduschek (203), Department of Biology, University of California, San Diego, La Jolla, California 92093

Thomas J. Gryczan (307), Department of Microbiology, The Public Health Research Institute of The City of New York, Inc., New York, New York 10016

*Dennis J. Henner** (1), Department of Cellular Biology, Research Institute of Scripps Clinic, Scripps Clinic and Research Foundation, La Jolla, California 92037

James A. Hoch (1), Department of Cellular Biology, Research Institute of Scripps Clinic, Scripps Clinic and Research Foundation, La Jolla, California 92037

Junetsu Ito (203), Department of Molecular and Medical Microbiology, University of Arizona College of Medicine, Tucson, Arizona 85724

Richard Losick (179), The Biological Laboratories, Harvard University, Cambridge, Massachusetts 02138

Lars Rutberg (247), Department of Bacteriology, Karolinska Institute, Stockholm 60, Sweden

Issar Smith (111), Department of Microbiology, The Public Health Research Institute of The City of New York, Inc., New York, New York 10016

* *Present address*: Genentech, Inc., 460 Point San Bruno Boulevard, South San Francisco, California 94080

Noboru Sueoka (35), Department of Molecular, Cellular, and Developmental Biology, University of Colorado, Boulder, Colorado 80309

Scott Winston (35), Department of Molecular, Cellular, and Developmental Biology, University of Colorado, Boulder, Colorado 80309

Stanley A. Zahler (269), Division of Biological Sciences, Cornell University, Ithaca, New York 14853

Preface

Since the discovery of DNA-mediated transformation in *Bacillus subtilis* by Spizizen in 1958, this organism has emerged as a second major prokaryotic subject for the study of molecular biology, providing a valuable gram-positive counterpoint to *Escherichia coli*. Although universal aspects of biology have properly received the lion's share of attention to date, particularistic studies that stress differences between organisms will probably become increasingly important. In this regard the study of *B. subtilis* will undoubtedly assume a major role as a representative of a second major class of prokaryotes. Some problems, like the control of sporulation and the uptake and integration of transforming DNA, can obviously be approached readily in *B. subtilis*; and their study will be facilitated by the considerable body of knowledge that exists concerning its genetics, biochemistry, and physiology.

Bacilli are also of major importance in the fermentation industry since they elaborate a variety of useful enzymes and antibiotics. As a consequence of this, industrial-scale techniques for the cultivation of bacilli are relatively well developed. The ability of several bacilli to excrete large amounts of certain proteins into the extracellular space is a particularly attractive feature for commercial exploitation. Unfortunately, little is known about the molecular basis of protein excretion in the bacilli and the coupling, if any, of excretion and synthesis. It is likely that this and the intriguing problems associated with growth stage related regulation (see Chapter 11) will yield to investigation within the next few years. This optimism is partly based on the advent of molecular cloning in *B. subtilis*.

The present treatise is directed toward two overlapping groups of scientists: those who are concerned with the use of the *B. subtilis* system as a tool for the study of molecular biology and those who wish to increase the medical, veterinary, and industrial usefulness of this and related organisms. In selecting the subject matter of this volume, I have given primary attention to those areas of research that have been traditionally investigated in this organism (sporulation, defective bacteriophages, transformation), to those topics in which outstanding progress has been made in recent years, and to those subjects that are of interest to the two groups of scientists described above. There has been no attempt to be comprehensive. It is hoped that subsequent volumes, when warranted by progress in the field, will make up for any omissions.

I would like to thank Eugenie Dubnau and Issar Smith for many discussions during the course of this work and for help with some of the editing. To Annabel Howard, my special appreciation for much of the secretarial work associated with preparing this volume.

David A. Dubnau

1

The Genetic Map of *Bacillus subtilis*

DENNIS J. HENNER AND JAMES A. HOCH

Introduction

Studies have focused on many different aspects of *Bacillus subtilis*, including biochemical and morphological changes during sporulation, germination, and outgrowth; mechanisms of genetic exchange in transformation and transduction; cell wall synthesis, cell growth, and division; enzymatic characterization of anabolic and catabolic pathways; and a host of others. The isolation of mutants defective in these processes with subsequent genetic analysis has resulted in our contemporary genetic map. As the most thoroughly characterized gram-positive microorganism, *B. subtilis* allows an evolutionary comparison of chromosome organization with that of other, distantly related, microorganisms (*10*). In addition, a genetic analysis of the primitive differentiation cycle this mi-

The Molecular Biology
of the Bacilli

ISBN 0-12-222701-8

croorganism undergoes, i.e., sporulation–germination, permits an assessment of unique aspects of chromosome structure and gene organization that may be attributed to developmental gene expression.

The advent of molecular cloning with its precise definitions of physical distances between mutations within genes or between genes themselves dictates that some relationship be formulated between observed genetic distance (i.e., cotransfer frequency) in transformation or transduction and actual physical distance between mutations. An initial attempt to define this relationship is presented below. Although imprecise at this time, the analysis allows an estimation of the physical distance from the genetic distance. A more detailed rationale for some of the assumptions used can be found in an earlier work (*66*).

A. Conversion of Genetic Cotransfer Frequency to Physical Distance between Markers

The frequency of genetic cotransfer between various linked markers is characteristic of the system used for genetic analysis. In general, the larger the piece of DNA transferred in the system, the higher the probability that two linked markers will be cotransferred. In *B. subtilis,* we have two useful systems of genetic analysis that differ greatly in the size of the piece of DNA transferred: PBS1 transduction and transformation. PBS1 transduction may allow fragments as large as 10% of the entire chromosome to be transferred, whereas transformation with naked DNA probably occurs with fragments $\frac{1}{10}$–$\frac{1}{20}$ the size of the transducing fragment (*66*). In order to compute the physical distance between markers on the transforming or transducing DNA fragment from genetic cotransfer frequencies obtained in each system, one must derive an expression that relates genetic values to physical distance.

The relationship between the frequency of cotransduction of markers and the physical distance between the markers on a transducing piece of DNA has been the subject of several mathematical models (*91,198*). The basic features of these models include the possibility that factors other than recombinational events will affect the apparent frequency of recombination between any two markers. In order to be cotransduced the two markers must be contiguous on a single fragment of transducing DNA and, as the distance between them increases, their chance of being on the same transducing fragment decreases. Furthermore, as the markers become closer to the end of the transducing fragment, the size of the pieces of DNA outside the markers in which crossing-over might occur becomes limited, reducing the number of cotransductants. The models also make certain assumptions about the transduction process.

They assume that recombination is linearly proportional to the length of DNA in which the recombination can occur, and that transducing fragments are of uniform length and have random end points. None of these assumptions has been verified in either the PBS1 transduction system or the transformation system, and there is evidence in some transduction systems for violation of certain of these assumptions, i.e., marker-specific effects, nonreciprocal crosses, preferred end points (*91*). However, the theoretical predictions of the models have shown good agreement with observed distances in P22 transduction in *Salmonella typhimurium* (*32,91*).

Kemper's formula (*91*), $C = 1 - t + t\,(\ln t)$, where C is the frequency of cotransduction of two markers and t is the fractional length of the transducing fragment separating the markers, has been used to convert cotransduction frequencies to physical distances along a transducing or transforming DNA fragment. A plot of this equation is shown in Fig. 1. An inspection of this plot shows that the cotransduction frequency falls quickly as the physical distance between two markers increases. For example, two markers, A and B, separated by half the length of a transducing fragment ($0.5t$), will show only 15% cotransduction, not 50%. This model predicts that recombination values will not be additive. If a marker, C, is halfway between the markers A and B above, it will be $0.25t$ from each and show 60% recombination from each, but the two outside markers, A and B, will show only 85% recombination, not 120%. Thus, in composing the final genetic map of *B. subtilis* the genetic distances from PBS1 transduction were converted to physical distances by

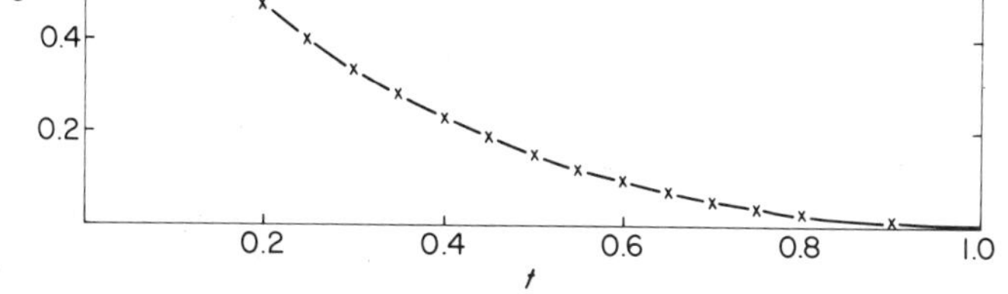

Fig. 1. Cotransfer frequency (C) in genetic analysis as a function of physical distance between markers (t) on donor DNA molecule plotted according to the equation $C = (1 - t) + t\,(\ln t)$.

the curve in Fig. 1. The validity of using this formula to correct PBS1 transduction values has been verified with a number of markers (*66*).

B. Relationship between the Computed Physical Distance and the Amount of DNA between Markers

Application of the Kemper formula to transduction or transformation data allows one to reduce these data to a value t that gives the fractional length of the transducing or transforming DNA fragment between any two markers. If the average length of these fragments were known, in terms of base pairs, an estimate of the number of base pairs between markers could be obtained. This relationship is simply $D = t(T)$, where D is the distance between markers in base pairs, t is the fractional length of the transducing or transforming DNA fragment, and T is the average size of this fragment in base pairs. In PBS1 transduction we have calculated that T is approximately 150×10^6 daltons or 250 kbp of DNA (*66*). Thus any set of genetic data for cotransfer between two markers may be converted to kilobase pairs of DNA between the markers by application of these equations. Such information is useful in many contexts, such as cloning experiments where decisions on what the chances for success are of overlapping hybridization experiments in moving from one marker to another if only the genetic distance between the markers is known. As more segments of the *B. subtilis* genome become available as cloned fragments, we shall have the opportunity to determine more precisely DNA distances between markers and their relationship to genetic cotransfer values.

Using a T value for PBS1 transduction of 150×10^6 daltons, we have estimated that the *B. subtilis* genome has a size of approximately 2×10^9 daltons (*66*), a value in agreement with physical determinations (*11,53*).

C. Detailed *Bacillus subtilis* Map

The genetic map of *B. subtilis* is circular and uninterrupted. We have chosen to divide this circle into 360° and to locate individual markers by a degree designation. Unfortunately, some regions of the chromosome are too crowded to permit a detailed genetic map in a circular format. Thus, a circular map with landmark loci (Fig. 2) may be used for orientation with the detailed map presented in Fig. 3. The circular map has been divided into 10 segments of 36° each. We have chosen to use the *guaA* locus as 0° because it is close to the origin of replication and is

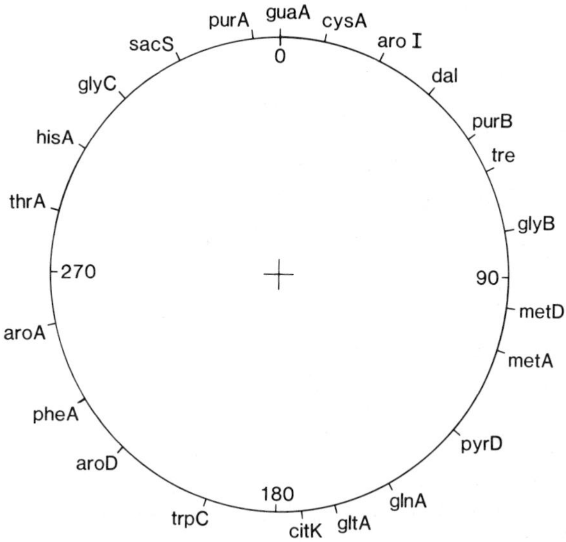

Fig. 2. Abbreviated genetic map of *B. subtilis* showing the landmark genetic markers.

an easily selectable marker in this region that has been used for a large number of genetic crosses.

A variety of new markers have been placed on this *B. subtilis* map that were not present on the previous version (*66*). These have been localized by their cotransduction with known genetic markers and positioned accordingly. No attempt was made to recalibrate the entire map using the new cotransduction data. One region, comprising the *metD–recA* interval, has been reanalyzed and some major changes made in both the order of markers and the intervals between markers. The order of the *cysC* and *pyr* cluster has been reversed in orientation from the previous map. There are a variety of conflicting reports regarding the order of these loci (*203;* R. Buxton, personal communication; K. Yamane, personal communication), but some recent mapping data favor a reversal of the previously published order (Ordal and Hoch, in preparation). Several other loci ordered with respect to *cysC* and *pyr* have also been moved. Many intervals between loci in this region have been changed, with a significant reduction in the overall size of the region. It would not be too startling if there proved to be an inversion of the *cycC–pyr* region in some strains of *B. subtilis*.

Since *B. subtilis* 168 is the most widely used strain of *B. subtilis,* we had difficulty in deciding whether to place markers introduced from

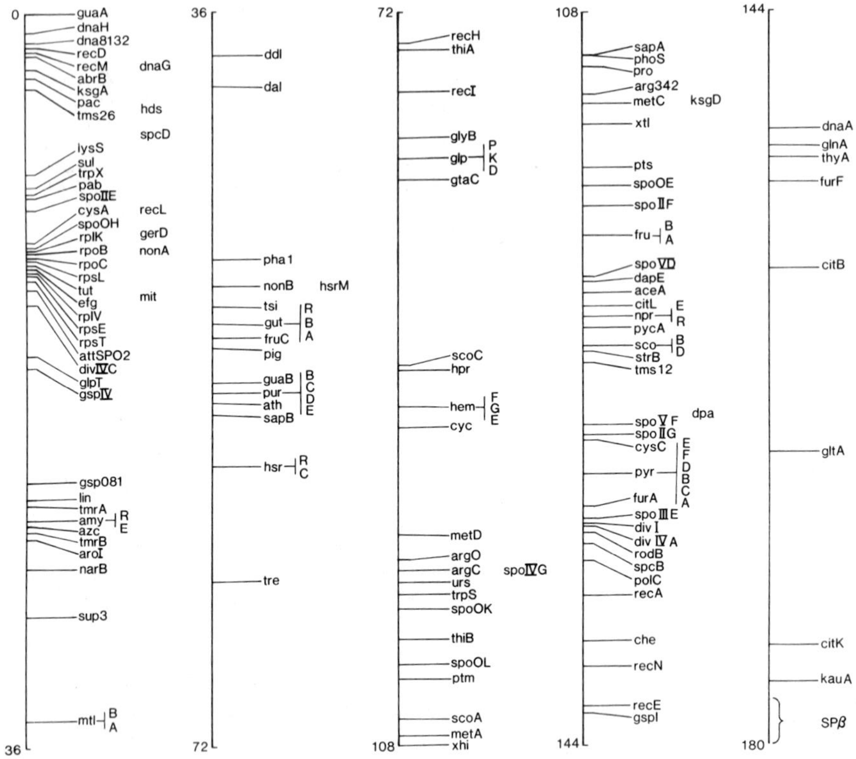

Fig. 3. Detailed genetic map of *B. subtilis*.

related strains on the genetic map. For example, the restriction–modification systems from various related strains may be introduced by transformation into the 168 strain and mapped by PBS1 transduction (*164*). We have placed these markers on the present map, since we feel that there must be a nearly homologous region in *B. subtilis* 168 in order for these markers to be introduced into the chromosome.

Many markers placed on this map have never been ordered with respect to their neighbors but mapped only with respect to landmark loci in the region. We hope that the publication of this genetic map will encourage workers to order new (and old) loci with respect to nearby markers on the map. Until such crosses are performed, the order and position of loci must be regarded as approximate and the original literature should be consulted.

The cloning of various genes of *B. subtilis* will certainly be useful in the near future in generating an accurate map of *B. subtilis*. Exact dis-

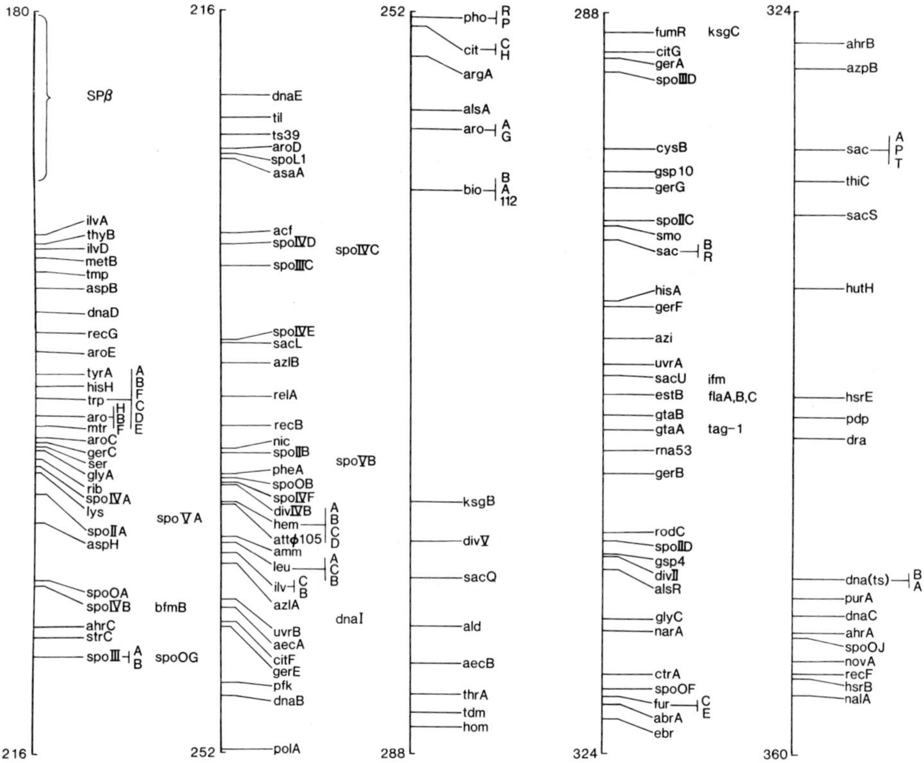

Fig. 3 Continued.

tances between loci will be determined, and the experimental relationship between physical distances and cotransformation and cotransduction determined. Conversely, the genetic map should prove to be of use in cloning experiments. For example, a marker that is not easily selectable could be cloned by selecting for a known nearby marker. Also, if the general chromosome organization of the various *B. subtilis* strains is similar, one might be able to screen for the desired gene from the related strain by hybridization with the similar cloned segment from the 168 strain.

D. Genetically Silent Regions

Several features of chromosome organization are common to both *B. subtilis* and *E. coli* that may indicate some fundamental structural constraints on gene distribution. In both organisms there is a tendency for the known loci to cluster in groups separated by relatively large genetically silent regions. The major clusters contain the genes for maintenance functions such as ribosomal components and amino acid synthesis, whereas the loci that appear in sparser regions seem of a less essential nature, e.g., sugar utilization enzymes. It has been suggested that these silent regions might be less available to RNA-synthesizing machinery because of a structural role in the folded and condensed bacterial nucleoid (*10,66*). One striking similarity between *B. subtilis* and *E. coli* is the large silent region near the terminus of replication, although a substantial proportion of this region is occupied by the temperate bacteriophage SPβ in *B. subtilis*. Whether silent regions are an integral part of the three-dimensional structure of bacterial genomes or whether they simply are less active coding regions for other reasons should be answered as molecular cloning makes these regions available for study.

Acknowledgments

We are grateful to our colleagues for unpublished mapping data. This work was supported, in part, by grants GM19416 and GM25891 awarded by the National Institute of General Medical Sciences, PHS/DNHS.

Appendix: *Bacillus subtilis* Genetic Loci

Gene symbol	Mnemonic	Map position[a]	Phenotype, enzyme deficiency, or other characteristics	References[b]
abrA	Antibiotic resistance	325	Partial suppressor of stage-0 mutant phenotypes	A
abrB	Antibiotic resistance	5	Partial suppressor of stage-0 mutant phenotypes	*184*
absA	Antibiotic sensitivity	—	Partial suppressor of stage-0 mutant phenotypes; see *abrB*	*83, 84*
absB	Antibiotic sensitivity	—	Partial suppressor of stage-0 mutant phenotypes; see *abrB*	*83, 84*
ace	Acetate	122	Pyruvate dehydrogenase defect, linked to *pycA*	C
acf	Acriflavine	235	Acriflavine resistance	*13, 82*

Appendix: (Continued)

Gene symbol	Mnemonic	Map position[a]	Phenotype, enzyme deficiency, or other characteristics	References[b]
aecA	Aminoethylcysteine	250	Aminoethylcysteine resistance, regulation of aspartokinase II	B, *116*
aecB	Aminoethylcysteine	290	Aminoethyl cysteine resistance, structural gene aspartokinase II	B, *116*
ahrA	Arginine hydroxamate	353	Arginine hydroxamate resistance, linked to *cysA*	*64, 127*
ahrB	Arginine hydroxamate	220	Arginine hydroxamate resistance	*64, 127*
ahrC	Arginine hydroxamate	209	Arginine hydroxamate resistance	A, *127*
ald	Alanine	285	L-Alanine dehydrogenase	*106*
alsA	Acetoin	265	Acetolactate synthase	*207*
alsR	Acetoin	320	Constitutive acetolactate synthase	*207*
amm	Ammonia	250	Glutamate requirement	*38*
amt	—	—	3-Aminotyrosine resistance, part of or very close to *tyrA* locus, see *tyrA*	*150*
amyB	Amylase	—	Control of amylase synthesis, probably identical to *sacQ* and *pap;* see *sacQ*	*162, 177*
amyE	Amylase	25	Amylase structural gene, also called *amyA*	*205*
amyR	Amylase	25	Control of amylase synthesis, also called *amyH*	*201, 204*
argA	Arginine	260	Arginine requirement	*17, 113*
argC	Arginine	100	Arginine or citrulline requirement	E, *113*
argO	Arginine	100	Arginine, ornithine, or citrulline requirement	E, *113*
arg342	Arginine	115	Arginine, ornithine or citrulline requirement	*72*
aroA	Aromatic	265	3-Deoxy-D-arabino-heptalosonic-7-phospate synthase	*73, 131*
aroB	Aromatic	210	Dehydroquinate synthase	*131*

(Continued)

Appendix: (Continued)

Gene symbol	Mnemonic	Map position[a]	Phenotype, enzyme deficiency, or other characteristics	References[b]
aroC	Aromatic	210	Dehydroquinate dehydratase	*73, 131*
aroD	Aromatic	230	Shikimate dehydrogenase	*3, 72, 73*
aroE	Aromatic	210	3-Enolpyruvyl-shikimate 5-phosphate synthase	*131, 134*
aroF	Aromatic	210	Chorismate synthase	*73, 131*
aroG	Aromatic	270	Chorismate mutase, isozyme 3	*73, 131*
aroH	Aromatic	210	Chorismate mutase, isozyme 1,2	111, 131
aroI	Aromatic	25	Shikimate kinase	*106*
aroJ	Aromatic	—	Tyrosine and phenylalanine; see *hisH*	*133*
asaA	Arsenate	230	Arsenate resistance	*1, 2*
aspA	Aspartate	—	Pyruvate carboxylase; see *pycA*	*25, 72*
aspB	Aspartate	200	Aspartate aminotransferase	*72*
aspH	Aspartate	215	Constitutive aspartase	*78*
ath	Adenine-thiamine	55	Adenine–thiamine requirement	*206*
attSPβ	Attachment	190	Integration site for SPβ	*208*
attSP02	Attachment	10	Integration site for SP02	*81, 169*
attϕ3T	Attachment	—	Integration site for ϕ3T, probably maps between *kuaA* and SPβ	E, *196*
attϕ105	Attachment	245	Integration site for ϕ105	*146, 160*
azc	—	25	L-Azitidin-2-carboxylic acid resistance	*47*
azi	Azide	310	Resistance to sodium azide	A
azlA	Azaleucine	250	4-Azaleucine resistance, derepressed leucine biosynthetic enzymes	*192*
azlB	Azaleucine	240	4-Azaleucine resistance	*192, 193*
azpA	Azopyrimidine	—	Resistance to azopyrimidines, alteration of DNA polymerase III; see *polC*	*34*
azpB	Azopyrimidine	330	Resistance to azopyrimidines	*28*

Appendix: (Continued)

Gene symbol	Mnemonic	Map position[a]	Phenotype, enzyme deficiency, or other characteristics	References[b]
bfmA	Branched fatty acid	UC	Lacks branched-chain ketoacid dehydrogenase	C
bfmB	Branched fatty acid	209	Requires branched-chain fatty acid or valine or isoleucine, maps between *strC* and *lys*	C
bioA	Biotin	270	7-KAP:DAP aminotransferase	*143*
bioB	Biotin	270	Biotin synthetase	*143*
bio112	Biotin	270	Early defect in biotin synthesis	*143*
bry	Bryamycin	—	Bryamycin resistance, maps in ribosomal protein cluster	*62, 169*
bsr	—	—	Restriction, modification by strain *B. subtilis* R; see *hsrR*	
but	—	NM	5-Bromouracil-tolerant	*19*
catA	—	—	Hyperproduction of extracellular proteases; see *hpr*	*106*
cdd	—	NM	Deoxycytidine-cytidine deaminase	*155*
che	Chemotaxis	140	Lacks chemotaxis to all attractants	K
citB	Citric acid cycle	165	Aconitate hydratase	*159, 207*
citC	Citric acid cycle	260	Isocitrate dehydrogenase	*74, 159*
citF	Citric acid cycle	255	Succinate dehydrogenase	*74, 159*
citG	Citric acid cycle	295	Fumarate hydratase	*159*
citH	Citric acid cycle	260	Malate dehydrogenase	A, *50*
citK	Citric acid cycle	185	α-Ketoglutarate dehydrogenase complex, enzyme E1	*71, 159*
citL	Citric acid cycle	125	Lipoamide dehydrogenase (NADH), E3 component of both pyruvate and α-ketoglutarate dehydrogenase complexes	*71*

(Continued)

Appendix: (Continued)

Gene symbol	Mnemonic	Map position[a]	Phenotype, enzyme deficiency, or other characteristics	References[b]
cml	Chloramphenicol	—	Caused by mutations in at least five 50 S ribosomal proteins, all mapping in the ribosomal protein cluster; see ribosomal protein cluster	*140*
crk	—	NM	Cytidine kinase	*155*
css	Cysteine	—	Cysteine sensitivity; see *cysA*	*88*
ctrA	Cytidine requirement	325	Requirement for cytidine in the absence of ammonium ion	*106, 207*
cyc	Cycloserine	94	Cycloserine resistance	L
cym	Cysteine–methionine	—	Requirement of cysteine or methionine; see *cysA*	*88, 147*
cysA	Cysteine	10	Cysteine requirement, serine transacetylase, a complex locus comprised of *css* (cysteine sensitivity), *cym* (cysteine or methionine), *hts* (hydrogen sulfide excretion), and *cysA*	*88*
cysB	Cysteine	300	Cysteine requirement	*38*
cysC	Cysteine	140	Cysteine, methionine, sulfite, or sulfide requirement	*72, 203*
dal	D-Alanine	40	D-Alanine requirement, DL-alanine racemase	*40, 45*
dapE	Diaminopimelic acid	122	*N*-Acetyl-LL-diaminopimelic acid ligase	Q
dck	—	NM	Deoxycytidine kinase	*155*
ddd	—	NM	Deoxycytidine kinase	*155*
ddl	—	40	D-Alanyl-D-alanine ligase	Q
divI	Division	130	Temperature-sensitive cell division, formerly *divD*	*188*
divII	Division	320	Temperature-sensitive cell division, formerly *divC*	*188*

Appendix: (Continued)

Gene symbol	Mnemonic	Map position[a]	Phenotype, enzyme deficiency, or other characteristics	References[b]
divIVA	Division	130	Minicell production	*153*
divIVB	Division	245	Minicell production	*153*
divIVC	Division	15	Minicell production, formerly *divA*	*188*
divV	Division	285	Temperature-sensitive cell division, formerly *divB*	*188*
dnaA	DNA	160	DNA synthesis, ribonucleotide reduction—probably not ribonucleotide reductase	*9, 15, 89, 118, 154*
dnaB	DNA	255	DNA synthesis, initiation of chromosome replication	*80, 89, 118, 119*
dnaC	DNA	355	DNA synthesis	*8, 89, 118*
dnaD	DNA	200	DNA synthesis, initiation of chromosome replication	*89, 118*
dnaE	DNA	230	DNA synthesis	*89, 118*
dnaF	DNA	—	DNA synthesis, DNA polymerase III; see *polC*	*89, 118*
dnaG	DNA	0	DNA synthesis	*89, 118*
dnaH	DNA	0	DNA synthesis	*89, 118, 185*
dnaI	DNA	250	DNA synthesis	*89, 118*
dna(ts)A	DNA	350	DNA synthesis	*60, 118*
dna(ts)B	DNA	350	DNA synthesis	*60, 118*
dna8132	DNA	0	DNA synthesis, initiation of chromosome replication	*60, 63, 185*
dpa	Dipicolinic acid	UC	Requires dipicolinic acid for heat-resistant spores, linked to *pyrA*	*12*
dra	—	346	Deoxyriboaldolase	M
D-tyr	D-Tyrosine	—	Resistance to D-tyrosine, maps within the *tyrA* locus; see *tyrA*	*30*
ebr	Ethidium bromide	325	Ethidium bromide resistance	*18*
efg	Elongation factor G	10	Elongation factor G	*4, 39, 96*

(Continued)

Appendix: (Continued)

Gene symbol	Mnemonic	Map position[a]	Phenotype, enzyme deficiency, or other characteristics	References[b]
ery	Erythromycin	—	Erythromycin resistance, ribosomal protein L22; see *rplV*	*180, 181*
estB	Esterase	310	Esterase B defect	*68*
flaA	Flagella	315	Defect in flagellar synthesis	*57*
flaB	Flagella	315	Defect in flagellar synthesis	*57*
flaC	Flagella	315	Defect in flagellar synthesis	*57*
fruA	Fructose	120	Fructose transport	*52*
fruB	Fructose	120	Fructose-1-phosphate kinase	*52*
fruC	Fructose	50	Fructokinase	0, *52*
fumR	Fumarase	295	Fumarate hydratase	A
furA	5-Fluorouracil	135	Resistance to 5-fluorouracil	*38*
furC	5-Fluorouracil	325	Resistance to 5-fluorouracil in the presence of uracil	E
furE	5-Fluorouracil	325	Resistance to 5-fluorouracil in the presence of uracil	E
furF	5-Fluorouracil	160	Resistance to 5-fluorouracil in the presence of uracil	*35*
fus	Fusidic acid	—	Fusidic acid resistance; see *efg*	*96*
gca	Glucosamine	NM	L-Glutamine-D-fructose-6-phosphate aminotransferase	*45*
gerA	Germination	295	Germination-defective	*124, 167*
gerB	Germination	315	Germination-defective	*124, 167*
gerC	Germination	210	Germination-defective	*124, 167*
gerD	Germination	10	Germination-defective	*124, 167*
gerE	Germination	255	Germination-defective	*124, 167*
gerF	Germination	305	Germination-defective	*124, 167*
gerG	Germination	300	Germination-defective	*124, 167*
glnA	Glutamine	160	Glutamine synthetase structural gene	*21, 35, 43*
glpD	Glycerol phosphate	75	Glycerol-3-phosphate dehydrogenase	*108*
glpK	Glycerol phosphate	75	Glycerol kinase	*108*
glpP	Glycerol phosphate	75	Pleiotropic glycerol mutant	*108*

Appendix: (Continued)

Gene symbol	Mnemonic	Map position[a]	Phenotype, enzyme deficiency, or other characteristics	References[b]
glpT	Glycerol phosphate	15	Fosfomycin resistance, glycerol phosphate transport defect	*107*
gltA	Glutamate	175	Glutamate or aspartate requirement, glutamine-2-ketoglutarate aminotransferase	*35, 72*
glyA	Glycine	210	Glycine requirement	*92–94*
glyB	Glycine	75	Glycine requirement	*63*
glyC	Glycine	320	Glycine requirement	*28*
gspl	Germination	150	Outgrowth-defective	*46*
gsp4	Germination	320	Outgrowth-defective	*46*
gsp10	Germination	300	Outgrowth-defective	*46*
gsp081	Germination	20	Outgrowth-defective	*5*
gspIV	Germination	20	Outgrowth-defective	*46*
gtaA	—	310	Glucosylation of teichoic acid, lacks UDP-glucose–poly(glycerol phosphate) α-glucosyltransferase	*106, 202*
gtaB	—	310	Glucosylation of teichoic acid	*202*
gtaC	—	75	Glucosylation of teichoic acid, lacks phosphoglucomutase	*202*
guaA	Guanine	0	IMP dehydrogenase	C, *185*
guaB	Guanine	50	Guanine requirement	A, C
gutA	Glucitol	50	D-Glucitol permease	0, *29*
gutB	Glucitol	50	D-Glucitol dehydrogenase	0, *29*
gutR	Glucitol	50	Constitutive synthesis of D-glucitol permease and dehydrogenase	0, *29*
hag	Flagella	315	Flagellar antigen, maps in the *flaA, flaB, flaC* region	*57, 115*
hds	—	5	Pleiotropic extragenic suppressors of DNA mutations	*165*
hemA	Heme biosynthesis	245	δ-Aminolevulinate synthetase	*95*
hemB	Heme biosynthesis	245	Porphobilinogen synthetase	*17*

(Continued)

Appendix: (Continued)

Gene symbol	Mnemonic	Map position[a]	Phenotype, enzyme deficiency, or other characteristics	References[b]
hemC	Heme biosynthesis	245	Porphobilinogen deaminase	*17*
hemD	Heme biosynthesis	245	Uroporphyrinogen III cosynthase	*121*
hemE	Heme biosynthesis	90	Uroporphyrinogen decarboxylase	*120*
hemF	Heme biosynthesis	90	Coproporphyrinogen oxidase	*120*
hemG	Heme biosynthesis	90	Ferrochelatase	*120*
hisA	Histidine	305	Histidine requirement, probable location of all the histidine enzymes except *hisH*	*24, 38, 41, 130*
hisH	Histidine	205	Histidinol phosphate aminotransferase, tyrosine and phenylalanine aminotransferase	*133*
homL	Homoserine	290	Threonine and methionine requirement, deletion lacking homoserine dehydrogenase	F
hpr	Protease	90	Overproduction of proteases	*69*
hsrB	Host-specific restriction	356	Host-specific restriction and modification of *B. subtilis* IAM 1247	*163, 164*
hsrC	Host-specific restriction	58	Host-specific restriction and modification of *B. subtilis* 1247 II	*163, 164*
hsrE	Host-specific restriction	342	Host-specific restriction and modification of *B. subtilis* IMA 1231	*163*
hsrM	Host-specific restriction	45	Host-specific restriction and modification of *B. subtilis* Marburg, probably identical to *nonB*	*164*
hsrR	Host-specific restriction	58	Host-specific restriction and modification of *B. subtilis* R, also called *bsr*	*79, 183*
hts	Hydrogen sulfide		Excretion of hydrogen sulfide; see *cysA*	*88*

Appendix: (Continued)

Gene symbol	Mnemonic	Map position[a]	Phenotype, enzyme deficiency, or other characteristics	References[b]
hutH	Histidine utilization	345	Histidine degradation	G, *31*
ifm	Flagella	UC	Increased flagella and motility, maps near *flaA, flaB, flaC,* cluster	*57*
ilvA	Isoleucine–valine	200	Threonine dehydrastase	*14, 71*
ilvB	Isoleucine–valine	250	Condensing enzyme	*14, 192*
ilvC	Isoleucine–valine	250	α-Hydroxy-β-ketoacid reductoisomerase	*14, 192*
ilvD	Isoleucine–valine	200	Dihydroxyacid dehydratase	*14*
inh	Inhibition by histidine	—	Inhibition by histidine, probably within *tyrA* locus; see *tyrA*	*131, 134*
kan	Kanamycin	10	Kanamycin resistance, maps in the ribosomal protein cluster	*55*
kauA	Ketoacid uptake	185	Branched-chain α-ketoacid transport	*54*
kir	Kiromycin	—	Probable mutation in the structural gene for EF-Tu; see *tuf*	*168*
ksgA	Kasugamycin	5	High-level kasugamycin resistance	A, *182*
ksgB	Kasugamycin	280	Low-level kasugamycin resistance	*182*
ksgC	Kasugamycin	UC	Fumarase-hydratase-defective, kasugamycin resistance	A
ksgD	Kasugamycin	31	Kasugamycin resistance	N
leuA	Leucine	250	α-Isopropylmalate synthase	*14, 38, 192, 193*
leuB	Leucine	250	Isopropylmalate isomerase	*192*
leuC	Leucine	250	β-Isopropylmalate dehydrogenase	*14, 192, 193*
lin	Lincomycin	25	Lincomycin resistance	*56, 62*
lpm	Lipiarmycin	—	Lipiarmycin resistance, RNA polymerase; see *rpoC*	*174*
lys	Lysine	210	Lysine requirement, diaminopimelate decarboxylase	*78, 92*
lysS	Lysine	5	Lysyl-tRNA synthetase	*152*

(Continued)

Appendix: (Continued)

Gene symbol	Mnemonic	Map position[a]	Phenotype, enzyme deficiency, or other characteristics	References[b]
lyt	Lytic	NM	Autolytic enzymes	*42*
mdh	—	—	Malate dehydrogenase; see *citH*	*50*
metA	Methionine	115	Responds to methionine, cystathionine, or homocysteine	*176*
metB	Methionine	200	Responds to methionine or homocysteine	*7*
metC	Methionine	115	Responds to methionine	*38*
metD	Methionine	95	Responds to methionine	*202*
mic	Micrococcin	—	Resistance to micrococcin; see *rplC*	*171*
mit	Mitomycin	UC	Resistance to mitomycin C, maps near *rplV*	*86*
mtlA	Mannitol	UC	Lacks mannitol transport, maps near *mtlB*	0
mtlB	Mannitol	35	Mannitol-l-phosphate dehydrogenase	*106*
mtr	5-Methyltryptophan	210	Resistance to 5-methyltryptophan, derepression of the tryptophan biosynthetic pathway	*75, 77*
nalA	Nalidixic acid	355	Resistance to nalidixic acid	*63, 185*
narA	Nitrate	320	Inability to use nitrate as a nitrogen source	*106, 207*
narB	Nitrate	30	Inability to use nitrate as a nitrogen source	*106, 207*
nea	Neamine	—	Neamine resistance; see ribosomal protein cluster	*56*
neo	Neomycin	—	Neomycin resistance; see ribosomal protein cluster	*56, 62*
nic	Nicotinic acid	245	Nicotinic acid requirement	*70, 92*
nonA	Nonpermissive	UC	Permissive for bacteriophage SP10 and ϕNR2, closely linked to *rfm*	*161*

Appendix: (Continued)

Gene symbol	Mnemonic	Map position[a]	Phenotype, enzyme deficiency, or other characteristics	References[b]
nonB	Nonpermissive	45	Permissive for bacteriophage SP10 and ϕNR2	*161*
novA	Novobiocin	0	Resistance to novobiocin	*62, 185*
novB	Novobiocin	UC	Resistance to novobiocin, linked to *thr*	*178*
nprE	Neutral protease	125	Structural gene for neutral protease	*187*
nprR	Neutral protease	125	Regulatory gene for neutral protease	*187*
ole	Oleandomycin	—	Oleandomycin resistance; see ribosomal protein cluster	*56, 62*
pab	*p*-Aminobenzoic acid	10	*p*-Aminobenzoic acid requirement, subunit A of the *p*-amino-benzoate synthase enzyme	*87*
pac	Pactomycin	5	Resistance to pactomycin	*62, 185*
pap	—	—	Hyperproduction of proteases and amylase; see *sacQ*	*177, 200*
pdp	—	342	Pyrimidine nucleoside phosphorylase	M
pfk	—	250	Phosphofructokinase	0, *52*
pgk	—	NM	3′-Phosphoglycerol-kinase	C
phal	—	50	Resistance to phage SP01	*106*
pheA	Phenylalanine	245	Phenylalanine requirement, prephenate dehydratase	*14*
phoP	Phosphatase	260	Regulation of alkaline phosphatase and alkaline phosphodiesterase	*101, 199*
phoR	Phosphatase	260	Regulation of alkaline phosphatase	*100, 101, 122*

(Continued)

Appendix: (Continued)

Gene symbol	Mnemonic	Map position[a]	Phenotype, enzyme deficiency, or other characteristics	References[b]
phoS	Phosphatase	110	Constitutive alkaline phosphatase	*149*
pig	Pigment	50	Sporulation-associated pigment	*158*
polA	Polymerase	260	DNA polymerase A	*50, 99*
polC	Polymerase	145	DNA polymerase III, azopyrimidine resistance	*9, 34, 112 156*
pro	Proline	112	Proline requirement	*26, 47*
ptm	—	105	Pyrithymine resistance	H
pts	—	120	Phosphoenolpyruvate phosphotransferase	*51, 136*
purA	Purine	355	Adenine requirement	*142*
purB	Purine	55	Adenine, guanine, or hypoxanthine requirement	*110, 142*
purC	Purine	55	Adenine or hypoxanthine requirement	*110*
purD	Purine	55	Adenine or hypoxanthine requirement	*110*
purE	Purine	55	Adenine requirement	*110*
purH	Purine	UC	Maps in the *purB* region	A, C
pycA	—	145	Pyruvate carboxylase	*25, 72*
pyrA	Pyrimidine	135	Carbamyl phosphate synthetase	*151*
pyrB	Pyrimidine	135	Aspartate transcarbamylase	*151*
pyrC	Pyrimidine	135	Dihydroorotase	*151*
pyrD	Pyrimidine	135	Dihydroorotate dehydrogenase	*151*
pyrE	Pyrimidine	135	Orotate phosphoribosyl transferase, also called *pyrX*	I, *151*
pyrF	Pyrimidine	135	Orotidine-5′-phosphate decarboxylase	*151*
pyrG	Pyrimidine	NM	Cytidine-5′-triphosphate synthetase	*155*
recA	Recombination	145	Genetic recombination and radiation resistance	*70, 118*
recB	Recombination	245	Genetic recombination and radiation resistance	*70, 118*

Appendix: (Continued)

Gene symbol	Mnemonic	Map position[a]	Phenotype, enzyme deficiency, or other characteristics	References[b]
recC	Recombination	—	Genetic recombination, indirect effect of SP02 lysogeny; see *attSP02*	*37, 49, 118*
recD	Recombination	5	Genetic recombination and radiation resistance	*61, 118, 185*
recE	Recombination	150	Genetic recombination and radiation resistance, ATP-dependent nuclease	*36, 37, 117, 118*
recF	Recombination	355	Genetic recombination and radiation resistance	*61, 118, 185*
recG	Recombination	205	Genetic recombination and radiation resistance	*59, 117, 118*
recH	Recombination	70	Genetic recombination and radiation resistance, ATP-dependent nuclease	*118, 132*
recI	Recombination	70	Genetic recombination and radiation resistance	*118, 132*
recL	Recombination	UC	Genetic recombination and radiation resistance, linked to *cysA*	*118, 117, 37*
recM	Recombination	5	Genetic recombination and radiation resistance	*37, 117, 118*
recN	Recombination	150	Genetic recombination and radiation resistance	A, *117, 118*
relA	Relaxed	65	ATP:GTP 3′-phosphotransferase	*172, 179*
relC	Relaxed	—	See *tsp*	*172*
rfm	Rifampicin	—	Rifampicin resistance. RNA polymerase; see *rpoB*	*58, 173*
rib	Riboflavin	210	Riboflavin biosynthetic pathway	P, *93*
rna53	RNA synthesis	315	Temperature-sensitive RNA synthesis	*157*
rodB	—	130	Cell wall-defective	*90*
rodC	—	320	Cell wall-defective	*90*

(Continued)

Appendix: (Continued)

Gene symbol	Mnemonic	Map position[a]	Phenotype, enzyme deficiency, or other characteristics	References[b]
rplA	Ribosomal protein, large	10	Ribosomal protein BL1, chloramphenicol resistance II	*140, 141*
rplC	Ribosomal protein, large	10	Ribosomal protein BL3, probable micrococcin resistance	*141, 171*
rplK	Ribosomal protein, large	10	Ribosomal protein BL11, thiostrepton resistance	*141, 145, 195*
rplL	Ribosomal protein, large	10	Ribosomal protein B12, chloramphenicol resistance VI	*140, 141*
rplO	Ribosomal protein, large	10	Ribosomal protein BL15, chloramphenicol resistance III	*140, 141*
rplV	Ribosomal protein, large	10	Ribosomal protein BL22, erythromycin resistance	*141, 180*
rpoB	RNA polymerase	10	β Subunit of RNA polymerase, rifampin resistance	*58, 173*
rpoC	RNA polymerase	10	β′ Subunit of RNA polymerase, streptolydigin resistance	*59, 173*
rpsE	Ribosomal protein, small	10	Ribosomal protein S5, spectinomycin resistance	*62, 85, 141*
rpsL	Ribosomal protein, small	10	Ribosomal protein S12, streptinomycin resistance	*56, 141*
rpsT	Ribosomal protein, small	10	Ribosomal protein S20	*139, 141*
Ribosomal protein cluster	—	—	Cluster of ribosomal proteins, including: S3, S5, S8, S12, S17, S19, S20, BL1, BL2, BL4, BL5, BL6, BL8, BL12, BL14, BL15, BL16, BL17, BL22, BL23, and BL25	*141*

Appendix: (Continued)

Gene symbol	Mnemonic	Map position[a]	Phenotype, enzyme deficiency, or other characteristics	References[b]
rRNA cluster	—	15	Probable site of five to seven rRNA gene clusters and many tRNA genes	J, *125, 126, 138, 170*
sacA	Sucrose	335	β-Fructofuranosidase	*103, 104*
sacB	Sucrose	305	Levansucrase	*103*
sacL	Sucrose	240	Levanase	*98*
sacP	Sucrose	335	Sucrose transport	*104*
sacQ	Sucrose	285	Hyperproduction of levansucrase and proteases	*97, 103, 162*
sacR	Sucrose	305	Constitutive β-fructofuranosidase production	*103*
sacS	Sucrose	335	Constitutive β-fructofuranosidase production	*103*
sacT	Sucrose	335	Constitutive β-fructofuranosidase production	*103*
sacU	Sucrose	310	Regulatory gene for levansucrase	*97, 103, 105*
sapA	—	110	Alkaline phosphatase	*149*
sapB	—	55	Alkaline phosphatase	*149*
scoA	—	110	Sporulation control	*123*
scoB	—	130	Sporulation control	*123*
scoC	—	90	Sporulation control	*123*
scoD	—	130	Sporulation control	*123*
ser	Serine	210	Requirement for serine	*75*
smo	Smooth	305	Smooth–rough colony morphology	*57*
spcA	Spectinomycin	—	Spectinomycin resistance; see *rpsE*	*27, 63, 85*
spcB	Spectinomycin	140	Resistance to spectinomycin	*62, 112*
spcD	Spectinomycin	UC	Spectinomycin dependence, maps between *cysA* and *purA*	*65*
spg	Sporangiomycin	—	Sporangiomycin resistance, 50 S ribosome alteration; see ribosomal protein cluster	*16*

(Continued)

Appendix: (Continued)

Gene symbol	Mnemonic	Map position[a]	Phenotype, enzyme deficiency, or other characteristics	References[b]
spoCM	Sporulation	355	Stage-0 sporulation, possibly identical to *spo0J*	*23*
spoL1	Sporulation	230	"Decadent" sporulation	*13*
spo0A	Sporulation	215	Stage-0 sporulation	*148*
spo0B	Sporulation	245	Stage-0 sporulation	*148*
spo0E	Sporulation	120	Stage-0 sporulation	*148*
spo0F	Sporulation	325	Stage-0 sporulation	*148*
spo0G	Sporulation	220	Stage-0 sporulation	*148*
spo0H	Sporulation	10	Stage-0 sporulation	*148*
spo0J	Sporulation	355	Stage-0 sporulation	*148*
spo0K	Sporulation	105	Stage-0 sporulation	*148*
spo0L	Sporulation	105	Stage-0 sporulation	A
spoIIA	Sporulation	210	Stage II sporulation	*148*
spoIIB	Sporulation	245	Stage II sporulation	*148*
spoIIC	Sporulation	300	Stage II sporulation	*148*
spoIID	Sporulation	320	Stage II sporulation	*148*
spoIIE	Sporulation	10	Stage II sporulation	*148*
spoIIF	Sporulation	120	Stage II sporulation	*148*
spoIIG	Sporulation	130	Stage II sporulation	*148*
spoIIIA	Sporulation	215	Stage III sporulation	*148*
spoIIIB	Sporulation	215	Stage III sporulation	*148*
spoIIIC	Sporulation	235	Stage III sporulation	*148*
spoIIID	Sporulation	295	Stage III sporulation	*148*
spoIIIE	Sporulation	140	Stage III sporulation	*148*
spoIVA	Sporulation	210	Stage IV sporulation	*148*
spoIVB	Sporulation	215	Stage IV sporulation	*148*
spoIVC	Sporulation	235	Stage IV sporulation	*148*
spoIVD	Sporulation	235	Stage IV sporulation	*148*
spoIVE	Sporulation	240	Stage IV sporulation	*148*
spoIVF	Sporulation	245	Stage IV sporulation	*148*
spoIVG	Sporulation	95	Stage IV sporulation	*148*
spoVA	Sporulation	210	Stage V sporulation	*148*
spoVB	Sporulation	245	Stage V sporulation	*148*
spoVC	Sporulation	10	Stage V sporulation	*148*
spoVD	Sporulation	125	Stage V sporulation	*148*
spoVE	Sporulation	130	Stage V sporulation	*148*
spoVF	Sporulation	NM	Stage V sporulation	*148*
sprA	—	NM	Derepression of homoserine kinase, homoserine dehydrogenase, and the minor threonine dehydratase (*tdm*)	*189*

Appendix: (Continued)

Gene symbol	Mnemonic	Map position[a]	Phenotype, enzyme deficiency, or other characteristics	References[b]
sprB	—	290	Partial suppression of isoleucine requirement, allows threonine dehydratase *sprA* mutants to grow in minimal medium, maps near *tdm* locus; see *tdm*	*190*
sspl	UV repair	NM	Endonuclease excising spore photoproducts	*129*
std	Streptolydigin	—	Streptolydigin resistance, RNA polymerase; see *rpoC*	*59, 173*
strA	Streptomycin	—	Resistance to streptomycin; see *rpsL*	*56*
strB	Streptomycin	130	Streptomycin resistance	*175*
strC	Streptomycin	220	Streptomycin resistance	*175*
sul	Sulfonilamide	10	Resistance to sulfonilamide	*81, 87*
sup3	Suppressor	30	Suppressor tRNA	*67*
tag	—	UC	Cell wall synthesis, near *hisA*	*22*
tdm	—	290	Minor threonine dehydratase	*190*
thiA	Thiamine	70	Blocked in synthesis of pyrimidine ring of thiamine	E, *93*
thiB	Thiamine	100	Blocked in synthesis of thiazole ring of thiamine	E, *73*
thiC	Thiamine	335	Blocked in fusion of the two rings of thiamine	E
thrA	Threonine	290	Threonine requirement, homoserine kinase	*38, 187, 189*
thyA	Thymidine	160	Thymidylate synthetase A	*7, 135, 197*
thyB	Thymidine	200	Thymidylate synthetase B	*7, 135, 197*
til	Tilerone	230	Tilerone resistance	A
tmp	Trimethoprim	200	Trimethoprim resistance	*191*
tmrA	Tunicamycin	25	Tunicamycin resistance, hyperproductivity of extracellular α-amylases	*137*

(Continued)

Appendix: (Continued)

Gene symbol	Mnemonic	Map position[a]	Phenotype, enzyme deficiency, or other characteristics	References[b]
tmrB	Tunicamycin	25	Tunicamycin resistance	*137*
tms12	—	130	Temperature-sensitive cell division	A, *33*
tms26	—	5	Temperature-sensitive cell division	*53, 183*
tolA	Tolerance	—	Tolerance to bacteriophage; see *abrB*	*83, 84*
tolB	Tolerance	NM	Tolerance to bacteriophage	*83, 84*
tre	Trehalose	60	Trehalose	*106*
trpA	Tryptophan	205	Tryptophan synthase α	*6, 76, 194*
trpB	Tryptophan	205	Tryptophan synthase β	*6, 76, 194*
trpC	Tryptophan	205	Indoleglycerolphosphate synthase	*6, 76, 194*
trpD	Tryptophan	205	Anthranilate phosphoribosyl-transferase	*6, 76, 194*
trpE	Tryptophan	205	Anthranilate synthase	*6, 76, 194*
trpF	Tryptophan	205	*N*-(5′-Phosphoribosyl) anthranilate isomerase	*76, 194*
trpS	Tryptophan	100	Tryptophanyl-tRNA synthase	*176*
trpX	Tryptophan	10	Glutamine-binding protein common to anthranilate synthetase and *p*-aminobenzoate synthetase	*87*
tsi	Temperature-sensitive induction	50	Temperature-sensitive induction of PBSX	*166*
tsp	Thiostrepton		Resistance to thiostrepton, 50 S subunit, maps in ribosomal protein cluster	*145, 172*
ts39	—	230	Temperature-sensitive synthesis of phosphatidylethanolamine	*109*
tuf	Tu factor	10	Elongation factor Tu	*39, 168*
tyrA	Tryosine	205	Tyrosine requirement, prephenate dehydrogenase	*131, 134*
urg	—	NM	*N*-Glycosidase	*114*

Appendix: (Continued)

Gene symbol	Mnemonic	Map position[a]	Phenotype, enzyme deficiency, or other characteristics	References[b]
urs	Uracil-sensitive	100	Uracil sensitivity, arginine specific carbamyl phosphate synthase	A, *144*
uvrA	UV repair	310	Excision of UV-induced pyrimidine dimers in DNA	*70, 106, 128*
uvrB	UV repair	250	Excision of UV-induced pyrimidine dimers in DNA	*128*
vas	—	—	Valine sensitivity, maps within threonine dehydratase locus; see *ilvA*	*102*
xhd	PBSX, headless	UC	Induces PBSX prophage, lack heads, maps near *xhi* and *xtl*	*186*
xhi	PBSX, heat-inducible	108	Heat-inducible PBSX prophage	*24, 26*
xin	—	UC	Noninducible PBSX prophage, maps near *xhi* and *xtl*	*186*
xtl	PBSX, tailless	110	Induced PBSX, lack tails	*48*

[a] UC, Map position not fully defined; NM, not mapped.

[b] References: Unpublished data from: A. J. Hoch; B. W. Steinberg; C. E. Freese; D. S. Baumberg; E. S. Zahler; F. C. Anagnostopoulos; G. S. Fisher; H. A. Galizzi; I. R. Switzer; J. W. Steinberg and D. Henner; K. G. Ordal; L. M. Polsinelli; M. J. Newhard; N. K. Bott; O. P. Gay; P. A. Stepanov; Q. R. Buxton.

References

1. Adams, A. (1971). *In* "Informative Molecules in Biological Systems" (L. Ledoux, ed.), pp. 418–428. North-Holland Publ., Amsterdam.
2. Adams, A. (1972). *Mol. Gen. Genet.* **118,** 311–322.
3. Adams, A., and Oishi, M. (1972). *Mol. Gen. Genet.* **118,** 295–310.
4. Aharonowitz, Y., and Ron, E. Z. (1972). *Mol. Gen. Genet.* **119,** 131–138.
5. Albertini, A. M., and Galizzi, A. (1975). *J. Bacteriol.* **124,** 14–25.
6. Anagnostopoulos, C., and Crawford, I. P. (1961). *Proc. Nat. Acad. Sci. USA* **47,** 378–390.

7. Anagnostopoulos, C., and Schneider-Champagne, A. M. (1966). *C. R. Acad. Sci. Paris Ser. D* **262,** 1311–1314.
8. Andersen, J. J., and Ganesan, A. T. (1975). *J. Bacteriol.* **121,** 173–183.
9. Attolini, C., Mazza, G., Fortunato, A., Ciarrochi, G., Mastromei, G., Riva, S., and Falaschi, A. (1976). *Mol. Gen. Genet.* **148,** 9–17.
10. Bachmann, B. J., and Brooks, L. K. (1980). *Microbiol. Rev.* **44,** 1–56.
11. Bak, A. L., Christiansen, C., and Stenderup, A. (1970). *J. Gen. Microbiol.* **64,** 377–380.
12. Balassa, G., Milhaud, P., Raulet, E., Silva, M. T., and Sousa, J. F. C. (1979a). *J. Gen. Microbiol.* **110,** 365–379.
13. Balassa, G., Milhaud, P., Sousa, J. C. F., and Silva, M. T. (1979b). *J. Gen. Microbiol.* **110,** 381–392.
14. Barat, M., Anagnostopoulos, C., and Schneider, A. M. (1965). *J. Bacteriol.* **90,** 357–369.
15. Bazill, W., and Karamata, D. (1972). *Mol. Gen. Genet.* **117,** 19–29.
16. Bazzicalupo, M., Parisi, B., Pirali, G., Polsinelli, M., and Sala, F. (1975). *Antimicrob. Agents Chemother.* **8,** 651–656.
17. Berek, I., Miczak, A., and Ivanovics, G. (1974). *Mol. Gen. Genet.* **132,** 233–239.
18. Bishop, P. E., and Brown, L. R. (1973). *J. Bacteriol.* **115,** 1077–1083.
19. Bishop, R. J., and Sueoka, N. (1972). *J. Bacteriol.* **112,** 870–876.
20. Borenstein, S., and Ephrati-Elizur, E. (1969). *J. Mol. Biol.* **45,** 137–152.
21. Bott, K. F., Reysset, G., Gregoire, J., Islert, D., and Aubert, J-P. (1977). *Biochem. Biophys. Res. Commun.* **79,** 996–1003.
22. Boylan, R. J., Mendelson, N. H., Brooks, D., and Young, F. E. (1972). *J. Bacteriol.* **110,** 281–290.
23. Bramucci, M. G., Keggins, K. M., and Lovett, P. S. (1977). *J. Virol.* **24,** 194–200.
24. Buxton, R. S. (1976). *J. Virol.* **20,** 22–28.
25. Buxton, R. S. (1978). *J. Gen. Microbiol.* **105,** 175–185.
26. Buxton, R. S. (1980). *J. Gen. Virol.* **46,** 427–437.
27. Cannon, J. G., and Bott, K. F. *Mol. Gen. Genet.* **174,** 149–162.
28. Canosi, U., Siccardi, A. G., Falaschi, A., and Mazza, G. (1976). *J. Bacteriol.* **126,** 180–121.
29. Chalumeau, H., Delobbe, A., and Gay, P. (1978). *J. Bacteriol.* **134,** 920–928.
30. Champney, W. S., and Jensen, R. A. (1969). *J. Bacteriol.* **98,** 205–214.
31. Chasin, L. A., and Magasanik, B. (1968). *J. Biol. Chem.* **243,** 5165–5178.
32. Chelala, C. A., and Margolin, P. (1974). *Mol. Gen Genet.* **131,** 97–112.
33. Copeland, J. C., and Marmur, J. (1968). *Bacteriol. Rev.* **32,** 302–312.
34. Cozzarelli, N. R., and Low, R. L. (1973). *Biochem. Biophys. Res. Commun.* **51,** 151–159.
35. Dean, D. R., Hoch, J. A., and Aronson, A. I. (1977). *J. Bacteriol.* **313,** 981–987.
36. Doly, J., Sasarman, E., and Anagnostopoulos, C. (1974). *Mutat. Res.* **22,** 15–23.
37. Dubnau, D., and Circgliano, C. (1974). *J. Bacteriol.* **117,** 488–493.
38. Dubnau, D., Goldthwaite, C., Smith, I., and Marmur, J. (1967). *J. Mol. Biol.* **27,** 163–185.
39. Dubnau, D., Pifko, S., Sloma, A., Cabane, K., and Smith, I. (1976). *Mol. Gen. Genet.* **147,** 1–12.
40. Dul, M. J., and Young, F. E. (1973). *J. Bacteriol.* **115,** 1212–1214.
41. Ephrati-Elizur, E., Srinivasan, P. R., and Zamenhof, S. (1961). *Proc. Nat. Acad. Sci. USA* **47,** 56–63.
42. Fein, J. E., and Rogers, H. J. (1976). *J. Bacteriol.* **127,** 1427–1442.
43. Fisher, S. H., and Sonenshein, A. L. (1977). *Biochem. Biophys. Res. Commun.* **79,** 987–995.

44. Freese, E., Park, S. W., and Chasel, M. (1964). *Proc. Nat. Acad. Sci. USA* **51,** 1164–1172.
45. Freese, E. B., Cole, R. M., Klofat, W., and Freese, E. (1970). *J. Bacteriol.* **101,** 1046–1062.
46. Galizzi, A., Siccardi, A. G., Albertini, A. M., Amileni, A. R., Meneguzzi, G., and Polsinelli, M. (1975). *J. Bacteriol.* **121,** 450–454.
47. Gallori, E., Bazzicalupo, M., Parisi, B., Peduggi, G., and Polsinelli, M. (1978). *Biochem. Biophys. Res. Commun.* **85,** 1518–1525.
48. Garro, A. J., Leffert, H., and Marmur, J. (1970). *J. Virol.* **6,** 340–343.
49. Garro, A. J., Sprouse, C., and Wetmur, J. G. (1976). *J. Bacteriol.* **126,** 556–568.
50. Gass, K. B., and Cozzarelli, N. R. (1973). *J. Biol. Chem.* **248,** 7688–7700.
51. Gay, P., Cordier, P., Marquet, M., and Delobbe, A. (1973). *Mol. Gen. Genet.* **121,** 355–368.
52. Gay, P., and Delobbe, A. (1977). *Eur. J. Biochem.* **79,** 363–373.
53. Gillis, M., DeLey, J., and DeCleene, M. (1970). *Eur. J. Biochem.* **12,** 143–153.
54. Goldstein, J. B., and Zahler, S. A. (1976). *J. Bacteriol.* **127,** 667–670.
55. Goldthwaite, C., and Smith, I. (1972). *Mol. Gen. Genet.* **114,** 181–189.
56. Goldthwaite, E., Dubnau, D., and Smith, I. (1970). *Proc. Nat. Acad. Sci. USA* **65,** 96–103.
57. Grant, G. F., and Simon, M. I. (1969). *J. Bacteriol.* **99,** 116–124.
58. Halling, S. M., Burtis, K. C., and Doi, R. H. (1977). *J. Biol. Chem.* **252,** 9024–9031.
59. Halling, S. M., Burtis, K. C., and Doi, R. H. (1978). *Nature (London)* **272,** 837–839.
60. Hara, H., and Yoshikawa, H. (1973). *Nature (London) New Biol.* **244,** 200–203.
61. Harford, N. (1974). *Mol. Gen. Genet.* **129,** 269–274.
62. Harford, N., and Sueoka, N. (1970). *J. Mol. Biol.* **51,** 267–286.
63. Harford, N., Lepesant-Kejzlarova, J., Lepesant, J-A, Hamers, R., and Dedonder, R. (1976). *In* "Microbiology, 1976" (D. Schlessinger, ed.), pp. 28–34. American Society for Microbiology, Washington, D. C.
64. Harwood, C. R., and Baumberg, S. (1977). *J. Gen. Microbiol.* **100,** 177–188.
65. Henkin, T. M., Campbell, K. M., and Chambliss, G. H. (1979). *J. Bacteriol.* **137,** 1452–1455.
66. Henner, D. J., and Hoch, J. A. (1980). *Microbiol. Rev.* **44,** 57–82.
67. Henner, D., and Steinberg, W. (1979). *J. Bacteriol.* **139,** 668–670.
68. Higerd, T. B. (1977). *J. Bacteriol.* **129,** 973–977.
69. Higerd, T. B., Hoch, J. A., and Spizizen, J. (1972). *J. Bacteriol.* 1026–1028.
70. Hoch, J. A., and Anagnostopoulos, C. (1970). *J. Bacteriol.* **103,** 295–301.
71. Hoch, J. A., and Coukoulis, J. H. (1978). *J. Bacteriol.* **133,** 265–269.
72. Hoch, J. A., and Mathews, J. (1972). *In* "Spores V" (H. O. Halvorson, R. Hanson, and L. L. Campbell, eds.), pp. 113–116. American Society for Microbiology, Washington, D. C.
73. Hoch, J. A., and Nester, E. W. (1973). *J. Bacteriol.* **116,** 59–66.
74. Hoch, J. A., and Spizizen, J. (1969). *In* "Spores IV" (L. L. Campbell, ed.), pp. 112–120. American Society for Microbiology, Washington, D. C.
75. Hoch, S. O. (1974). *J. Bacteriol.* **117,** 315–317.
76. Hoch, S. O., Anagnostopoulos, C., and Crawford, I. P. (1969). *Biochem. Biophys. Res. Commun.* **34,** 838–844.
77. Hoch, S. O., Roth, C. W., Crawford, I. P., and Nester, E. W. (1971). *J. Bacteriol.* **105,** 38–48.
78. Iijima, T., Diesterhaft, M. D., and Freese, E. (1977). *J. Bacteriol.* **129,** 1440–1447.
79. Ikawa, D., Shibata, T., and Ando, T. (1979). *Mol. Gen. Genet.* **170,** 123–127.
80. Imada, S., Carroll, L. E., and Sueoka, N. (1980). *Genetics* **94,** 809–823.

81. Inselberg, J. W., Eremenko-Volpe, T., Greenwald, L., Meadow, W. L., and Marmur, J. (1969). *J. Virol.* **3,** 627–628.
82. Ionesco, H., Michel, J., Cami, B., and Schaeffer, P. (1970). *J. Appl. Bacteriol.* **33,** 13–24.
83. Ito, J. (1973). *Mol. Gen. Genet.* **124,** 97–106.
84. Ito, J., Mildner, G., and Spizizen, J. (1971). *Mol. Gen. Genet.* **112,** 104–109.
85. Itoh, T. (1976). *Mol. Gen. Genet.* **144,** 39–42.
86. Iyer, V. N. (1966). *J. Bacteriol.* **92,** 1663–1669.
87. Kane, J. F. (1977). *J. Bacteriol.* **132,** 419–425.
88. Kane, J. F., Goode, R. L., and Wainscott, J. (1975). *J. Bacteriol.* **121,** 204–211.
89. Karamata, D., and Gross, J. D. (1970). *Mol. Gen. Genet.* **108,** 277–287.
90. Karamata, D., McConnell, M., and Rogers, H. J. (1972). *J. Bacteriol.* **111,** 73–79.
91. Kemper, J. (1974). *J. Bacteriol.* **117,** 94–99.
92. Kelly, M. S. (1967a). *Mol. Gen. Genet.* **99,** 333–349.
93. Kelly, M. S. (1967b). *Mol. Gen. Genet.* **99,** 350–361.
94. Kelly, M. S., and Pritchard, R. H. (1963). *Heredity* **17,** 598–603.
95. Kiss, I., Berek, I., and Ivanovics, G. (1971). *J. Gen. Microbiol.* **66,** 153–159.
96. Kobayashi, H., Kobayashi, K., and Kobayashi, Y. (1977). *J. Bacteriol.* **132,** 262–269.
97. Kunst, F., Pascal, M., Lepesant-Kejzlarova, J., Lepesant, J-A, Billault, A., and Dedonder, R. (1974). *Biochimie* **56,** 1481–1489.
98. Kunst, F., Steinmetz, M., Lepesant, J-A, and Dedonder, R. (1977). *Biochimie* **59,** 287–292.
99. Laipis, J. P., and Ganesan, A. T. (1972). *J. Biol. Chem.* **247,** 5867–5871.
100. LeHegarat, J-C, and Anagnostopoulos, C. (1969). *C. R. Acad. Sci. Paris* **269,** 2048–2050.
101. LeHegarat, J-C, and Anagnostopoulos, C. (1973). *Eur. J. Biochem.* **39,** 525–539.
102. Leibovici, J., and Anagnostopoulos, C. (1969). *Bull. Soc. Chim. Biol.* **51,** 691–707.
103. Lepesant, J-A, Kunst, F., Lepesant-Kejzlarova, J., and Dedonder, R. *Mol. Gen. Genet.* **118,** 135–160.
104. Lepesant, J-A, Billault, A., Lepesant-Kejzlarova, J., Pascal, M., Kunst, F., and Dedonder, R. (1974). *Biochimie* **56,** 1465–1470.
105. Lepesant-Kejzlarova, J., Walle, J., Billault, A., Kunst, F., Lepesant, J-A, and Dedonder, R. (1974). *C. R. Acad. Sci. Paris* **278,** 1911–1914.
106. Lepesant-Kejzlarova, J., Lepesant, J-A, Walle, J., Billault, A., and Dedonder, R. (1975). *J. Bacteriol.* **121,** 823–834.
107. Lindgren, V. (1978). *J. Bacteriol.* **133,** 667–670.
108. Lindgren, V., and Rutberg, L. (1974). *J. Bacteriol.* **119,** 431–442.
109. Lindgren, V., Holmgren, E., and Rutberg, L. (1977). *J. Bacteriol.* **132,** 473–484.
110. Lopez, M. E., Ferrari, F. A., Siccardi, A. G., Mazza, G., and Polsinelli, M. *J. Bacteriol.* **126,** 533–535.
111. Lorence, J. H., and Nester, E. W. (1967). *Biochemistry* **6,** 1541–1553.
112. Love, E., D'Ambrosio, J., Brown, N. C., and Dubnau, D. (1976). *Mol. Gen. Genet.* **144,** 313–321.
113. Mahler, I., Neumann, J., and Marmur, J. (1963). *Biochim. Biophys. Acta* **72,** 69–70.
114. Makino, F., and Munakata, N. (1977). *J. Bacteriol.* **131,** 438–445.
115. Martinez, R. J., Ichiki, A. T., Lundh, N. P., and Tronick, S. R. (1968). *J. Mol. Biol.* **34,** 559–564.
116. Mattioli, R., Bazzicalupo, M., Federici, G., Gallori, E., and Polsinelli, M. (1979). *J. Gen. Microbiol.* **114,** 223–225.
117. Mazza, G., Fortunato, A., Ferrari, E., Canosi, U., Falaschi, A., and Polsinelli, M. (1975). *Mol. Gen. Genet.* **136,** 9–30.

118. Mazza, G., and Galizzi, A. (1978). *Microbiologica* **1,** 111–135.
119. Mendelson, N. H., and Gross, J. D. (1967). *J. Bacteriol.* **94,** 1603–1608.
120. Miczak, A., Berek, I., and Ivanovics, G. (1976). *Mol. Gen. Genet.* **146,** 85–87.
121. Miczak, A., Pragai, B., and Berek, I. (1979). *Mol. Gen. Genet.* **174,** 293–295.
122. Miki, T., Minimi, Z., and Ikeda, Y. (1965). *Genetics* **52,** 1093–1100.
123. Milhaud, P., Balassa, G., and Zucca, J. (1978). *Mol. Gen. Genet.* **163,** 35–44.
124. Moir, A., Lafferty, E., and Smith, D. A. (1979). *J. Gen. Microbiol.* **111,** 165–180.
125. Moran, C. P., Jr., and Bott, K. F. (1979a). *J. Bacteriol.* **140,** 99–105.
126. Moran, C. P., Jr., and Bott, K. F. (1979b). *J. Bacteriol.* **140,** 742–744.
127. Mountain, A., and Baumberg, S. (1980). *Mol. Gen. Genet.* **178,** 691–701.
128. Munakata, N. (1977). *Mol. Gen. Genet.* **156,** 49–54.
129. Munakata, N., and Ikeda, Y. (1968). *Biochem. Biophys. Res. Commun.* **33,** 469–475.
130. Nasser, D., Henderson, G., and Nester, E. W. (1969). *J. Bacteriol.* **98,** 44–50.
131. Nasser, D., and Nester, E. W. (1967). *J. Bacteriol.* **94,** 1706–1714.
132. Naumov, L. S., Savchenko, G. V., and Prozorov, A. (1974). *Genetika (USSR)* **10,** 126–131.
133. Nester, E. W., and Montoya, A. L. (1976). *J. Bacteriol.* **126,** 699–705.
134. Nester, E. W., Schafer, M., and Lederberg, J. (1963). *Genetics* **48,** 529–551.
135. Neuhard, J., Price, A. R., Schack, L., and Thomassen, E. (1978). *Proc. Nat. Acad. Sci. USA* **75,** 1194–1198.
136. Niaudet, B., Gay, P., and Dedonder, R. (1975). *Mol. Gen. Genet.* **136,** 337–349.
137. Nomura, S., Yamane, K., Sasaki, T., Yamasaki, M., Tamura, G., and Maruo, B. (1978). *J. Bacteriol.* **135,** 818–821.
138. Oishi, M., Oishi, A., and Sueoka, N. (1966). *Proc. Nat. Acad. Sci. USA* **55,** 1095–1103.
139. Osawa, S. (1976). *Mol. Gen. Genet.* **144,** 49–51.
140. Osawa, S., Takata, R., Tanaka, K., and Tamaki, M. (1973). *Mol. Gen. Genet.* **127,** 163–173.
141. Osawa, S., and Tukui, A. (1978). *Mol. Gen. Genet.* **164,** 113–129.
142. O'Sullivan, A., and Sueoka, N. (1967). *J. Mol. Biol.* **27,** 349–368.
143. Pai, C. H. (1975). *J. Bacteriol.* **121,** 1–8.
144. Paulis, T. J., and Switzer, R. L. (1979). *J. Bacteriol.* **137,** 82–91.
145. Pestka, S., Weiss, D., Vince, R., Wienen, B., Stoffler, G., and Smith, I. (1976). *Mol. Gen. Genet.* **144,** 235–241.
146. Peterson, A. M., and Rutberg, L. (1969). *J. Bacteriol.* **98,** 874–877.
147. Piggot, P. J. (1975). *J. Gen. Microbiol.* **89,** 371–374.
148. Piggot, P. J., and Coote, J. G. (1976). *Bacteriol. Rev.* **40,** 908–962.
149. Piggot, P. J., and Taylor, S. Y. (1977). *J. Gen. Microbiol.* **102,** 69–80.
150. Polsinelli, M. (1965). *J. Gen. Microbiol.* **13,** 99–110.
151. Potvin, B. W., Kelleher, R. J., Jr., and Gooder, H. (1975). *J. Bacteriol.* **123,** 604–615.
152. Racine, F. M., and Steinberg, W. (1974). *J. Bacteriol.* **120,** 384–389.
153. Reeve, J. N., Mendelson, N. H., Coyne, S. I., Hallock, L. L., and Cole, R. M. (1973). *J. Bacteriol.* **114,** 860–873.
154. Rima, B. K., and Takahashi, I. (1978a). *J. Gen. Microbiol.* **107,** 130–145.
155. Rima, B. K., and Takahashi, I. (1978b). *Can. J. Biochem.* **56,** 158–160.
156. Riva, S. *et al.* (1975). *Mol. Gen. Genet.* **137,** 185–202.
157. Riva, S., Villani, G., Mastromei, G., and Mazza, G. (1976). *J. Bacteriol.* **127,** 679–690.
158. Rogolsky, M. (1968). *J. Bacteriol.* **95,** 2426–2427.
159. Rutberg, B., and Hoch, J. A. (1970). *J. Bacteriol.* **104,** 826–833.
160. Rutberg, L. (1969). *J. Virol.* **3,** 38–44.
161. Saito, H., Shibata, T., and Ando, T. (1979). *Mol. Gen. Genet.* **170,** 117–122.

162. Sekiguchi, J., Takada, N., and Okada, H. (1975). *J. Bacteriol.* **121,** 688–694.
163. Shibata, T., Ikawa, S., Kim, C., and Ando, T. (1976). *J. Bacteriol.* **128,** 473–476.
164. Shibata, T., Ikawa, S., Komutsu, Y., Ando, T., and Siato, H. (1979). *J. Bacteriol.* **139,** 308–310.
165. Siccardi, A. G., Ottolenghi, S., Fortunato, A., and Mazza, G. (1976). *J. Bacteriol.* **138,** 174–181.
166. Siegel, E. C., and Marmur, J. (1969). *J. Virol.* **4,** 610–618.
167. Smith, D. A., Moir, A., and Sammons, R. (1978). *In* "Spores VII" (G. Chambliss and J. C. Vary, eds.), pp. 158–163. American Society for Microbiology, Washington, D. C.
168. Smith, I., and Paress, P. (1978). *J. Bacteriol.* **135,** 1101–1117.
169. Smith, I., and Smith, H. (1973). *J. Bacteriol.* **114,** 1138–1142.
170. Smith, I., Dubnau, D., Morell, P., and Marmur, J. (1968). *J. Mol. Biol.* **33,** 123–140.
171. Smith, I., Weiss, D., and Pestka, S. (1976). *Mol. Gen. Genet.* **144,** 231–233.
172. Smith, I., Paress, P., Cubane, K., and Dubnau, D. (1980). *Mol. Gen. Genet.* **178,** 271–279.
173. Sonenshein, A. L., Cami, B., Brevet, J., and Cote, R. (1974). *J. Bacteriol.* **120,** 253–265.
174. Sonenshein, A. L., Alexander, H. B., Rothstein, D. M., and Fisher, S. H. (1977). *J. Bacteriol.* **132,** 73–79.
175. Staal, S. P., and Hoch, J. A. (1972). *J. Bacteriol.* **110,** 202–207.
176. Steinberg, W., and Anagnostopoulos, C. (1971). *J. Bacteriol.* **105,** 6–19.
177. Steinmetz, M., Kunst, F., and Dedonder, R. (1976). *Mol. Gen. Genet.* **148,** 281–285.
178. Sugino, A., and Bott, K. F. (1980). *J. Bacteriol.* **141,** 1331–1339.
179. Swanton, M., and Edlin, G. (1972). *Biochem. Biophys. Res. Commun.* **46,** 583–588.
180. Tanaka, K., Tamaki, M., Osawa, S., Kimura, A., and Takata, R. (1973). *Mol. Gen. Genet.* **127,** 157–161.
181. Tipper, D. J., Johnson, C. W., Ginther, C. L., Leighton, T., and Wittmann, H. G. (1977). *Mol. Gen. Genet.* **150,** 147–159.
182. Tominago, A., and Kobayashi, Y. (1978). *J. Bacteriol.* **135,** 1149–1150.
183. Trautner, T. A., Pawlek, B., Bron, S., and Anagnostopoulos, C. (1974). *Biol. Aspects Mol. Gen. Genet.* **131,** 181–191.
184. Trowsdale, J., Shiflett, M., and Hoch, J. A. (1978). *Nature (London)* **272,** 179–180.
185. Trowsdale, J., Chen, S. M. H., and Hoch, J. A. (1979). *Mol. Gen. Genet.* **173,** 61–70.
186. Thurm, P., and Garro, A. J. (1975). *J. Virol.* **16,** 184–191.
187. Uehara, H., Yamane, K., and Maruo, B. (1979). *J. Bacteriol.* **139,** 583–590.
188. Van Alstyne, D., and Simon, M. I. (1971). *J. Bacteriol.* **108,** 1366–1379.
189. Vapnek, D., and Greer, S. (1971a). *J. Bacteriol.* **106,** 615–625.
190. Vapnek, D., and Greer, S. (1971b). *J. Bacteriol.* **106,** 983–993.
191. Wainscott, V. J., and Kane, J. F. (1976). *In* "Microbiology, 1976" (D. Schlessinger, ed.), pp. 208–213. American Society for Microbiology, Washington, D. C.
192. Ward, J. B., and Zahler, S. A. (1973a). *J. Bacteriol.* **116,** 719–726.
193. Ward, J. B., and Zahler, S. A. (1973b). *J. Bacteriol.* **116,** 727–735.
194. Whitt, D. D., and Carlton, B. C. (1968). *J. Bacteriol.* **96,** 1273–1280.
195. Wienen, B. *et al.* (1979). *J. Biol. Chem.* **254,** 8031–8041.
196. Williams, M. T., and Young, F. E. (1977). *J. Virol.* **21,** 522–529.
197. Wilson, M. C., Farmer, J. L., and Rotham, F. (1966). *J. Bacteriol.* **92,** 186–196.
198. Wu, T. T. (1966). *Genetics* **54,** 405–410.
199. Yamane, K., and Maruo, B. (1978). *J. Bacteriol.* **134,** 108–114.
200. Yoneda, Y., and Maruo, B. (1975). *J. Bacteriol.* **124,** 48–54.

201. Yoshikawa, M., Okuyama, A., and Tanaka, N. (1975). *J. Bacteriol.* **122,** 796–797.
202. Young, F. E., Smith, C., and Reilly, B. E. (1969). *J. Bacteriol.* **98,** 1087–1097.
203. Young, M. (1975). *J. Bacteriol.* **122,** 1109–1116.
204. Yuki, S. (1968). *Biochem. Biophys. Res. Commun.* **31,** 182–187.
205. Yuki, S. (1975). *Jpn. J. Genet.* **50,** 155–157.
206. Zahler, S. A. (1978). *J. Gen. Microbiol.* **197,** 199–201.
207. Zahler, S. A., Benjamin, L. G., Glatz, B. S., Winter, P. F., and Goldstein, B. J. (1976). *In* "Microbiology, 1976" (D. Schlessinger, ed.), pp. 35–43. American Society for Microbiology, Washington, D. C.
208. Zahler, S. A., Korman, R. Z., Rosenthal, R., and Hemphill, H. E. (1977). *J. Bacteriol.* **129,** 556–558.

2

DNA Replication in *Bacillus subtilis*

SCOTT WINSTON AND NOBORU SUEOKA

The Molecular Biology
of the Bacilli

ISBN 0–12–222701–8

I. Introduction

Investigations into the mechanism of DNA replication in *Bacillus subtilis* have greatly contributed to the basic understanding of this important process. The unique attributes of *B. subtilis* make this system amenable to further investigations into many of the unanswered questions surrounding the replication process. *Bacillus subtilis* is a gram-positive soil bacterium which is nonpathogenic, and studies on this organism can be directly compared and contrasted with those on the pathogenic gram-negative bacterium *Escherichia coli* [for *E. coli,* see recent reviews by Wickner (1978), Tomizawa and Selzer (1979), and Kolter and Helinski (1979)]. *Bacillus subtilis* can be grown and stored easily. A defined minimal medium is available, and this allows easy radioactive labeling of specific molecules and isolation and characterization of auxotrophic mutants. A circular map of *B. subtilis* has been constructed (Young and Wilson, 1975; Henner and Hoch, 1980) with over 200 well-defined loci. A number of mutations affecting DNA replication have been isolated (Mazza and Galizzi, 1978). Thus, biochemical and genetic studies can be performed on *B. subtilis*.

Bacillus subtilis can form endospores. The sporulation and germination processes of *B. subtilis* are most interesting in terms of control mechanisms governing gene expression and membrane biosynthesis. As described later, germinating spores have proven to be an excellent means of examining DNA replication and cell division in a synchronized system.

The most important tool in studies on DNA replication in *B. subtilis* has been genetic transformation (Sueoka, 1975). Transformation in *B. subtilis* routinely occurs at a high frequency (0.1–2%), with a minimal of cell manipulation involved (Spizizen, 1958). Transformation allows precise indentification of specific regions of DNA, which is extremely important in identifying and characterizing regulatory sites of DNA synthesis such as the origin and terminus of replication. The effective use of genetic transformation in DNA replication studies will become evident later.

Bacillus subtilis is a potential host for experiments involving genetic engineering, using both recombinant DNA technology and standard genetic manipulations such as transformation and transduction (Erickson and Young, 1978). A knowledge of plasmid DNA replication in *B. subtilis* will certainly aid the advancement of *B. subtilis* as a host–vector system.

II. Replication *in Vivo*

A. Size and Circularity of the Chromosome

The *B. subtilis* chromosome is a circular molecule of approximately 2–3 × 10^9 daltons. Direct evidence for the circularity of the chromosome has been provided by the autoradiographs of Wake (1973) and genetic linkage analysis (Kejzalarová-Lepesant *et al.*, 1975). The size of the chromosome has been determined by viscoelastometry (Klotz and Zimm, 1972), sedimentation (Kavenoff, 1972), and renaturation studies (Bak *et al.*, 1970). As seen by electron microscopy, the chromosome exists in a nucleoid structure (reviewed by Pettijohn, 1976) and is usually observed associated with mesosomes (Ryter, 1968).

B. Replication from a Fixed Origin

Replication of the *B. subtilis* chromosome begins at a fixed position or origin. This was first demonstrated by analyzing the frequency of genetic markers by transformation (Yoshikawa and Sueoka, 1963a). Yoshikawa and Sueoka (1963a) reasoned that sequential replication from a fixed origin would be reflected in the frequency of a particular marker depending on the marker's location with respect to the origin. Thus, a polarity of marker frequency should exist, which decreases with the increasing distance of the marker from the origin. The detection of polarity is dependent on three conditions: (1) The cells are not growing synchronously; (2) the rate of fork migration is constant; and (3) the sequence of replication of the markers is fixed (Sueoka, 1975). Since the efficiency of transformation is not equal for all markers, the absolute number of transformants must be normalized to standard DNA in which all markers are equally frequent. In stationary-phase cells of *B. subtilis*, the chromosome is in a nearly completed state of replication (Yoshikawa and Sueoka, 1963a). Spores also contain nonreplicating chromosomes in a completed state, and spores are particularly excellent sources of DNA in which all markers are equally represented (Yoshikawa *et al.*, 1964).

In their initial experiments, Yoshikawa and Sueoka (1963a) observed an *ade-6* (*purB6*)/*met-5* (*metB5*) marker ratio of 2, indicating that replication proceeded from an origin near *ade-6* to a terminus near *met-5*. These results were confirmed by following the sequential replication of markers using the density transfer technique of Meselson and Stahl (1958). The transfer of markers to the hybrid density was monitored during the outgrowth of stationary-phase cells, and the replication order coincided with that obtained from marker frequency analysis (Yoshikawa and Sueoka, 1963b). Density transfer analysis of DNA replication in germinating spores also demonstrates the sequential replication of the chromosome beginning at a single origin near *ade-16* (*purA16*) (Oishi *et al.*, 1964; O'Sullivan and Sueoka, 1967). Thus, stationary-phase cells and spores contain completed chromosomes which replicate from the origin upon outgrowth. Further information regarding the identification and isolation of the replication origin is discussed in Section V,B.

C. Dichotomous Replication

Sueoka and co-workers noted that the ratio of *purB6* to *metB5* reached 4 in spores that had been germinated in rich media and had a doubling time of 20 min (Yoshikawa *et al.*, 1964). The high *purB6/metB5* ratio was also observed in cells growing in rich media with a rapid growth rate. They explained this result as a consequence of a new initiation event occurring before the previously formed replication forks had reached the terminus. Further studies utilizing density transfer techniques and synchronous replication from the origin during spore germination demonstrated that the origin marker *ade* was replicated twice before any terminus markers were replicated (Oishi *et al.*, 1964).

This premature initiation event is termed dichotomous replication. Dichotomous replication allows a cell to shorten its generation time by increasing the frequency of initiation with no increase in the rate of DNA synthesis of the replication fork. General mathematical solutions for marker frequency and cell generation time as a function of initiation frequency have been presented (Sueoka and Yoshikawa, 1965). Initiation of replication has since been considered the determining factor in the rate of cellular division in bacteria (Maaløe and Kjeldgaard, 1966; Helmstetter and Cooper, 1968).

Perturbations in the elongation step of DNA replication can also cause dichotomous replication. After thymine starvation of *B. subtilis thy* cells, a single premature initiation event occurs upon the addition of thymine (Kallenbach and Ma, 1968). Some thymine-requiring strains of *B. subtilis* normally possess multiple replication forks during log-phase growth in defined media (Ephrati-Elizur and Borenstein, 1971; Gillin and Ganesan, 1975). Up to four initiation events prior to completion of the initial round of replication can be observed when *thy* spores are germinated in the presence of 5-bromouracil (Yoshikawa and Haas, 1968). Since protein synthesis continues normally during replication in the presence of bromouracil, Yoshikawa and Haas have proposed that a growth rate-dependent regulator molecule accumulates during such treatment, which causes dichotomous replication. This proposal is supported by the findings of Nagley and Wake (1969). They observed that bromouracil decreased the rate of fork movement without affecting RNA and/or protein synthesis. They concluded that dichotomous replication was not a result of initiation occurring earlier in time in relation to the previous initiation, but rather a result of the normal progression of the cells toward the next initiation event, whereas the rate of DNA replication was much slower. In the case of thymine starvation, the number of premature initiations is limited because of a decreased rate of protein synthesis during starvation. If the timing of initiation is directly related to the rate of protein synthesis, then the growth of cells in a rich medium should cause dichotomous replication, which does indeed happen. However, these results do not confine the nature or function of such a regulatory molecule or molecules to a specific mode of action. These initiator molecules have yet to be characterized.

Two possible configurations of initiation, symmetric and asymmetric, exist for chromosomes replicating dichotomously. In the symmetric configuration, both arms of the daughter chromosomes reinitiate, whereas initiation on only one arm of the daughter chromosomes gives the asymmetric configuration. The two modes of initiation can be differentiated by density transfer experiments. The asymmetric mode predicts that, as long as the initial round of replication has not terminated, half of the origin region will be density-labeled in cells that have been first radioactively pulse-labeled at the origin and then grown in the presence of a density label. On the other hand, 100% of the pulse-labeled DNA will be completely density-labeled if the dichotomy is symmetric.

To determine experimentally whether the dichotomy is symmetric or asymmetric, Quinn and Sueoka (1970) pulse-labeled and density-labeled DNA in germinating *B. subtilis* spores with [^{3}H]-5-bromouracil. After a short pulse time, the radioactively labeled bromouracil was replaced with bromouracil, and the transfer of the pulse-labeled DNA was followed in cesium chloride equilibrium gradients. Eighty-three percent of the pulse-labeled DNA was found to be completely density-labeled after 86% of the chromosomes had reinitiated prior to termination of the initial replication forks. Thus, dichotomous replication in germinating spores of *B. subtilis* has the symmetric configuration. These density transfer experiments were made possible by using a bromouracil-tolerant strain, as continuous incorporation of bromouracil is lethal to normal cells (Bishop and Sueoka, 1972).

As mentioned previously, thymine starvation induces dichotomous replication. Kallenbach and Ma (1968) observed that, after readdition of thymine to thymine-starved *B. subtilis* cells, the *ade/met* ratio increased only to 3, rather than to the expected value of 4. They interpreted the results to mean that premature initiation occurred on only one of two possible daughter chromosomes, which would be asymmetric. Kaufman (1976) investigated the effect of thymine starvation on germinating spores and exponentially growing cells using density transfer techniques. His results were opposite those of Kallenbach and Ma and suggest that the symmetric configuration exists after thymine starvation. Wake (1972) visualized dichotomous replication after thymine starvation using autoradiography. He observed a predominance of structures that were symmetric but, as he stated, there were many facets of the experiment that could have possibly biased the results in favor of either a symmetric or an asymmetric mode.

Kaufman (1976) did find, however, that nalidixic acid treatment induced the asymmetric mode of replication in *B. subtilis*. Nalidixic acid directly inhibits the *gyrA* subunit of DNA gyrase in *B. subtilis* (Sugino and Bott, 1980). The inhibition of DNA gyrase may affect the proper formation of initiation complexes, which causes asymmetric reinitiations. Thymine starvation affects only the DNA polymerization reaction by depriving it of a substrate and may not affect the normal mode of symmetric reinitiation.

D. Bidirectionality of Replication

The question of unidirectional or bidirectional replication in *B. subtilis* was not answered satisfactorily until 1975 (reviewed by Wake, 1975). Experimental evidence for circularity of the *B. subtilis* chromosome and bidirectionality of its replication were not available until then. The construction of the genetic map at a time at which there were a limited number of markers and linkage groups led to a linear map that was not contradictory to the replication order of markers derived from density transfer experiments (Dubnau, 1970). The revision of the circular map by Harford (1975) and Kejalarová-Lepesant *et al.* (1975) and the autoradiographic analysis of replicating chromosomes by Wake (1973) conclusively demonstrate the bidirectional replication of the entire *B. subtilis* chromosome.

The first and most convincing evidence for bidirectional replication was provided by Gyurastis and Wake (1973) and by Wake (1973). Their autoradiographs of replicating chromosomes show bidirectional replication over half the chromosome. Wake's (1974) further analysis of chromosomes in the process of termination shows two replication forks approaching each other at a similar velocity and to within 10% of the length of the chromosome. These results make the complete bidirectional replication of the *B. subtilis* chromosome the most likely mode of replication.

Genetic evidence for partial bidirectional replication was provided by Hara and Yoshikawa (1973) and O'Sullivan *et al.* (1975a). These data were in conflict with those of Harford (1975), who found by studying the replication order of markers in density transfer experiments that replication was completely bidirectional. However, after proper identification and mapping of mutants used in their previous studies, the most recent work of O'Sullivan *et al.* (1975b) and Hye *et al.* (1976) concurs with the conclusion of complete bidirectionality by Wake and Harford.

E. Discontinuity of Replication

The antiparallel structure of the complementary DNA strands necessitates sequential synthesis of daughter strands in both the $5' \rightarrow 3'$ and $3' \rightarrow 5'$ directions during semiconservative replication. In *B. subtilis,* as well as in all other systems that have been studied, the DNA polymerases have a capacity for chain elongation only in the $5' \rightarrow 3'$ direction. The synthesis in the $3' \rightarrow 5'$ direction has been found to occur by the polymerization of short DNA fragments in the $5' \rightarrow 3'$ direction in a retrograde fashion, followed by processing and joining until a complete DNA strand is synthesized (Okazaki *et al.,* 1968). DNA replication in *B. subtilis* occurs by this discontinuous mechanism.

In their initial studies on discontinuous synthesis in *B. subtilis,* Okazaki and co-workers found a majority of a short radioactive pulse label in DNA fragments 8 S in size. These fragments are eventually chased into DNA of a much greater size, but the joining of fragments is slower than that observed during similar experiments on *E. coli* (Okazaki *et al.,* 1968). The fact that almost all the pulse label was initially found in small fragments suggested that both DNA strands were synthesized discontinuously. However, in their later studies, approximately half of the pulse label appeared in fragments much larger than 8 S (Okazaki *et al.,* 1970). This result raises the possibility that one strand is synthesized continuously. Hybridization of the 8 S fragments almost exclusively with the separated L strand of the *B. subtilis* chromosome supports this idea (Okazaki *et al.,* 1970).

In contrast with the results of Okazaki, Sternglanz *et al.* (1975) have observed that the method of terminating the pulse label determines the percentage of pulse-labeled DNA in small, 8 S fragments. If the pulse is terminated with acetone at $-10°C$ rather than pyridine, the vast majority of the pulse label is found in fragments less than 16 S in size. Moreover, the pulse-labeled material hybridizes equally well with either the purified H or L strand. These data argue for both strands being synthesized in a discontinuous fashion. To resolve the different results of Okazaki and Sternglanz, these experiments should be repeated in stable L cells of *B. subtilis* in order to ensure rapid lysis of the cells and prevent either rapid joining of small fragments or endonucleolytic degradation of larger fragments. Andersen and Ganesan (1976) studied nascent fragment synthesis in protoplasts and used ethanol–phenol buffer to stop the pulse-labeling reaction. In these experiments the nascent fragments did not self-anneal to any measurable extent above background. This result favors discontinuous synthesis on only one strand.

To investigate whether the formation of Okazaki fragments in *B. subtilis* was an artifact due to the use of thymidine for pulse-labeling, Wang and Sternglanz (1974) compared the synthesis of replicative intermediates labeled with either thymine or thymidine. The majority of pulse label is found in small fragments regardless of which is used, but there is one difference in the labeling pattern. After pulse-labeling with thymine, a trichloroacetic acid-precipitable 2 S component is present. This component is 50–100 nucleotides in length, and its synthesis is inhibited by 6-(*p*-hydroxyphenylazo)uracil, a specific inhibitor of the replicative DNA polymerase in *B. subtilis* polymerase III (pol III). The 2 S component is chased slowly into higher-molecular-weight DNA, but it has not been conclusively shown that it is a true replicative intermediate and not a breakdown product.

Quite possibly, these 2 S DNA fragments may result from the excision of uracil from DNA, which occurs in *B. subtilis thy* strains. Tamanoi and Okazaki (1978) and Tamanoi *et al.* (1978) detected the accumulation of short DNA chains in thy^- but not thy^+ strains of *B. subtilis*. The *thy* mutation results in an increased incorporation of uracil into DNA. The uracil moiety is then excised by a uracil DNA glycosidase, causing chain cleavage. DNA from thy^+ cells incorporates very little uracil, so the repair of uracil incorporation in thy^+ cells does not hinder the analysis of *in vivo* replicative intermediates.

The possible incorporation of uracil does not explain the appearance of 2 S fragments only after pulse-labeling with thymine. It may, however, explain the inability of Sternglanz *et al.* (1975) to detect large pulse-labeled DNA fragments, since a thy^- strain was used. In a thy^+ strain, Tamanoi and Okazaki (1978) detected both large and small pulse-labeled fragments, but the shorter fragments hybridized with both the H and L strands of *B. subtilis*. This result differs from their previous results in which the small fragments hybridized with the L strand (Okazaki *et al.*, 1970). The question of discontinuous replication on both daughter strands thus remains unresolved.

RNA is covalently bound to the 5′-end of nascent DNA chains in *B. subtilis* (Tamanoi *et al.*, 1977). The RNA-linked DNA pieces accumulate when DNA polymerase I (pol I) is inactivated, suggesting that this enzyme may function during removal of the RNA primer (Tamanoi *et al.*, 1977). However, no 5′ → 3′ nuclease activity has been shown to be associated with *B. subtilis* DNA pol I. DNA pol III has a low level of intrinsic RNase activity and can extend an RNA primer, implicating pol III in the removal of nascent RNA fragments and primer extension (Low *et al.*, 1976b). Since DNA replication proceeds normally in the presence of rifampicin, an inhibitor of RNA polymerase, the RNA primer is probably synthesized by either a heretofore unidentified enzyme similar to the primase product of the *E. coli dnaG* loci or by a modified RNA polymerase molecule.

F. Protein and RNA Synthesis Requirements in Initiation

In *E. coli,* amino acid starvation inhibits new initiation but does not affect the ability of the cell to complete ongoing rounds of replication (Maaløe and Hanawalt, 1961; Lark *et al.*, 1963). This result provided the foundation for the concept of the regulation of chromosome replication at the initiation step. Although the inability to initiate without protein synthesis is generally observed in *B. subtilis,* the results on the completion of replication during amino acid starvation have been conflicting. This may stem from the fact that the *B. subtilis* transformation system allows one to examine critically the state of the chromosome without depending on pulse-labeling or quantitation of the amount of radioactive label incorporated.

Anraku and Landman (1968) found that, when starved for tryptophan, the 168 *trp thy* strain completed replication. This was also observed by Winston and Matsushita (1976). *Bacillus subtilis* 168 *trp his* and 168 *phe* also contain completed chromosomes after amino acid starvation (Gillin and Ganesan, 1975). However, *thy*$^-$ derivatives complete only half of the preexisting replication forks (Gillin and Ganesan, 1975). Replication in 168 *thy leuA8 metB5* starved for leucine appears to stop at the terminus, but new initiations occur during the amino acid starvation, and these forks stop primarily in the last half of the chromosome (Copeland, 1969).

Greater than 80% of W23 cells starved for leucine contain incomplete chromosomes (Copeland, 1971a). The replication forks are randomly distributed (Copeland, 1971a), although preferred stopping sites may exist (Copeland, 1971b). Copeland (1969, 1971a,b) has proposed that a second regulatory circuit exists which acts to control DNA replication and is independent of the initiation event that requires protein synthesis. Since the effects of amino acid starvation seem to differ from strain to strain and are dependent on the amino acid that is lacking, the results should be interpreted cautiously. Protein synthesis is certainly required for initiation (Yoshikawa, 1965), and in general, *B. subtilis* completes preexisting replication forks in the absence of protein synthesis.

RNA synthesis is also required for initiation (Laurent, 1973). The possibility of an RNA origin primer linked to DNA has been presented (Laurent, 1979) but remains unconfirmed. The product of the obligatory transcriptional step during initiation could play a structural role as well (Murakami *et al.,* 1976a). Elongation is insensitive to the RNA polymerase inhibitors rifampicin and streptolydigin, but the synthesis of nascent DNA chains is primed by small RNA oligonucleotides (Tamanoi *et al.,* 1977). The role of RNA during initiation is discussed more fully in Section III,A.

III. Genetic Analysis of Replication

A. Initiation-Defective Mutants

A list and brief description of the mutations that affect DNA replication in *B. subtilis* has recently been published (Mazza and Galizzi, 1978). Only more recent studies and studies that have contributed to our knowledge of the physiology of DNA replication will be considered here. The map positions of the mutations described in this chapter are presented in Fig. 1.

The fact that *de novo* protein synthesis is required for *B. subtilis* initiation has spurred the search for mutations in the genes coding for intitiation proteins (Yoshikawa, 1965). The most well-studied initiation mutations have been those that map in the *dnaB* locus between *polA* and *citF* (Karamata and Gross, 1970). Included in the *dnaB* group are the 168 mutants *dnaB151* and *dnaB134* isolated by Mendelson and Gross (1967), *dna-37* isolated by Laurent and Vannier (1973), *dna-1* and *dna-17* isolated by White and Sueoka (1973), *dna-103* isolated by Imada *et al.* (1980), NB62 from N. Brown's collection, and *dna-27* and 10 *dna ind* mutants isolated by Murakami *et al.* (1976a,b). A W23 mutant, *dna-20,* defective in the initiation of replication, also maps at the *dnaB* locus (Upcroft *et al.,* 1975). Three other loci that affect initiation have been found: *dna-808, dna-809,* and *dna-199* have been classified as initiation mutants and map between *purA* and *sacA* (Hara and Yoshikawa, 1973). On the basis of the amount of residual DNA synthesis at 45°C, the *dnaG*

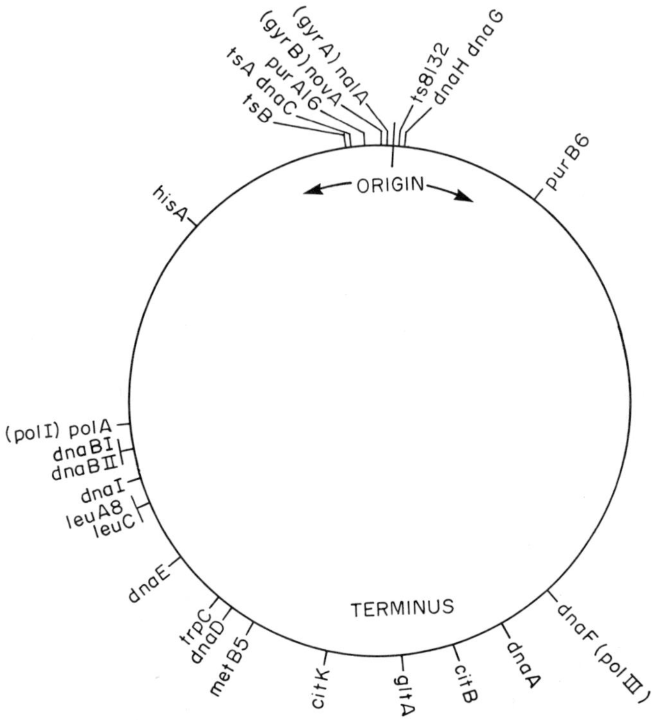

Fig. 1 Map positions of mutations affecting DNA replication in *B. subtilis*.

mutation, which maps near the origin, may also be initiation-defective (Shivakumar and Dubnau, 1978a). Another mutant originally described as *dnaP* (Riva *et al.*, 1975) affects only the initiation of DNA replication and maps in the gene coding for DNA pol III (Love *et al.*, 1976; Attolini *et al.*, 1976).

The general criteria for classification of *dna* mutants as initiation-defective is the continued incorporation of a radioactively labeled DNA precursor for a limited period of time after the cells have been shifted to the nonpermissive temperature. RNA and protein syntheses proceed normally. The residual DNA synthesis continues until all rounds of replication already in progress have been completed. The amount of increase in the total amount of radioactive label incorporated is dependent on the number of replication forks present at the time of temperature shift. With cells having two replication forks, the expected increase in the amount of incorporated label is 39%, and for dichotomously replicating chromosomes with six forks, the increase is 84% (Sueoka and Yoshikawa, 1965).

Since the residual synthesis of DNA at the nonpermissive temperature could be a result of anomalous replication or incomplete inhibition of DNA synthesis, more stringent experimental criteria have been developed for defining initiation mutants. Using classic density transfer experiments and genetic marker distribution, White and Sueoka (1973) developed a theoretical analysis of the behavior of chromosome replication and marker transfer after inhibition of initiation and after release from the inhibition of initiation. Basically, their theory states that an exponentially growing cell culture will transfer all markers equally to the hybrid region when a DNA density label is added to the growth medium. If the density label is added at the time of inhibition of initiation, a gradient of transfer will exist, with markers nearest the origin showing the least amount of transfer. The amount of transfer of an origin marker is strictly dependent on its location relative to the origin, assuming the inhibition is immediate upon shift to the nonpermissive temperature. The theoretical kinetics of density transfer with and without an initiation block, and the effect of a marker's position and the number of replication forks per chromosome on the transfer of that marker, have been presented (White and Sueoka, 1973, Appendix).

The density transfer test for initiation has been applied in the study of *dna-1* (White and Sueoka, 1973), *dna-134* (Callister and Wake, 1977), *dnaB19* (Winston and Sueoka, 1980a), *ts-37* (Laurent, 1973), *dna-20* (*ts*) (Upcroft *et al.*, 1975), *dna-27* (Murakami *et al.*, 1976a), and *dna-103* and *dna-62* (Imada and Sueoka, unpublished), and all these mutants are strictly initiation-defective. Genetic transformation alone is also quite useful for identifying initiation mutants. The ratio of origin markers to terminus markers in exponentially growing cells is at least 2:1 (Yoshikawa and Sueoka, 1963a). A decrease in this ratio to 1:1 indicates termination of all current rounds of replication with no new initiation. This criterion, along with the kinetics of DNA synthesis, has been used to classify *dna-199, dna-808,* and *dna-809* (Hara and Yoshikawa, 1973), and the 10 *dna ind* mutants (Murakami *et al.*, 1976a), as initiation-defective.

The inhibition of initiation in *dnaB134* and *dna-20* (*ts*) cells is not a result of incomplete termination (Burnett and Wake, 1977). At the nonpermissive temperature, normal division septation occurs in *dnaB134* and *dna-1* cells (Callister and Wake, 1974). The action of the *dnaB134* gene product is required for the initiation of replication during the germination of spores (Callister *et al.*, 1977), although the gene product is more temperature-resistant inside the spore. Increased rates of RNA synthesis have been observed at the nonpermissive temperature in four *dnaB* mutants, *dnaB134, dna-1, dna-20,* and *dnaB19* (Sargent, 1977). As a result of the increased RNA synthesis, the overall rate of protein synthesis also increased. These data clearly show that mutations at the *dnaB* locus directly affect the initiation process, and that the inhibition of initiation at the nonpermissive temperature is not a result of decreases in macromolecular synthesis or perturbations in other physiological processes associated with normal cell growth.

Two mutants, *dna-37* (Laurent, 1973) and *dna-27* (Murakami *et al.*, 1976a), have been used to demonstrate an obligatory transcriptional step during the initiation of replication. After incubation at the nonpermissive temperature, reinitiation in *dna-37* is sensitive to the RNA polymerase inhibitors streptolydigin and rifampicin, but not to chloramphenicol (Laurent, 1973). Identical results were obtained using *dna-27* (Murakami *et al.*, 1976a).

Murakami *et al.* (1976a) have concluded from their studies that the *dna-27* gene product helps to stabilize the transcriptional product required for initiation. They propose that the *dna-27* gene product is part of the primary initiation complex. This complex is a precursor of the secondary initiation complex in which an RNA molecule participates. Synthesis and proper insertion of the RNA moiety allow initiation to occur. The primary initiation complex is a stable intermediate in the process of accumulating the potential to initiate, while the RNA component turns over rapidly and becomes stable only after its proper insertion into the complex. Laurent (1974) has proposed a similar model for the buildup of initiation potential. In this model a second initiator component, which is synthesized at a specific time in the cell cycle, interacts with the first initiator component, which is synthesized continually before initiation occurs. Possibly this second component is the initiator RNA molecule, and it is this step that is rate-limiting for initiation.

De novo synthesis of the thermolabile product of *dna-37* is not required for reactivation of initiation after returning the cells to the permissive temperature (Laurent, 1973). Five other *dnaB* mutations—*dna-1, dna-103, dna-51, dna-60* (Imada and Sueoka, unpublished), and *dna-27* (Murakami *et al.*, 1976a)—also do not require *de novo* protein synthesis for initiation after shifting the cells to the permissive temperature. These five mutations are all at the same locus. The *dna-1, dna-103, dna-51,* and *dna-60* mutations have no detectable ($<10^{-5}$) recombination index among them, and all are quite close to, but distinct from, the *dna-27* mutation (Imada *et al.*, 1980). The precise map location of *dna-37* relative to these five mutations is not known, but it most probably maps in the same region.

Whether the *dnaB* locus is actually comprised of two separate genes or one large gene remains to be decided. Imada *et al.* (1980) have divided the *dnaB* locus into *dnaBI* and *dnaBII* on the basis of the recombination frequency (0.25) between mutations that map at the two subloci. The mutations *dna-1, dna-27, dnaB134,* and seven others have been designated *dnaBI,* whereas *dnaB19* is located in *dnaBII.* The two subloci can be distinguished by their physiological effects. Although both inhibit initiation at the *B. subtilis* chromosome, replication of the *Staphylococcus aureus* plasmid pUB110 is inhibited only by the *dna-1* mutant and not by the *dnaB19* mutant (Shivakumar and Dubnau, 1978a; Winston and Sueoka, 1980a). This is discussed in more detail in Section II, F,1. Murakami *et al.* (1976b) have found that the 10 *dna ind* mutations are closely linked (recombination indexes of approximately 0.2) to *dna-27.* The *dna ind* mutants are phenotypically different from *dna-27.* All the mutants are defective in chromosomal initiation at 47°C, but SP02 prophage induction at 47°C by HPura is affected only in *dna ind* mutants and not in *ts-27.* Either there are two separate loci in the *dnaB* region that are involved in the regulation of initiation, or a multifunctional *dnaB* protein can have one or both of its functions altered depending on the location of the mutation. The development of an *in vivo* complementation system in *B. subtilis* is essential for distinguishing between these two possibilities.

B. Elongation-Defective Mutants

Although many mutants that affect DNA replication have been isolated, only three of the proteins involved in the elongation process have been identified (Mazza and Galizzi, 1978). These three are ribonucleoside reductase (*dnaA*), DNA pol III (*polC*), and DNA gyrase (*nalA* and *novA*).

DNA (pol III is the major replicative enzyme in *B. subtilis* (Brown *et al.,* 1972; Cozzarelli and Low, 1973; MacKenzie *et al.,* 1973; Gass and Cozzarelli, 1973; Bazill and Gross, 1973). This enzyme has been extensively studied because of its specific inhibition by 6-(arylhydrazino)pyrimidines such as HPura (Brown, 1971; Neville and Brown, 1972; Bazill and Gross, 1972; Cozzarelli *et al.,* 1975). The mechanism of action of arylhydrazinopyrimidines has recently been reviewed (Cozzarelli, 1977).

Genetic mapping of mutations that have differential effects on pol III have proven most interesting in terms of function–structure relationships. The *dnaF* mutation originally isolated by Karamata and Gross (1970) maps in the *pyr–thyA* region of the chromosome. The *dnaP* mutation isolated by Riva *et al.* (1975) maps very close to *dnaF* (Attolini *et al.,* 1976). At nonpermissive temperatures these mutants have decreased levels of pol III, inhibit initiation but not elongation, maintain pol III activity *in vitro*, and are sensitive to HPura. The mapping data and phenotypes of the mutants indicate that they are at the same locus, so the *dnaP* mutation is now designated *dnaF69* (Attolini *et al.*, 1976). Since *dnaF* mutants affect the level of pol III in cells and are closely linked to the structural gene for pol III (Love *et al.*, 1976); Attolini *et al.* have proposed that the *dnaF* locus represents one region of the pol III polypeptide. This region of pol III may interact specifically with the initiation complex, because mutations in this region do not affect the polymerization activity of pol III.

During spore germination the amount of pol III in the cell increases from no detectable amount to levels at the time of initiation that are about one-sixth the amount found in log-phase cultures (Ciarrocchi *et al.*, 1977). The levels of pol I are high in spores and remain constant through germination.The coincident increase in the amount of pol III and the ability to initiate in spores suggests that pol III may be one of the limiting factors during initiation. Thus, pol III is probably responsible for the synthesis of the first deoxyribonucleotides at the origin.

The *dnaF* mutation is closely linked to four other mutations, *polC25*, *polC26*, *azp-12*, and *mut-1*, that effect the levels and activities of pol III (Love *et al.*, 1976). The first three mutants are resistant to arylhydrazinopyrimidines and are conditionally lethal at the nonpermissive temperature because of a lack of DNA synthesis (Cozzarelli and Low, 1973). The *polC25* and *polC26* mutants significantly decrease the expression of pol III activity (Gass and Cozzarelli, 1973). The *mut-1* mutant has a mutator phenotype at the permissive temperature (Bazill and Gross, 1973), and the pol III activity in crude extracts from this strain is low compared to wild-type activity, as well as being temperature-sensitive *in vitro*. The *polC25* locus is also highly mutagenic. The *azp-12* mutation maps in the middle of the *polC* locus with *dnaF*, *polC25*, *polC26*, and *mut-1* on either end. The active polymerization site is represented by the *azp-12* mutation, whereas the *polC25*, *polC26*, and *mut-1* mutations may represent the region of the *polC* locus involved in the regulation of gene expression and possibly proofreading functions (Attolini *et al.*, 1976), although the $3' \rightarrow 5'$ exonuclease associated with pol III appears normal in terms of general single-stranded exonuclease activity (Low *et al.*, 1976a).

In protoplasts, nascent fragment synthesis is slowed when pol III activity is partially inhibited by HPura (Andersen and Ganesan, 1976). Complete inhibition of pol III by HPura stops elongation of nascent fragments. Neither DNA pol I nor DNA polymerase II (pol II) can extend nascent DNA fragments. In addition, in the absence of active pol III enzyme, there is no gap filling and ligation of nascent fragments to form high-molecular-weight DNA. Thus, unlike that of *E. coli,* pol I of *B. subtilis* may play no role in discontinuous synthesis. Studies on DNA replication *in vitro* using *polA5* cells that lack pol I demonstrate that semiconservative replication occurs in the absence of pol I and that the DNA synthesized is biologically active (Ganesan *et al.*, 1973b). However, DNA replication *in vivo* and *in vitro* is inhibited in another *polA ts* mutant isolated by Bazill (Ganesan *et al.,* 1973a). Also, accumulation of RNA-linked DNA fragments has been seen in *polA ts* mutants of *B. subtilis* (Tamanoi *et al.,* 1977). Possibly these contrary results come from the different pol I mutants used. The *polA5* cells have no detectable activity of pol I, but the other *polA* mutants may still contain enough active enzyme for DNA replication. The question of whether pol I plays a role in DNA replication in *B. subtilis* remains unresolved.

Nalidixic acid and novobiocin inhibit DNA replication in *B. subtilis.* Recently, these two inhibitors have been shown to affect directly the two subunits of DNA gyrase in *E. coli* (see the review of Cozzarelli, 1980), and these results have been extended to *B. subtilis* (Sugino and Bott, 1980). The *nalA* and *novB* loci code for the *gyrA* and *gyrB* subunits, respectively. The DNA gyrase of *B. subtilis* maintains activities nearly identical to those ascribed to the *E. coli* enzyme. These include DNA supercoiling, relaxation of DNA coiling, double-stranded DNA cleavage in the presence of oxolinic acid, and DNA-binding properties. In *B. subtilis, nalB* and *novB* map quite closely, whereas in *E. coli* they are separated by a fairly large distance.

As mentioned previously, *dna199, dna808,* and *dna809* (*tsB* group) isolated by Hara and Yoshikawa (1973) map near the origin (close to *purA16*) and are initiation-defective. Another series of mutants classified in the *tsA* group, *dna604, dna56, dna810,* and *dna8133,* stop DNA synthesis immediately at the nonpermissive temperature (Hara and Yoshikawa, 1973). Transduction and transformation mapping (Karamata and Gross, 1970; Andersen and Ganesan, 1975) indicate that the *dnaC* locus is close to *purA16,* and most probably the *dnaC* loci and *ts8133* are mutations at the same loci. What may prove to be interesting is that the recombination frequency between the *dnaC tsA* markers and the *tsB* markers is from 0.083 to 0.455 (Hara and Yoshikawa, 1973). On the basis of recombination frequencies obtained by Carleton (1966) during fine mapping of the tryptophan operon, there may be either one large-sized gene or two closely linked genes in the *tsA–tsB* region. The *dnaC tsA-tsB* mutants may produce a product like the *dnaB* gene product of *E. coli* (Wickner, 1978; Tomizawa and Selzer, 1979). In *E. coli, dnaB* mutants can either be initiation- or replication-defective (Zyskind and Smith, 1977), because this protein is an integral part of both processes.

Careful analysis of *dnaC* mutants has been made (Andersen and Ganesan, 1975), and these mutants are replication-defective. Further mapping and characterization of the other *dna* mutants in this region are in order before it can be definitely decided whether this region near the origin contains one or two genes that function during both initiation and replication. Once again, an *in vivo* complementation system is needed in *B. subtilis*.

One mutant, *ts8132,* cannot be strictly classified as either an initiation or replication mutant (Hara and Yoshikawa, 1973). This mutant continues to synthesize DNA at high temperature up to 50% of the initial amount of DNA, but the *purA/metB* ratio remains high during growth at 48°C. Cells containing the *ts8132* mutation are sensitive to radiation and chemical damage of DNA, and the cells have a high spontaneous reversion frequency (Sadaie and Narui, 1976). DNA precursor synthesis is normal. The *ts8132* gene product may interact with the replicative and repair machinery of the cells. Possibly the phenotype of the mutant could be a result of the inability of these cells to repair DNA damaged by radiation and error-prone replication. This mutation maps very close to the origin of replication and is in close proximity to *dnaH, dnaG, recD,* and *recG* (Trowsdale *et al.,* 1979).

C. Termination of Replication

The latest replicating markers during a synchronous round of DNA replication in *B. subtilis* map in the *citK–gltA–citB* region of the chromosome (Harford, 1975; O'Sullivan *et al.,* 1975b; Hye *et al.,* 1976). Termination of replication most probably occurs in this area of the chromosome. Wake (1973) has shown by autoradiography that symmetrical bidirectional replication occurs over 90% of the length of the chromosome.

It is not known whether termination in *B. subtilis* is a result of the meeting of two replication forks that begin at the same time and travel with equal rates, or if there exists a specific termination site at which forks cease movement. O'Sullivan and Anagnostopoulus (personal communication) have used merodiploid strains of *B. subtilis* to determine whether large DNA inserts into the chromosome change the position of the terminus. Density transfer experiments were performed with a synchronized merodiploid strain constructed by Anagnostopoulus and Trowsdale (1976), which contains a duplicated region of the chromosome extending from *thr* to *aro*. This duplication increases the size of the chromosome by about 20%. In this strain, the latest marker to be replicated was *gltA,* which was previously shown to be the last marker replicated in nonmerodiploid strains (Harford, 1975; Hye *et al.,* 1976).

Thus, these preliminary results indicate that a fixed termination site exists near the *gltA* marker.

There are no known mutants that specifically affect termination. Although the kinetics of DNA replication in a termination mutant would be similar to that of an initiation mutant, spores offer the unique opportunity to differentiate between the two types of mutants. Termination mutants could initiate and complete the initial round of replication during germination of spores at the nonpermissive temperature, whereas initiation mutants could not (Mendelson, 1968). The *dnaG* mutant should be examined in this manner, since it may affect termination (Shivakumar and Dubnau, 1978a). However, this screening procedure would select for segregation and cell division mutants as well.

As a means of keeping replication in time with cell division, termination of replication could signal the onset of septum formation. This does not seem to be the case in *B. subtilis*. Inhibition of replication did not prevent cell division in *B. subtilis,* and enucleated cells resulted (Donachie *et al.,* 1971). Enucleate cells are formed after replication is inhibited in *dnaB134* cells at 45°C by thymine starvation or nalidixic acid treatment (Sargent, 1975b). Inhibition of replication by HPura after 70% of the chromosome has been replicated in synchronized *dnaB134* and *dna-1* cells also does not inhibit central division septation (McGinness and Wake, 1979). Inhibition of replication before that time inhibits septation. The absence of termination and the formation of the septum in a central position require the septum to pass through the partially replicated chromosome (McGinness and Wake, 1979), and this has been observed by electron microscopy (Van Iterson and Aten, 1976). It can be concluded from these results that termination of replication is not the signal for cell division. Rather, a developmental sequence leading to cell division and synchronized with DNA replication commences at an earlier time than termination (Sargent, 1975b; McGinness and Wake, 1979).

IV. Replication *in Vitro*

A. Toluene-Treated Cells

The study of DNA replication *in vitro* has greatly contributed to current understanding of the DNA elongation process. In *B. subtilis,* the most well-studied *in vitro* system has been toluene-treated cells. This system was first developed in *E. coli* (Moses and Richardson, 1970) and adapted to *B. subtilis* by Matsushita *et al.* (1971).

The basic criteria for an *in vitro* system that closely resembles the *in vivo* state are (1) semiconservative DNA replication, (2) ATP-dependent synthesis, (3) normal movement of the replication fork, (4) synthesis of biologically active DNA, and (5) a normal effect of *dna* mutants on the replication process (Matsushita *et al.*, 1971). Basically, all these criteria have been met in toluene-treated *B. subtilis*.

Density transfer experiments show that in the presence of all four triphosphates (BrdUTP replacing dTTP) and ATP, DNA replication is semiconservative and completely dependent on ATP. On the basis of transformation assays, the ATP-dependent DNA synthesized *in vitro* is biologically active. When synchronized cells are toluene-treated and allowed to replicate, replication proceeds sequentially from the origin. From the amount of the origin marker, *purA16,* transferred in toluene-treated synchronized cells, it appears that the number of forks that are elongated is equal to the number that had been replicating in a parallel *in vivo* sample. Although these studies indicate that replication proceeds normally, the *in vitro* rate of synthesis is approximately one-tenth the normal rate, and only a small amount of synthesis equal to about 10% of the size of the whole chromosome is achieved. This limitation has not been improved upon in more recent studies.

DNA synthesis *in vivo* is resistant to chloramphenicol and/or rifampicin treatment. Replication *in vitro* is also resistant to these macromolecular inhibitors (Sueoka *et al.*, 1973). Ribonucleoside triphosphates (other than ATP) are not required *in vitro,* suggesting that either endogenous pools of nucleoside triphosphates are sufficient or they are regenerated *in vitro*. Addition of nucleoside triphosphates increased synthesis to a small extent at later times of incubation. DNA synthesis *in vitro* is sensitive to HPura (Brown *et al.*, 1972). DNA synthesis both *in vivo* and *in vitro* is thus carried out by pol III. Replication is also inhibited *in vitro* by *p*-chloromercurial benzoate (PCMB) and *N*-ethylmaleimide (NEM) (Sueoka *et al.*, 1973).

A localization test has been devised to distinguish between cell-associated DNA synthesis and synthesis that may occur free of the cell (Matsushita and Sueoka, 1974). Over 80% of the ATP-dependent incorporation of radioactive thymine is cell-associated. If toluene-treated cells are subjected to Brij-58 treatment, ATP-dependent incorporation is decreased and is no longer cell-associated. Thus, the integrity of the replication complex must be maintained *in vitro* for proper replication. Toluene-treated cells are permeable to DNAse I (MW 31,000) but not to pol I (Matsushita and Sueoka, 1974).

No new initiation of replication forks can be detected in toluene-treated cells (Sueoka *et al.*, 1973; Winston and Matsushita, 1975). Removal of toluene does not result in the recovery of initiation, so toluene does not directly inhibit initiation (Winston and Matsushita, 1975). Protein synthesis in toluene-treated cells is only 10% of the *in vivo* rate. Since protein synthesis is required for initiation *in vivo,* its decrease *in vitro* may contribute to the lack of initiation *in vitro* (Winston and Matsushita, 1975).

B. Azide Detergent-Treated Cells

Another *in vitro* replication system using *B. subtilis* has been developed by Ganesan (1971; Ganesan *et al.,* 1973a). Azide-poisoned cells become nonviable within 20 min. Treatment of these cells with Brij-58 results in a loss of 60% of the proteins into the supernatant. A substantial amount (90%) of the DNA pol I is also released from the cell. After the addition of ATP and the four dNTPs, DNA replication commences. The DNA is synthesized semiconservatively and is sensitive to the sulfhydryl inhibitors NEM and PCMB. Replication proceeds sequentially along the chromosome, and the DNA synthesized has transforming activity. However, on the basis of density transfer experiments, up to 70% of the incorporation observed *in vitro* is attributable to repair synthesis. The high amount of DNA repair *in vitro* is greatly decreased by using a pol I mutant, *polA5*. With the *polA5* mutant, it can be shown that up to 15% of the chromosome is replicated in a semiconservative fashion. As in toluene-treated *B. subtilis,* the semiconservative DNA replication is ATP-dependent. γ-Labeled ATP is not incorporated into DNA, so it probably functions as an energy source for unwinding DNA during replication. Both pronase and DNase can enter the permeable cells. The synthesis is also stimulated by the addition of all ribonucleoside triphosphates. Nalidixic acid inhibits replication *in vitro*. Initiation of new rounds of DNA synthesis does not occur in this system.

C. Freeze-Thawed Cells

Bacillus subtilis cells that have been frozen in liquid nitrogen and then thawed incorporate TMP, TDP, TTP, as well as thymidine (Billen *et al.,* 1971a). The incorporation of thymidine represents semiconservative replication (Billen *et al.,* 1971b). The incorporation of deoxyribonucleotides is indicative of DNA repair. The nucleotides do not compete with thymidine for incorporation during replication, and this suggests that the replication apparatus is compartmentalized and inaccessible to phosphylated compounds (Billen *et al.,* 1971b). However, it has not been demonstrated that cells that incorporate thymidine are also capable of incorporating nucleotides. Intact viable cells may be the only cells in the population that incorporate thymidine and replicate DNA. Thymine is incorporated in minimal amounts by freeze-thawed cells (Billen *et al.*, 1971a). Since thymidine is rapidly converted to thymine in *B. subtilis* (Mastromei and Riva, 1975), it is not readily apparent why thymine is not incorporated.

V. Plasmid Replication

A. Replication of *Staphylococcus aureus* plasmids in *Bacillus subtilis*

Although a number of naturally occurring plasmids in *B. subtilis* have been identified (Lovett and Dramucci, 1975; Tanaka and Koshikawa, 1977; Lehegarat and Anagnostopolous, 1977), very little is known about the replication of these plasmids in *B. subtilis*. The only plasmids whose replication in *B. subtilis* have been examined are *S. aureus* plasmids (Ehrlich, 1977) and chromosomal-origin-containing plasmids constructed using recombinant DNA technology.

The most well-studied plasmid in *B. subtilis* is the *S. aureus* plasmid pUB110. This plasmid maintains a copy number of 40 in *B. subtilis* and expresses resistance to kanamycin or neomycin (Gryczanetal, 1978). The effect of various *dna ts* mutations on pUB110 replication has been examined. The plasmid does not require the *dnaA13, dnaB19, dnaC6, dnaD23, dnaE20,* and *dnaI102* gene products (Shivakumar and Dubnau, 1978a; Winston and Sueoka, 1980a). The replication of pUB110 is dependent on DNA pol III, as plasmid replication is inhibited at the nonpermissive temperature in *polC*-defective strains (Shivakumar and Dubnau, 1978a) or in the presence of HPura (Winston, unpublished observations). Pol I-deficient cells do not prevent plasmid replication. The plasmid fails to replicate in *dna-1* (Winston and Sueoka, 1980a) *dnaG,* and *dnaH* cells (Shivakumar and Dubnau, 1978a) at the nonpermissive temperature. In *dnaG* cells at 45°C, pUB110 accumulates in a molecular form with a much lower electrophoretic mobility, suggesting that this mutant may affect termination (Shivakumar and Dubnau, 1978a). The kinetics of DNA synthesis in this mutant are not inconsistent with this possibility, but further characterization of this mutant is necessary. The *dnaH* mutation affects elongation of the host chromosome as well as pUB110 replication. The *dnaH* gene product has not yet been identified. This mutation maps close to the replication origin (Trowsdale *et al.*, 1979).

The *dnaA* gene codes for ribonucleotide reductase in *B. subtilis* (Bazill and Karamata, 1972). Hydroxyurea inhibits this enzyme. Neither the *dnaA ts* mutation nor hydroxyurea affects pUB110 replication in the presence of deoxyadenosine and thymidine, but chromosomal replication is rapidly inhibited. Since DNA replication in toluene-treated cells is also inhibited by hydroxyurea in the presence of all the required precursors, the reductase must function in some role during chromosomal replication other than precursor synthesis, and this function is not required for pUB110 synthesis (Shivakumar and Dubnau, 1978b).

Replication of pUB110 in *B. subtilis* is random (Winston *et al.*, 1980a,b). As defined by Rownd (1969), the mode of replication is random when the DNA molecule to be replicated is randomly selected from the entire population regardless of whether it has been previously replicated during that generation. By the end of one generation, the theoretical distribution of plasmid in cesium chloride gradients after a density transfer experiment is 25% of the plasmid in LL (parental), 50% in HL (hybrid), and 25% in HH (newly replicated). This distribution is observed for pUB110 isolated from exponentially growing *B. subtilis* cells. The random mode is common for high-copy number plasmids in bacteria.

B. Isolation and Replication of *oriC*-Containing Plasmids

Plasmids make good model systems for studying the initiation and replication of DNA. To gain insight into bacterial DNA replication, the most useful plasmids are those that use all the host *dna* gene products. In terms of initiation, plasmids that contain and use the chromosomal origin for initiation are most worthwhile to study. To this end, *B. subtilis* *oriC*-containing plasmids have been constructed. It should be noted that, at the time of writing, three separate groups have identified what appear to be different *Eco*RI restriction fragments which may contain the *B. subtilis* origin.

Bott (1980) has isolated a self-replicating *Eco*RI restriction fragment that confers nalidixic acid resistance (Nal^r). The Nal^r phenotype is not satisfactory for primary selection, so this 5-kb fragment has been cloned into a plasmid, pCS540, that carries a chloramphenicol resistance marker (Cm^r) and lacks a functional origin for replication in *B. subtilis*. The Nal^r fragment confers on pCS540 the ability to replicate in *B. subtilis* and is more stable than the circularized Nal^r fragment by itself. This fragment hybridizes to an *Eco*RI fragment of identical size from an *Eco*RI digest of whole *B. subtilis* 168 DNA from which it was derived (Bott, 1980). Surprisingly, the 5-kb fragment hybridizes to different size *Eco*RI fragments from different *B. subtilis* strains (Bott, 1980). For example, it hybridizes with an *Eco*RI fragment of approximately 20 kb from the 168 *trp thy* strain. Quite possibly, the many genetic manipulations by transformation and transduction in *B. subtilis* have resulted in localization of the replication origin in distinctly different microenvironments in various strains, although the location of the origin relative to more distant genetic markers remains approximately the same (Bott, 1980).

The restriction fragment containing the replication origin has been identified by pulse-labeling the DNA synthesized during synchronous initiation *in vivo*. During a short pulse-labeling of DNA from either germinating spores synchronized by thymine deprivation or synchronized initiation-defective cells the origin fragment will have the highest specific radioactivity. A restriction map of the *B. subtilis* origin has been constructed using DNA of pulse-labeled origin isolated from 168 *leu trp thy* cells (Seiki *et al.*, 1979). An *Eco*RI fragment of 1.7 kb has the highest specific radioactivity, indicating that this fragment contains the replication origin (Ogasawara *et al.*, 1979). Using 168 *trp thy* cells, Winston and Sueoka (unpublished) found an *Eco*RI fragment of 9.4 kb to have the highest specific radioactivity. These two fragments have not yet been directly compared. It is not known whether the 1.7-kb fragment has self-replicating properties. The DNA fragment carrying the replication origin of *B. subtilis* chromosome has not been unequivocally identified yet. It is likely that one of the three candidates described above will turn out to be the true origin. *B. subtilis* isolation of *ori C*-carrying fragment as a self replicating unit has not been accomplished in spite of the fact that several laboratories, including our own, have made extensive efforts. This may result from a strong interference between the chromosome and the *ori C*-carrying plasmid, resulting in strict incompatability.

VI. DNA–Membrane Association

A. Characterization of DNA–Membrane Complexes

One of the postulates put forth by Jacob *et al.* (1963) in their replicon hypothesis was that DNA–membrane attachment served to regulate replication and ensure proper segregation of daughter chromosomes. This idea gained early support with the demonstration by electron microscopy of the association of the nuclear body in *B. subtilis* with membrane invaginations termed "mesosomes" (Ryter and Jacob, 1963). Further evidence for DNA–membrane association in *B. subtilis* was the finding of nascent DNA bound to membranous particles by analyzing the amount of pulse-labeled [^{3}H]thymidine found in the membrane fraction (Ganesan and Lederberg, 1965).

The idea of controlling initiation via DNA–membrane associations gained further support with the discovery that DNA near the origin of chromosomal replication was specifically enriched in DNA–membrane fractions isolated from *B. subtilis* (Sueoka and Quinn, 1968; Snyder and Young, 1969). By assaying DNA that was either membrane-bound or membrane-free for transforming activity of markers at various locations on the chromosome, it was determined that the amount of transforming activity for an origin marker (*purA16*) was enriched in the membrane fraction relative to other markers such as *leuA8*. DNA pulse-labeled at the origin was also enriched in the DNA–membrane fraction and remained so upon chasing. DNA–membrane fractions isolated by centrifugation of *B. subtilis* lysates in renografin density gradients are also enriched for DNA near the replication origin (Ivarie and Pené, 1970).

Low-salt sucrose gradients were first used in the separation of membrane-bound and membrane-free DNA (Ganesan and Lederberg, 1965; Sueoka and Quinn, 1968). More recently, a *high salt-resistant* DNA–membrane complex in *B. subtilis* has been isolated (Sueoka and Hammers, 1974). Centrifugation of *B. subtilis* cell lysates in cesium chloride (5–0.5 *M*)–sucrose (20–25%) linear gradients produces two DNA–membrane fractions. These fractions, M_2 and M_1, are both identically enriched in markers near the chromosomal origin and terminus. M_2 and M_1 differ in their protein/DNA ratios, and this is probably the reason for their separation. The fact that M_2 has a lower protein/DNA ratio and appears to be comprised of a subset of proteins found in M_1 suggests that M_2 represents a more specific fraction of the membrane complexing with DNA than the whole membrane (M_1).

The M_2 fraction is sensitive to detergents such as Sarkosyl, Tween 80, and sodium deoxycholate but is not affected by RNase, phospholipase A, or Triton X-100 (Imada *et al.*, 1975). This fraction can also be iodinated *in vivo*, making it likely that some part of the M_2 fraction is exposed to the outside of the membrane (Imada *et al.*, 1975). The proteins iodinated in M_2 are also iodinated in M_1, further supporting the notion that M_2 is derived from the bulk membrane M_1 fraction.

Several proteins are found to be enriched in the M_2 DNA–membrane fraction isolated from *B. subtilis* by cesium chloride–sucrose gradient centrifugation (Sueoka and Hammers, 1974; Imada *et al.*, 1976). Harmon and Taber (1977a) also observed several unique proteins in DNA–membrane complexes purified from *B. subtilis*. However, these complexes contain many other DNA fragments different from origin-region fragments, which has made it quite difficult to establish specific DNA–protein interactions. Toyoda *et al.* (1978) used a sucrose gradient free-flow electrophoresis technique to purify a specific DNA–membrane complex from the bulk membrane fragments of *B. subtilis*. By using the *B. subtilis* transformation system, they were able to identify and further purify a complex that contained predominantly *purA16*$^+$ DNA and five unique proteins, three of which were tightly bound to the DNA. These proteins have not yet been correlated with any specific functions or loci involved in DNA initiation or replication. Complexes such as these are promising substrates for *in vitro* reconstitution experiments and for genetic analysis of proteins that interact with specific DNA sequences near the replication origin.

Yamaguchi and Yoshikawa (1977) have identified a *salt-sensitive* DNA–protein complex highly enriched for DNA near the origin in *B. subtilis*. Their lysis and centrifugation procedure results in the isolation of a bulk DNA–membrane complex (M), which is enriched for all markers near the origin, and a soluble complex (S). The M complex is *resistant* to high salt and gives rise to M_2 and M_1 fractions in cesium chloride–sucrose gradients (Sueoka and Hammers, 1974).

In contrast, the S complex is denatured at salt concentrations greater than 0.5 *M* and is enriched only for the *purA16* marker. The *purA16* marker is located approximately 3–4 $\times$ 10^7 daltons away from the *B. subtilis* origin (O'Sullivan and Sueoka, 1967). The S complex contains 4 $\times$ 10^7 daltons of unique DNA, but the replication origin is not found in the S complex (Yoshikawa *et al.*, 1978). It is not known whether the S complex is actually part of the bulk DNA–membrane complex *in vivo* or whether there are two different types of origin-region associations near the *purA16* marker. The amount of the S complex isolated is not dependent on the presence of detergents during lysis, which suggests that the S complex is not derived from the membrane fraction (Yamaguchi and Yoshikawa, 1977). On the other hand, many of the proteins that comigrate with the S complex are qualitatively similar to those found in the M complex, although the direct association of these proteins with the DNA has not been demonstrated (Yamaguchi and Yoshikawa, 1977; McKenzie, Winston and Sueoka, unpublished). Phospholipids and RNA are also associated with the S complex. Thus, the origin region exists in two states or consists of two components, one that is membrane-associated and salt-resistant, and another that is associated with protein, RNA, and phospholipid, is salt-sensitive, and contains only DNA in the *purA16* region. The relationship between M and S complexes and their role during initiation is further discussed in Section VI,B.

B. Initiation and DNA–Membrane Association

To demonstrate a direct role for origin membrane complexes during initiation, the effect of *dna ts* mutants on initiation and DNA–membrane association has been studied (Winston and Sueoka, 1980a). Along with examining the membrane association of the chromosomal origin marker *purA16,* the membrane association of pSL103 was also examined. This plasmid is a chimera containing the *S. aureus* vector pUB110 and an *Eco*RI restriction fragment that can complement a defect in tryptophan biosynthesis (*trpC*) in *B. subtilis* (Keggins *et al.,* 1978). Replication of pSL103 is via the pUB110 vector, which also can transform neomycin-sensitive *B. subtilis* to neomycin resistance. The majority of pSL103 is found in either the M fraction after cesium chloride–sucrose gradient centrifugation or in the M and S complexes after sucrose gradient centrifugation (Winston and Sueoka, 1980b; Yoshikawa *et al.,* 1980). The high enrichment of pSL103 in the membrane fraction and its easy detection by transformation make it an excellent molecule for studying DNA–membrane associations.

In *dna-1* (pSl103) cells at nonpermissive temperatures, both chromosomal and plasmid initiation are inhibited (Winston and Sueoka, 1980a). When DNA–membrane association was subsequently examined using both cesium chloride–sucrose and sucrose gradient centrifugation, it was observed that both *purA16* and pSL103 were greatly decreased in their membrane association. Both initiation and membrane association of the *B. subtilis* chromosome and pSL103 recover when the cells are returned to the permissive temperature. The *dnaB19* mutant has a differential effect on chromosomal and plasmid initiation and membrane association. In these cells, only initiation of the *B. subtilis* chromosome is inhibited, and only the *purA16* marker is decreased in its membrane enrichment. The effect of high temperature on M fractions and M and S complexes purified from mutant cells at the permissive temperature was also examined. At 45°C, *purA16* and pSL103 DNA are selectively released from complexes purified from *dna-1* cells. Only *purA16* DNA is released at 45°C from complexes purified from *dnaB19* cells. The consistent coupling between initiation and membrane association and the selective effect of initiation mutants on the origin DNA–membrane association are strong evidence for the critical role of origin DNA–membrane association during initiation in *B. subtilis*.

Proper conformation of DNA near the origin is probably important during initiation. Electron micrographs of loop structures in the S complex suggest that the S complex contains superhelical structures (Yamaguchi and Yoshikawa, 1977). Ethidium bromide markedly decreases the amount of DNA in the M and S complexes (Yoshikawa *et al.*, 1978, 1980). It destroys both the initiation potential in germinating *B. subtilis* and the M and S complexes at identical concentrations of intercalating dye. Replication and membrane association of pSL103 are similarly inhibited, although higher concentrations of dye are required. The effect of ethidium bromide is reversible, but *de novo* RNA synthesis and an active *ts27* (*dnaBI*) gene product are necessary for initiation. In the presence of the DNA gyrase inhibitor, novobiocin, a limited amount of initiation and elongation occurs (Ogasawara *et al.*, 1979). The amount of initiation is decreased if novobiocin is present before the initiation potential has formed. Thus, DNA gyrase is important in the formation of initiation potential and subsequent elongation from the origin. Novobiocin has a similar effect on Col E1 replication *in vitro* (Gellert *et al.*, 1976). The M and S complexes appear to be formed *de novo* in parallel with the cell's ability to initiate during spore germination (Yoshikawa *et al.*, 1978). After initiation, the S complex then begins to become more rapidly sedimenting, possibly because of its association with the cell membrane.

As described above, a general picture of the origin structure is beginning to emerge, but additional information will be required before the precise relationship among the M complex, the S complex, and initiation can be determined. Formation of the M and S complexes appears to be crucial for initiation, and the conformation of the M and S complexes and their interaction may serve to regulate initiation.

The effect of temperature on membrane proteins in *dna ts* mutants has also been analyzed. As mentioned previously, *dna-1* cells do not initiate new rounds of DNA synthesis at the nonpermissive temperature (White and Sueoka, 1973). A single protein of 35,000 MW is missing from the *dna-1* membrane fraction purified by cesium chloride–sucrose centrifugation at the nonpermissive temperature (Imada *et al.*, 1976). A protein of similar molecular weight is also missing from M and S complexes isolated from *dna-1* cells (McKenzie, Winston and Sueoka, unpublished). It has not yet been conclusively proven that the missing protein band represents the primary gene product of the *dna-1* locus.

Changes in the membrane protein patterns of initiation mutants must be interpreted cautiously in terms of a role of the membrane during initiation. Alterations of membrane proteins also occur in *B. subtilis dna-151 ts* cells at the nonpermissive temperature (Harmon and Taber, 1977b). The DNA–membrane fractions studied (M bands) were isolated on the basis of their affinity for Mg^{2+}–Sarkosyl crystals (Tremblay *et al.*, 1969). The M band contained three proteins that were not found in the whole membrane. One of these proteins (mac-2) accumulates in the M band at the nonpermissive temperature, but the accumulation occurs 3–4 hr after DNA synthesis has ceased. After mac-2 accumulates, it can be detected in the whole membrane. The protein is thus a membrane protein but is probably not involved in DNA replication. A 30,000-MW protein that is in the greatest quantity in both the M band and whole membrane decreases in relative concentration in *dna-151* cells at high temperature. From the kinetics of the disappearance of the protein, it is likely that it functions during segregation rather than initiation (Harmon and Taber, 1977a).

If the membrane association of origin in *B. subtilis* is required only for initiation, then the origin association may be cyclic, with the highest enrichment occurring at the time of initiation. However, DNA pulse-labeled at the origin remains associated with the membrane during successive generations (Sueoka and Quinn, 1968). The enrichment of a marker near the origin, *purA16,* has been examined during one synchronous round of replication (Beeson and Sueoka, 1979). No concomitant changes in the *purA16* enrichment with initiation are observed during the replication cycle. There is also no change in the amount of DNA that is membrane-associated during the replication cycle (Harmon and Taber, 1977b), but these experiments are not sufficiently sensitive in detecting site-specific changes in DNA–membrane association. The experiments of Sueoka and co-workers support a permanent association of the replication origin with the membrane, but this point should be further investigated using origin-containing plasmids.

C. *Membrane Association of the Replication Fork and Other Regions of DNA*

The first evidence for association in the replication fork with the membrane in *B. subtilis* was presented by Ganesan and Lederberg (1965). They found that a 2-sec pulse of [^{3}H]thymidine was enriched in a rapidly sedimenting complex over DNA uniformly labeled with [^{14}C]thymine. The enrichment of the pulse decreases with longer pulse times, and the pulse is removed from the membrane fraction by chasing with cold thymidine. The ^{3}H pulse in the membrane fraction is released by pronase. From these results, it appears that DNA synthesis occurs on the membrane and that, once replicated, the DNA moves away from the membrane-bound replication complex (Ganesan and Lederberg, 1965).

The membrane association at the replication fork may be different from the replication origin association. Yamaguchi *et al.* (1971) isolated both complexes in a membrane-bound form from cells lysed by lysozyme and Brij-58, but they found only the origin to be membrane-associated when membrane fractions were prepared from protoplasts. Protoplasts lack mesosomes, so this structure may be important to the integrity of the replication complex.

The enrichment of pulse-labeled DNA in the membrane fraction in *B. subtilis* is usually low (Sueoka and Quinn, 1968; Yamaguchi *et al.*, 1971). This may be due to the loss of Okazaki fragments from the membrane fraction during lysis and shearing (Hye *et al.*, 1976). The amount of enrichment is inversely related to the amount of single-stranded fragments released into the free fraction. The release of Okazaki fragments from membrane-bound T4 during isolation of the complex has also been reported (Leibowitz and Schaehter, 1975). The previous reports of pulse enrichment may then be underestimates of the actual amount of replication forks that are membrane-associated.

Membrane enrichment of the replication fork was not seen during one synchronous round of replication in *B. subtilis* (Beeson and Sueoka, 1979). In these experiments, the membrane fractions were isolated in cesium chloride–sucrose gradients, which should have eliminated DNA–membrane association mediated by ionic interactions. The replication fork binding is probably transient, so it is unlikely that, even if the replication fork is salt-resistant, the synchrony was good enough to detect binding of a genetic marker as it was being replicated. However, Fujita *et al.* (1973) reported that markers in *B. subtilis* showed increased enrichment in the low-salt membrane fraction as they were replicated.

Although the genetic transformation data in *B. subtilis* demonstrates the enrichment of origin and terminus markers in the membrane fraction, transforming activity for all markers are detected in the membrane fraction. Thus, there may be a number of nonspecific DNA–membrane sites. In fact, the number of attachment sites in *B. subtilis* has been estimated at 70–90 (Ivarie and Pene, 1973). Some of the DNA–membrane attachment sites may be related to transcription. Ryter and Chang (1975) detected a high percentage of transcriptional activity on the membranes of *B. subtilis* by using electron microscope autoradiography. Possible functions for the other sites of attachment include segregation of DNA during cell division or locations of DNA repair (Ivarie and Pene, 1973).

D. Membrane Association of the Replication Terminus

Sueoka and Quinn (1968) observed that a marker near the terminus of *B. subtilis* was also enriched in the DNA–membrane complex. The levels of enrichment were not as high as those seen with markers near the origin. The enrichment of the terminus was confirmed by Snyder and

Young (1969) and Ivarie and Pene (1970). The terminus marker used in all these studies was *metB5*, which is located at least 2×10^8 daltons away from the terminus (Rutberg and Hoch, 1970).

In light of the bidirectional mode of *B. subtilis* replication, Yamaguchi and Yoshikawa (1975) have examined the terminus region to determine if markers on both sides of the terminus are membrane-associated. Markers on either side of the terminus are found to be membrane-associated, and the region of the terminus associated with the membrane is approximately 4×10^8 daltons. The terminus region that is membrane-associated is then almost 10 times larger than the origin region that is membrane-associated. If the terminus is associated at a specific site, then a gradient of enrichment extending bidirectionally from this point may exist. Yamaguchi and Yoshikawa (1975) observed that all terminus markers were equally enriched. This supports their hypothesis that the large region of terminus association region is a result of a number of equally probable attachment sites within a large defined region of DNA. Hye *et al.* (1976) found that some *B. subtilis* terminus markers were more enriched than others, but they could not strictly correlate an increase in enrichment with the markers' relative position to the terminus. This may, in fact, stem from the difficulty in precisely locating the terminus. Horowitz *et al.* (1979) also noted that markers closer to the *B. subtilis* terminus were more enriched, but only a few markers were studied. Thus, specific sites of attachment close to the terminus are also possible, with other variable regions of association distal to the actual site of termination.

The role of the terminus–membrane association is much less well defined than that of the origin–membrane association. Jacob *et al.* (1963) proposed that chromosome segregation could be achieved by growth between the attachment sites of two daughter chromosomes. The most likely site for such growth would be between the origin–membrane attachment sites (Imada *et al.*, 1975). From the observed pattern of nuclear segregation in *B. subtilis* (Sargent, 1974), it was proposed (Sargent, 1975a) that surface growth and chromosome replication were coordinated by growth of the cell envelope between the replication origins. The termination event involves release of the terminus from the previous growth zone to the new growth zones of the daughter cells, with the previous growth zone serving as the site of septum formation. In support of Sargent's (1975a) model, it has been shown that, after the termination of replication, protein synthesis is required for release of the terminus from the membrane (Winston and Matsushita, 1976). This suggests that the origin, rather than the terminus–membrane binding, functions during segregation. However, there is no experimental data concerning the terminus–membrane association during actual segregation and cell division.

E. DNA–Cell Wall Association

DNA is tightly associated with the cell wall in *B. subtilis* (Brown *et al.*, 1976). DNA near the origin and terminus is selectively enriched in cell wall preparations, and the chromosomal area of enrichment is greater than that seen in membrane–DNA complexes (Doyle *et al.*, 1980). The cell wall–DNA fractions are composed of unique fatty acids and proteins. The presence of proteins and lipid in the cell wall fractions suggests that membranous particles may be involved in DNA–cell wall association. These specific complexes containing DNA, membrane, and cell wall could function as a segregation apparatus (Doyle *et al.*, 1980). In stable L forms of *B. subtilis*, which completely lack cell walls, the origin, but not the terminus, is membrane-bound (Horowitz *et al.*, 1979). The L forms characteristically undergo aberrant cell divisions, so there may be a relationship between cell division and terminus–membrane association. L forms can be simulated in wild-type *B. subtilis* cells by growth in high salt (Streips *et al.*, 1980). Under these conditions, plasmolysis occurs, and abnormal cell division is induced. Further work is needed to substantiate the role, if any, of the terminus–membrane association during termination and cell division.

References

Anagnostopoulos, C., and Trowsdale, J. (1976). *In* "Microbiology, 1976" (D. Schlessinger, ed.), pp. 44–57. American Society for Microbiology, Washington, D.C.

Andersen, J. J., and Ganesan, A. T. (1975). *J. Bacteriol.* **121,** 173–183.

Andersen, J. J., and Ganesan, A. T. (1976). *J. Mol. Biol.* **106,** 285–295.

Anraku, N., and Landman, O. E. (1968). *J. Bacteriol.* **95,** 1813–1827.

Attolini, C., Mazza, G., Fortunato, A., Ciarrocchi, G., Mastromei, G., Riva, S., and Falaschi, A. (1976). *Mol. Gen. Genet.* **148,** 9–17.

Bak, A. L., Christiansen, C., and Stenderup, A. (1970). *J. Gen. Microbiol.* **64,** 377–380.

Bazill, G. W., and Gross, J. D. (1972). *Nature (London) New Biol.* **240,** 82–83.

Bazill, G. W., and Gross, J. D. (1973). *Nature (London) New Biol.* **243,** 241–243.

Bazill, G. W., and Karamata, D. (1972). *Mol. Gen. Genet.* **117,** 19–29.

Beeson, J., and Sueoka, N. (1979). *J. Bacteriol.* **139,** 911–916.

Billen, D., Carreira, L. B., Hadden, C. T., and Silverstein, S. (1971a). *J. Bacteriol.* **108,** 1250–1256.

Billen, D., Carreira, L. B., and Silverstein, S. (1971b). *Biochem. Biophys. Res. Commun.* **43,** 1150–1157.

Bishop, B., and Sueoka, N. (1972). *J. Bacteriol.* **112,** 870–876.

Bott, K. (1980). *In* "Journal Supra Structure Supplement 4" (C. F. Fox, ed.), p. 326. Alan Liss, New York.
Brown, N. C. (1971). *J. Mol. Biol.* **59,** 1–17.
Brown, N. C., Wisseman, C., and Matsushita, T. (1972). *Nature (London)* **237,** 72–74.
Brown, N. C., Doyle, R. J., and Streips, U. N. (1976). *Prep. Biochem.* **6,** 479–482.
Burnett, L., and Wake, R. G. (1977). *J. Bacteriol.* **130,** 538–539.
Callister, H., and Wake, R. G. *J. Bacteriol.* **120,** 579–582.
Callister, H., and Wake, R. G. (1977). *J. Mol. Biol.* **117,** 71–84.
Callister, H., LeMesurier, S., and Wake, B. G. (1977). *J. Bacteriol.* **130,** 1030–1037.
Carelton, B. C. (1966). *J. Bacteriol.* **91,** 1795–1803.
Ciarrocchi, G., Attolini, C., Cobianchi, F., Riva, S., and Falaschi, A. (1977). *J. Bacteriol.* **131,** 776–783.
Copeland, J. C. (1969). *J. Bacteriol.* **99,** 730–736.
Copeland, J. C. (1971a). *Science* **172,** 159–161.
Copeland, J. C. (1971b). *J. Bacteriol.* **105,** 595–603.
Cozzarelli, N. R. (1977). *Ann. Rev. Biochem.* **46,** 641–668.
Cozzarelli, N. R. (1980). *Science* **207,** 953–960.
Cozzarelli, N. R., and Low, R. L. (1973). *Biochem. Biophys. Res. Commun.* **51,** 151–157.
Cozzarelli, N. R., Low, R. L., Rashbaum, S. A., and Peebles, C. L. (1975). *In* "DNA Synthesis and its Regulation" (M. Goulian, P. Hanawalt, and C. F. Fox, eds.), pp. 14–37. Benjamin, New York.
Donachie, W. D., Martin, D. T. M., and Begg, K. J. (1971). *Nature (London) New Biol.* **231,** 274–276.
Doyle, R. J., Streips, U. N., Imada, S., Fan, U. S. C., and Brown, W. C. (1980). *J. Bacteriol.* **144,** 957–966.
Dubnau, D. (1970). *In* "Handbook of Biochemistry" (H. Saber and R. A. Harte, eds.), 2nd ed., pp. 39–45. Chemical Rubber Co., Ohio.
Ehrlich, S. D. (1977). *Proc. Nat. Acad. Sci. USA* **74,** 1680–1682.
Ephrati-Elizur, E., and Borenstein, S. (1971). *J. Bacteriol.* **106,** 58–64.
Erickson, R. J., and Young, F. E. (1978). *Dev. Ind. Microbiol.* **19,** 245–257.
Fujita, H., Komano, T., and Maruyama, Y. (1973). *Biochem. Biophys. Res. Commun.* **52,** 1361–1367.
Ganesan, A. T. (1971). *Proc, Nat. Acad. Sci. USA* **68,** 1296–1300.
Ganesan, A., and Lederberg, J. (1965). *Biochem. Biophys. Res. Commun.* **18,** 824–835.
Ganesan, A. T., Laipis, P. J., and Yehle, C. O. (1973a). *In* "DNA Synthesis *in vitro*" (R. D. Wells and R. B. Inman, eds.), pp. 405–436. University Park Press, Baltimore, Maryland.
Ganesan, A. T., Yehle, C. O., and Yu, C. C. (1973b). *Biochem. Biophys. Res. Commun.* **50,** 155–163.
Gass, K. B., and Cozzarelli, N. R. (1973). *J. Biol. Chem.* **248,** 7688–7700.
Gellert, M., O'Dea, M., Itoh, T., and Tomizawa, J. (1976). *Proc. Nat. Acad. Sci. USA* **73,** 4474–4478.
Gillen, F. D., and Ganesan, A. T. (1975). *J. Bacteriol.* **123,** 1055–1067.
Gryczan, T. J., Contente, S., and Dubnau, D. (1978) *J. Bacteriol.* **134,** 318–329.
Gyurasitis, E., and Wake, R. (1973). *J. Mol. Biol.* **73,** 55–63.
Hara, H., and Yoshikawa, H. (1973). *Nature (London) New Biol.* **244,** 200–203.
Harford, N. (1975). *J. Bacteriol.* **121,** 835–847.
Harmon, J., and Taber, H. W. (1977a). *J. Bacteriol.* **129,** 789–795.
Harmon, J., and Taber, H. W. (1977b). *J. Bacteriol.* **130,** 1224–1233.
Helmstetter, C. E., and Cooper, S. (1968). *J. Mol. Biol.* **31,** 507–518.

Henner, D. J., and Hoch, J. A. (1980). *Microbiol. Rev.* **44,** 57–82.

Horowitz, S., Doyle, R. J., Young, F. E., and Streips, U. N. (1979). *J. Bacteriol.* **138,** 915–922.

Hye, R. J., O'Sullivan, A., Howard, K., and Sueoka, N. (1976). *In* "Microbiology, 1976" (D. Schlessinger, ed.), pp. 83–90. American Society for Microbiology, Washington, D.C.

Imada, S., Carroll, L. E., and Sueoka, N. (1975). *In* "DNA Synthesis and its Regulation" (M. Goulian, P. Hanawalt, and C. F. Fox, eds.), pp. 187–200. Benjamin, New York.

Imada, S., Carroll, L. E., and Sueoka, N. (1976). *In* "Microbiology, 1976" (D. Schlessinger, ed.), pp. 116–122. American Society for Microbiology, Washington, D.C.

Imada, S., Carroll, L., and Sueoka, N. (1980). *Genetics* **94,** 809–823.

Ivarie, R. D., and Pene, J. (1970). *J. Bacteriol.* **104,** 839–850.

Ivarie, R. D., and Pene, J. (1973). *J. Bacteriol.* **114,** 571–576.

Jacob, F., Brenner, S., and Cuzin, F. (1963). *Cold Spring Harbor Symp. Quant. Biol.* **28,** 329–348.

Kallenbach, N. R., and Ma, R. I. (1968). *J. Bacteriol.* **95,** 304–309.

Karamata, D., and Gross, J. D. (1970). *Mol. Gen. Genet.* **108,** 277–287.

Kaufman, K. (1976). Ph.D. Thesis, Princeton Univ., New Jersey.

Kavenoff, R. (1972). *J. Mol. Biol.* **72,** 801–806.

Keggins, K., Lovett, P., and Duvall, E. J. (1978). *Proc. Nat. Acad. Sci. USA* **75,** 1423–1427.

Kejzalarová-Lepesant, J., Lepesant, V., Walle, J., Billault, H., and Dedonder, R. (1975). *J. Bacteriol.* **121,** 823–834.

Klotz, L., and Zimm, B. (1972). *J. Mol. Biol.* **72,** 779–800.

Kolter, R., and Helinski, D. (1979). *Ann. Rev. Genet.* **13,** 355–391.

Lark, K., Repko, T., and Hoffman, E. (1963). *Biochem. Biophys. Acta* **76,** 9–24.

Laurent, S. (1973). *J. Bacteriol.* **116,** 141–145.

Laurent, S. (1974). *J. Bacteriol.* **117,** 329–336.

Laurent, S. (1979). *Control of Replication and Partitioning of Bacterial Chromosomes and Plasmids. EMBO Meeting Abstract, p. 47, Lancaster, England.*

Laurent, S. J., and Vannier, F. (1973). *J. Bacteriol.* **114,** 474–489.

Lehegarat, A., and Anagnostopolous, C. (1977). *Mol. Gen. Genet.* **157,** 167–174.

Leibowitz, P. J., and Schaechter, M. (1975). *Int. Rev. Cytol.* **41,** 1–28.

Love, E., D'Ambrosio, J., Brown, N. C., and Dubnau, D. (1976). *Mol. Gen. Genet.* **144,** 313–321.

Lovett, P., and Dramucci, M. G. (1975). *J. Bacteriol.* **124,** 484–490.

Low, R. L., Peebles, C. L., Rashbaum, S. A., and Cozzarelli, N. R. (1976a). *In* "Microbiology, 1976" (D. Schlessinger, ed.), pp. 185–193. American Society for Microbiology, Washington, D.C.

Low, R. L., Rashbaum, S. A., and Cozzarelli, N. R. (1976b). *J. Biol. Chem.* **251,** 1311–1325.

Maaløe, O., and Hanawalt, P. (1961). *J. Mol. Biol.* **3,** 144–155.

Maaløe, O., and Kjeldgaard, N. O. (1966). *In* "Control of Macromolecular Synthesis" (B. Davis, ed.). Benjamin, New York.

Mackenzie, J. M., Neville, M. N., Wright, G. E., and Brown, N. C. (1973). *Proc. Nat. Acad. Sci. USA* **70,** 512–516.

Mahler, I., and Halvorson, H. O. (1977). *J. Bacteriol.* **131,** 374–377.

Mastromei, G., and Riva, S. (1975). *Biochem. Biophys. Res. Commun.* **63,** 522–528.

Matsushita, T., and Sueoka, N. (1974). *J. Bacteriol.* **118,** 974–979.

Matsushita, T., White, K., and Sueoka, N. (1971). *Nature (London) New Biol.* **232,** 111–114.

Mazza, G., and Galizzi, A. (1978). *Microbiologica* **1,** 111–135.

McGinnes, T., and Wake, R. G. (1979). *J. Mol. Biol.* **134,** 251–264.
Mendelson, M. (1968). *Cold Spring Harbor Symp. Quant. Biol.* **33,** 313–316.
Mendelson, N., and Gross, J. (1967). *J. Bacteriol.* **94,** 1603–1608.
Meselson, M., and Stahl, F. (1958). *Proc. Nat. Acad. Sci. USA* **44,** 671–711.
Moses, R. F. and Richardson, C. C. (1970). *Proc. Nat. Acad. Sci. USA* **67,** 674–681.
Murakami, S., Inuzaka, N., Yamaguchi, M., Yamaguchi, K., and Yoshikawa, H. (1976a). *J. Mol. Biol.* **108,** 683–701.
Murakami, S., Murakami, S., and Yoshikawa, H. (1976b). *Nature (London)* **259,** 215–218.
Nagley, P., and Wake, R. G. (1969). *J. Mol. Biol.* **43,** 619–630.
Neville, M. N., and Brown, N. C. (1972). *Nature (London) New Biol.* **240,** 80–81.
Ogasawara, N., Seiki, M., and Yoshikawa, H. (1979). *Nature (London)* **281,** 702–704.
Oishi, M., Yoshikawa, H., and Sueoka, N. (1964). *Nature (London)* **204,** 1069–1073.
Okazaki, R. *et al.* (1968) *Cold Spring Harbor Symp. Quant. Biol.* **33,** 129–143.
Okazaki, R., Sugimoto, K., Okazaki, T., Imae, Y., and Sugino, A. (1970). *Nature (London)* **228,** 223–226.
O'Sullivan, A., Howard, K., and Sueoka, N. (1975a). *J. Mol. Biol.* **91,** 15–38.
O'Sullivan, A., Howard, K., and Sueoka, N. (1975b). *J. Mol. Biol.* **99,** 347–348.
O'Sullivan, A., and Sueoka, N. (1967). *J. Mol. Biol.* **27,** 349–368.
Pettijohn, D. (1976). *Crit. Rev. Biochem.* **4,** 175–203.
Quinn, W., and Sueoka, N. (1970). *Proc. Nat. Acad. Sci. USA* **67,** 717–723.
Riva, S. *et al.* (1975). *Mol. Gen. Genet.* **137,** 185–202.
Rownd, R. (1969). *J. Mol. Biol.* **44,** 387–402.
Rutberg, B., and Hoch, J. A. (1970). *J. Bacteriol.* **104,** 826–833.
Ryter, A. (1968). *Bacteriol. Rev.* **32,** 39–54.
Ryter, A., and Chang, A. (1975). *J. Mol. Biol.* **98,** 797–810.
Ryter, A., and Jacob, F. (1963). *C. R. Acad. Sci. Paris* **257,** 3060–3063.
Sadaie, Y., and Narui, K. (1976). *J. Bacteriol.* **126,** 1037–1041.
Sargent, M. (1974). *Nature (London)* **250,** 252–254.
Sargent, M. (1975a). *J. Bacteriol.* **123,** 7–19.
Sargent, M. (1975b). *J. Bacteriol.* **123,** 1218–1234.
Sargent, M. G. (1977). *Mol. Gen. Genet.* **155,** 329–338.
Seiki, M., Ogasawara, N., and Yoshikawa, H. (1979). *Nature (London)* **281,** 699–701.
Shivakumar, A. G., and Dubnau, D. (1978a). *Plasmid* **1,** 405–416.
Shivakumar, A. G., and Dubnau, D. (1978b). *J. Bacteriol.* **136,** 1205–1207.
Snyder, R., and Young, F. (1969). *Biochem. Biophys. Res. Commun.* **35,** 354–362.
Spizizen, J. (1958). *Proc. Nat. Acad. Sci. USA* **44,** 1072–1078.
Sternglanz, R., Wang, H. F., and Donegan, J. J. (1975). *In* "DNA Synthesis and its Regulation" (M. Goulian, P. Hanawalt, and C. F. Fox, eds.), pp. 309–321. Benjamin, New York.
Streips, U., Horowitz, S., and Doyle, R. J. (1980). *In* "Microbiology, 1980" (D. Schlessinger, ed.) American Society for Microbiology, Washington, D.C. pp. 284–287.
Sueoka, N. (1975). *Stadler Genet. Symp.* **7,** 71–90.
Sueoka, N., and Hammers, S. (1974). *Proc. Nat. Acad. Sci. USA* **71,** 4787–4791.
Sueoka, N., and Quinn, W. (1968). *Cold Spring Harbor Symp. Quant. Biol.* **33,** 695–705.
Sueoka, N., and Yoshikawa, H. (1965). *Genetics* **52,** 747–757.
Sueoka, N., Matsushita, T., Ohi, S., O'Sullivan, A., and White, K. (1973). *In* "DNA Synthesis *in vitro*" (R. D. Wells, and R. B. Inman, eds.), pp. 385–404. University Park Press, Baltimore, Maryland.
Sugino, A., and Bott, K. (1980). *J. Bacteriol.* **141(3),** 1331–1339.
Tamanoi, F., and Okazaki, T. (1978). *Proc. Nat. Acad. Sci. USA* **75,** 2195–2199.

Tamanoi, F., Okazaki, T., and Okazaki, R. (1977). *Biochem. Biophys. Res. Commun.* **77,** 290–297.

Tamanoi, F., Machida, Y., and Okazaki, T. (1978). *Cold Spring Harbor Symp. Quant. Biol.* **43,** 239–242.

Tanaka, T., and Koshikawa, T. (1977). *J. Bacteriol.* **131,** 699–701.

Tomizawa, J., and Selzer, G. (1979). *Ann. Rev. Biochem.* **48,** 999–1034.

Toyoda, H., Yamaguchi, K., and Yoshikawa, H. (1978). *J. Bacteriol.* **136,** 799–802.

Tremblay, G. Y., Daniels, M. J., and Schaechter, M. (1969). *J. Mol. Biol.* **40,** 65–76.

Trowsdale, T., Chen, S. H., and Hoch, J. A. (1979). *Mol. Gen. Genet.* **173,** 61–70.

Upcroft, P., Dyson, H. J., and Wake, R. G. (1975). *J. Bacteriol.* **121,** 121–127.

Van Iterson, W., and Aten, J. A. (1976). *In* "Microbiology, 1976" (D. Schlessinger, ed.), pp. 128–140. American Society for Microbiology, Washington, D.C.

Wake, R. G. (1972). *J. Mol. Biol.* **68,** 501–509.

Wake, R. G. (1973). *J. Mol. Biol.* **77,** 569–574.

Wake, R. G. (1974). *J. Mol. Biol.* **86,** 223–231.

Wake, R. G. (1975). *In* "DNA Synthesis and its Regulation" (M. Goulian, P. Hanawalt, and C. F. Fox, eds.), pp. 650–676. Benjamin, New York.

Wang, H. F., and Sternglanz, R. (1974). *Nature (London)* **248,** 147–150.

White, K., and Sueoka, N. (1973). *Genetics* **73,** 185–214.

Wickner, S. (1978). *Ann. Rev. Biochem.* **47,** 1163–1192.

Winston, S., and Matsushita, T. (1975). *J. Bacteriol.* **123,** 921–927.

Winston, S., and Matsushita, T. (1976). *In* "Microbiology, 1976" (D. Schlessinger, ed.), pp. 123–167. American Society for Microbiology, Washington, D.C.

Winston, S., and Sueoka, N. (1978). *In* "DNA Synthesis: Present and Future" (I. Molineux and M. Kohiyama, eds.), pp. 33–39. Plenum Press, New York.

Winston, S., and Sueoka, N. (1980a). *Proc. Nat. Acad. Sci. USA* **77,** 2834–2838.

Winston, S., and Sueoka, N. (1980b). *J. Bacteriol.* **142(1),** 339–343.

Winston, S., Korn, R., and Sueoka, N. (1980a). *In* "Mechanistic Studies of DNA Replication and Genetic Recombination" (*ICN-UCLA Symp. Mol. Cell. Biol.*) (B. Alberts and C. F. Fox, eds.). Academic Press, New York pp. 207–219.

Yamaguchi, K., and Yoshikawa, H. (1975). *J. Bacteriol.* **124,** 1030–1033.

Yamaguchi, K., and Yoshikawa, H. (1977). *J. Mol. Biol.* **110,** 219–253.

Yamaguchi, K., Murakami, S., and Yoshikawa, H. (1971). *Biochem. Biophys. Res. Commun.* **44,** 1559–1565.

Yoshikawa, H. (1965). *Proc. Nat. Acad. Sci. USA* **53,** 1476–1483.

Yoshikawa, H., and Haas, M. (1968). *Cold Spring Harbor Symp. Quant. Biol.* **33,** 843–855.

Yoshikawa, H., and Sueoka, N. (1963a). *Proc. Nat. Acad. Sci. USA* **49,** 806–813.

Yoshikawa, H., and Sueoka, N. (1963b). *Proc. Nat. Acad. Sci. USA* **49,** 559–566.

Yoshikawa, H., O'Sullivan, A., and Sueoka, N. (1964). *Proc. Nat. Acad. Sci. USA* **52,** 973–980.

Yoshikawa, H., Yamaguchi, K., Seiki, M., Ogasawara, N., and Toyoda, H. (1978). *Cold Spring Harbor Symp. Quant. Biol.* **43,** 569–576.

Yoshikawa, H., Ogasawara, N., and Seiki, M. (1980). *Mol. Gen. Genet.* **179,** 265–272.

Young, F. E., and Wilson, G. A. (1975). *In* "Spores VI" (P. Gerhardt, R. N. Costilow, and H. L. Sadoff, eds.), pp. 596–614. American Society for Microbiology, Washington, D.C.

Zyskind, J., and Smith, D. W. (1977). *J. Bacteriol.* **129,** 1476–1486.

3

RNA Polymerase of *Bacillus subtilis*

ROY H. DOI

The Molecular Biology
of the Bacilli

ISBN 0-12-222701-8

I. Introduction

The DNA-dependent RNA polymerase (RPase) of *Bacillus subtilis* plays a major role in gene expression during growth and sporulation of the organism, since the primary transcripts for tRNA, rRNA, and mRNA are all synthesized by this enzyme. In order to gain a better understanding of RPase function it is essential that the biochemical and genetic properties of the enzyme be well characterized. Therefore purification procedures, which result in the isolation of native enzyme, are necessary. The development of suitable procedures has not been an easy task with this organism. However, several purification protocols have been developed which now allow a relatively high degree of biochemical characterization of this vital enzyme.

The general properties of *B. subtilis* RPase are similar to those reported for a number of prokaryotic organisms (Burgess, 1976); however, significant differences have also been noted for *B. subtilis* RPase, which will be discussed in detail. These differences may be related to either the added complexity of transcription required for spore formation or to the nature of the evolution of gram-positive organisms. The latter point has taken on added significance as a result of the recent studies on the RPase of archaebacteria (Zillig *et al.*, 1978, 1979). The structure of the RPase from these bacteria is quite distinct from that found in gram-positive and gram-negative organisms, indicating that the transcription apparatus has evolved quite differently among prokaryotic organisms.

A major variation between the *Escherichia coli* and *B. subtilis* transcription machineries is the multiplicity of RPase forms found in *B. subtilis* compared to the single RPase form reported in *E. coli*. The various forms found in *B. subtilis* consist of RPase core (E) associated with σ factors and with other polypeptide factors. Since the primary form of the enzyme is $E\sigma^{55}$, and the other forms of the core occur in relatively smaller amounts, the minor forms may be overlooked or discarded inadvertently during the purification of $E\sigma^{55}$. Various purification procedures may also favor the yield of one form over the other. Although part of the difference in the number of enzyme forms found in *E. coli* and in *B. subtilis* might be explained by losses of minor *E. coli* forms during the purification of *E. coli* Eσ, other explanations are more likely. And these are based on the fact that *B. subtilis* is a gram-positive organism that sporulates. Gram-positive organisms, whether they are spore formers or not, appear to have more polypeptides associated with their core enzyme than the core enzyme of gram-negative organisms (Duffy and Geiduschek, 1975; Lill *et al.*, 1977; Pero *et al.*, 1975; Stetter, 1977; Stetter *et al.*, 1974). Thus it is possible that there is a basic difference in the transcription apparatus of gram-positive and gram-negative organisms.

Also, several new forms of RPase are observed in sporulating cells of *B. subtilis* when dramatic morphological changes are occurring. Sporulation is accompanied by a sequential expression of a number of genes that are not transcribed during the log phase of growth (DiCioccio and Strauss, 1973; Doi and Igarashi, 1964; Sumida-Yasumoto and Doi, 1974; Yamakawa and Doi, 1971). It is essential that the RPase have a high degree of gene selectivity to ensure that the sporulation process occurs correctly, since abortive sporulation results invariably in lysis and death of the cell. The occurrence of new developmental forms of RPase may therefore be necessary for recognition and expression of these sporulation-specific genes.

In addition studies on phage-infected *B. subtilis* have provided evidence for the concept that the promoter specificity of the enzyme can be regulated by the association of polypeptides other than σ^{55} with the core, i.e., that other σ-like polypeptides exist. The results indicate that different classes of promoters are present in the phage chromosome and that their recognition requires modification(s) of host enzyme specificity by the addition of phage-coded polypeptides (Burgess, 1976; Duffy *et al.*, 1975; Spiegelman and Whiteley, 1974b). Thus one or several early phage gene products appear to be transcription regulatory factors. Although earlier studies with phage-infected *E. coli* cells had demonstrated that the host enzyme was modified both covalently and by the addition of phage-coded polypeptides, they did not show that the promoter specificity of the enzyme had been altered (Doi, 1977a). Thus for this purpose the studies with phage-infected *B. subtilis* have been more fruitful.

This discussion on RPase will be concerned with current methods of purification of the enzyme, its structural and functional properties, and its genetics and transcriptional properties. Several earlier reviews relevant to this discussion include those by Losick and Pero (1976a,b), Doi (1977a,b), Losick (1972), Chamberlin (1974), Burgess (1976), and Yura and Ishihama (1979).

II. Purification of RPase

The primary consideration in the purification of RPase from vegetative and sporulating cells of *B. subtilis* is to isolate all forms of the enzyme in their native state. This has not been an easy task, since *B. subtilis* cells contain a high level of proteolytic enzymes and several forms of RPase, some of them present in very small amounts. The minor forms of the enzyme usually consist of RPase core enzyme associated with small polypeptides. It is possible that the current methods of purification and detection still result in loss of small polypeptides from the core, loss of complete minor enzyme forms, or plain lack of detection of forms present in very small quantities. In addition it has been necessary to establish experimental criteria for determining whether a RPase core-associated polypeptide is a true polypeptide factor that regulates core activity or an adventitious contaminant of the core. This has required the development of purification procedures for putative regulatory polypeptides and reassociation conditions to study their effect on core activity. A number of purification procedures have been reported (Losick *et al.*, 1970; Avila *et al.*, 1971; Maia *et al.*, 1971; Shorenstein and Losick, 1973a; Spiegelman and Whiteley, 1974a; Linn *et al.*, 1975; Pero *et al.*, 1975; Duffy and Geiduschek, 1975; Fukuda and Doi, 1977; Halling *et al.*, 1977a; Plevani *et al.*, 1977; Spiegelman *et al.*, 1978; Nakayama *et al.*, 1978b; Davison *et al.*, 1979). A typical purification scheme is illustrated in Table I.

A. Inhibition of Protease Activities

The inhibition of proteolytic activity is essential during enzyme purification, since partially cleaved RPase is still catalytically active and can confuse both structural and activity studies (Losick *et al.*, 1970; Leighton *et al.*, 1973; Orrego *et al.*, 1973). Most procedures have employed the serine protease inhibitors phenylmethylsulfonyl fluoride (PMSF) and diisopropyl fluorophospate (DFP) in the extraction and purification buffers at a final concentration of 1 m*M*. However, studies by Nakayama *et al.* (1977) showed that the use of these powerful serine protease inhibitors and metal chelators for metal-activated proteases blocked only 85% of the proteolytic activity in crude extracts derived from sporulating cells. In order to eliminate virtually all proteolytic activity in crude extracts they used a protease-affinity material, hemoglobin–Sepharose,

TABLE I Steps in RPase Purification[a;b]

1. Crude cell extract
2. Hemoglobin–Sepharose column chromatography
3. PEG–dextran phase extraction
4. DEAE–cellulose column chromatography
5. DNA–cellulose column chromatography
6. Glycerol gradient centrifugation

[a] From Halling *et al.* (1977a).
[b] See text for detailed discussion of purification procedures.

in addition to the chemical protease inhibitors. They first batch-treated crude supernatant extracts with hemoglobin–Sepharose (25 g wet weight of the material to 100 ml of crude extract) by slowly stirring the mixture at 4°C for 45 min. The proteolytic activity associated with the hemoglobin–Sepharose was removed by centrifugation. The supernatant was treated for the second time by passage through a hemoglobin–Sepharose column (1.5 × 25 cm). Complete recovery of the RPase activity was observed in the flow-through fraction of this column, and the enzyme in this crude state retained its activity and subunit structure for 30 days at 4°C (Nakayama *et al.*, 1977). This two-step treatment for removing proteases is highly recommended for extracts from sporulating cells; the single-batch treatment is usually sufficient for vegetative cell extracts when used in conjunction with serine protease inhibitors. This method for treating crude cell extracts has also resulted in a very stable system for *in vitro* protein synthesis in *B. subtilis* (Nakayama *et al.*, 1978a), indicating that it is a suitable general method for removing proteolytic activity from *B. subtilis* crude extracts.

B. Buffer

Buffer conditions to consider during the purification of *B. subtilis* RPase include the proper pH and ionic and glycerol concentrations. The requirement for proper glycerol and salt conditions is particularly evident as the enzyme becomes purer and is subjected to certain steps such as passage through a DNA–cellulose column. High salt concentrations should be avoided as the enzyme becomes purer, since inactivation of the enzyme by $(NH_4)_2SO_4$ precipitation has been noted (Davison *et al.*, 1979). Furthermore, salt concentrations greater than 0.5 *M* can result in the removal of polypeptide factors normally associated with the core (Duffy and Geiduschek, 1975). Pure RPase is quite stable at 4°C in buffer A (20 m*M* Tris · HCl, 100 m*M* KCl, 10 m*M* $MgCl_2$, 0.1 m*M* dithiothreitol, pH 7.9) plus 20% glycerol for 2–3 weeks. For longer storage periods the enzyme is kept at −20°C in buffer A containing 50% glycerol. When the stored enzyme is to be used, buffer should be added in small amounts to the enzyme to dilute the glycerol gradually; when enzyme in 50% glycerol is added quickly to the dilution buffer, significant inactivation of the enzyme occurs (Gonzalez *et al.*, 1977).

C. RPase Assay

The typical reaction mixture used during enzyme purification included in a final volume of 125 μl: 120 m*M* Tris · HCl, pH 7.9, 10 m*M* $MgCl_2$,

0.4 m*M* dithiothreitol, 150 m*M* KCl, 2 m*M* $MnSO_4$, 0.4 m*M* ATP, 0.2 m*M* [^{3}H]UTP (2 μCi), and 2–3 μg poly d(A-T). When natural templates were used, $MnSO_4$ was omitted, GTP and CTP at 0.4 m*M* were added, and the KCl concentration was optimized for each template, e.g., 50 m*M* for *B. subtilis* DNA, 100–150 m*M* for ϕe or ϕ29 DNA. The reactions were carried out for 10 min at 37°C and assayed as described by Leighton *et al.* (1973). Under these conditions, the Eδ form of vegetative RPase had little or no activity, whereas the Eσ forms had maximum activity. These forms are described in detail in the following sections. The specific activity of the enzyme is usually reported as nanomoles of mononucleotide incorporated per 10 min/mg protein or nanomoles of mononucleotide incorporated per 60 min/mg protein.

D. Purification of RPase from Vegetative Cells

Most purification procedures involve the separation of RPase from DNA in an early step. This has been accomplished by using the polyethylene glycol (PEG)–dextran phase separation method of Babinet (1967), the passage of crude extract through a heparin–agarose column (Sternbach *et al.,* 1975), or high-salt and high-speed centrifugation techniques (Duffy and Geiduschek, 1975). The PEG–dextran phase extraction method has been used effectively by a number of investigators (Spiegelman *et al.,* 1978; Halling *et al.,* 1977a; Fukuda and Doi, 1977; Shorenstein and Losick, 1973a); with a slight modification one can obtain removal of most of the DNA and good recovery of concentrated enzyme; it has been found that, with proper concentrations of salt and dextran, the RPase can be recovered in the small dextran phase (Fukuda and Doi, 1977). A heparin–agarose column can also be used to remove DNA from the RPase at a very early step in purification (Davison *et al.,* 1979). In all steps of purification one has to consider whether the conditions being used could result in the loss of any core-associated polypeptides. After removal of most of the DNA an effective subsequent step has been chromatography of the enzyme through a DEAE–cellulose column, which provides partial purification of the RPase from other proteins, but more importantly removal of remaining nucleic acids that interfere with binding of the RPase to the DNA affinity column. If the PEG–dextran step is used prior to the DEAE–cellulose column step, the dextran-containing RPase preparation is absorbed batchwise onto the DEAE–cellulose and then poured into a column, since dextran tends to plug up preformed DEAE–cellulose columns.

The use of DNA–cellulose (Litman, 1968) or DNA–agarose (Schaller *et al.,* 1972) column chromatography has been exceptionally effective in purifying nucleic acid-free RPase from other proteins, particularly when a linear salt gradient has been used to elute the enzyme from the column (Halling and Doi, 1978). The enzyme is dialyzed against a Mg^{2+}-free loading buffer containing 0.1 *M* KCl and 20% glycerol, since Mg^{2+} interferes with binding of the enzyme to the DNA–cellulose and 20% glycerol stabilizes the σ-containing form of the enzyme. After loading of the enzyme, the column is washed with 0.28 *M* KCl until no further A_{280}-absorbing material is eluted, and then a linear gradient is applied from 0.3 to 1.1 *M* KCl. The wash with 0.28 *M* KCl removes many DNA-binding proteins which tend to elute with the δ-containing form of the enzyme. A typical elution pattern for vegetative cell enzyme is illustrated in Fig. 1. The δ-containing core (Eδ) can be partially separated from the σ-containing cores ($E\sigma^{55}$ and $E\sigma^{37}$) as shown in Fig. 2. It is particularly important to have 20% glycerol present in the equilibration and elution buffers of the DNA–cellulose column, since at lower glycerol concentrations a significant fraction of $E\sigma^{55}$ is converted to E and free σ^{55} with

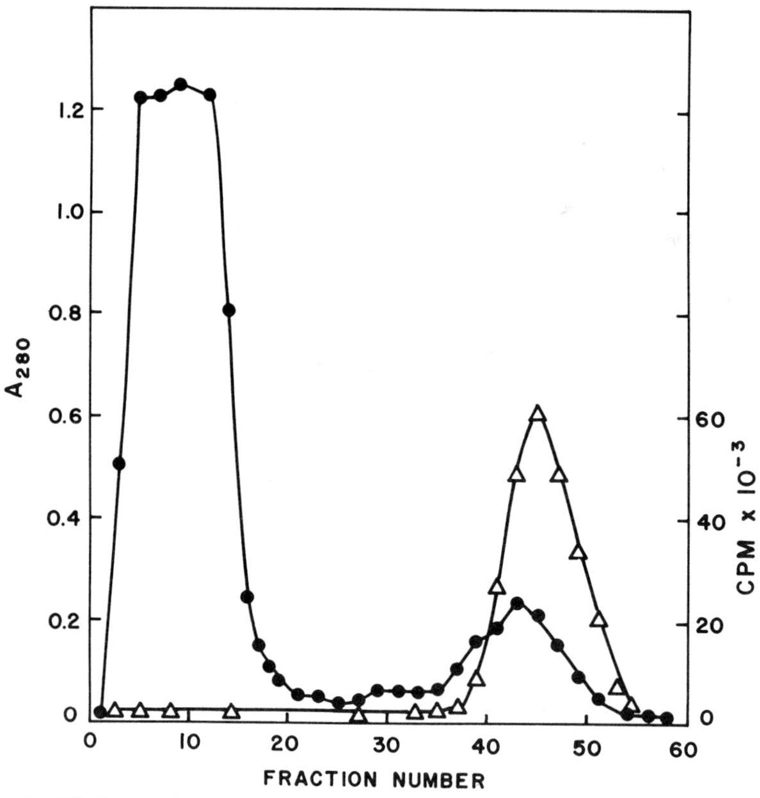

Fig. 1. Elution profile of vegetative RPase from a DNA–cellulose column. ●, A_{280}; △, activity.

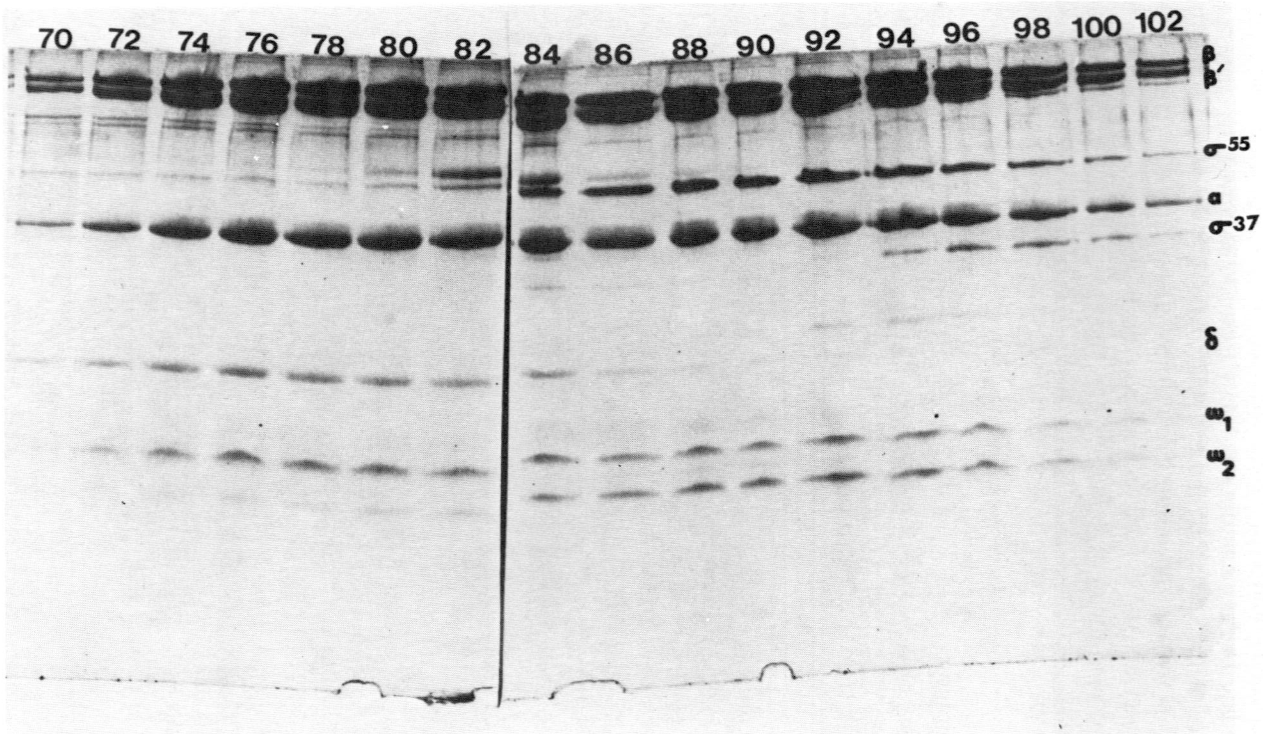

Fig. 2. SDS–polyacrylamide gel electrophoresis pattern of vegetative RPase fractions from a DNA–cellulose column.

a concomitant drastic loss in enzyme activity. Free σ^{55} does not bind to the column and appears in the flow-through fraction. A significant change in the A_{280} elution profile is also observed with an increase in the early eluting peak, since free E and Eδ elute at almost the same ionic concentrations (Plevani *et al.*, 1977; Halling and Doi, 1978). After elution of the enzyme from the DNA–cellulose column, Mg^{2+} is added as soon as possible to each fraction to a final concentration of 10 m*M* in order to stabilize the enzyme. The A_{280} profile indicates that Eσ^{55} comprises about 70% and Eδ about 30% of the total enzyme. No free E fraction was observed in any purification fractions when the method of Halling *et al.* (1977a) was used for RPase purification.

Polypeptides other than σ^{55} and δ have been found to be associated with the core fractions from the DNA–cellulose column, and their true affinity for the core can be tested or demonstrated by several methods. These include centrifugation through a 20–40% glycerol gradient in the presence of 0.5 *M* KCl (Duffy *et al.*, 1975), passage through a Bio-Gel A-5m column in the presence of 0.5 *M* KCl, and electrophoresis through a nondenaturing polyacrylamide gel system (Williamson and Doi, 1979). If the polypeptide(s) remains associated with the core after these treatments in a constant stoichiometric ratio, it suggests that the polypeptide is truly associated with the core and is not an adventitious contaminant. Other criteria will also have to be applied to confirm this possibility, and they will be discussed later.

The enzyme obtained from the DNA–cellulose column is greater than 98% pure as determined by sodium dodecyl sulfate (SDS)–polyacrylamide gel electrophoresis techniques, and the Eσ^{55} fractions have a specific activity varying from 500 to 6000 nmol UMP incorporated per milligram of protein per 10 min at 37°C with ϕ29 DNA as template. The varying activity of RPase preparations has been the subject of much discussion, but there is no firm explanation concerning this disturbing problem (Wiggs *et al.*, 1979). The primary reason proposed is that the purification procedures and assay conditions affect the native structure and the activity of the enzyme in a variable way. Eδ, on the other hand, consistently has low activity with all templates tested (Doi *et al.*, 1980) and therefore, if Eδ is mixed with Eσ^{55}, it could significantly lower the specific activity of the preparation. The variable presence of Eδ therefore could be a partial explanation for the observed variation in specific activity.

E. Purification of Core Subunits and Core-Associated Polypeptides

The purification scheme for RPase core subunits and core-associated polypeptides is illustrated in Fig. 3. This scheme has been used suc-

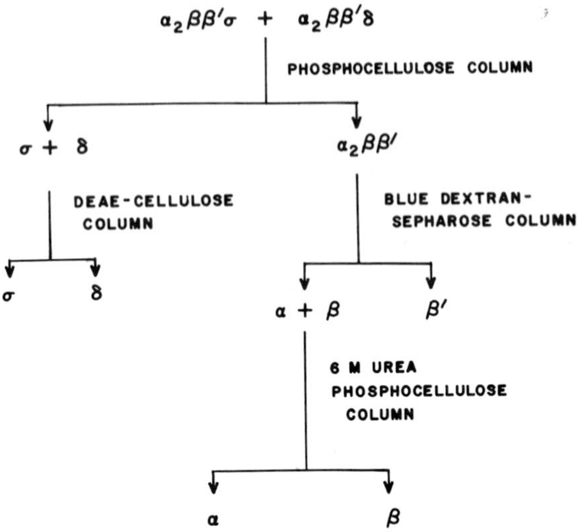

Fig. 3. Purification scheme for RPase subunits.

cessfully to purify each of the core subunits (α, β', β) and σ^{55} and δ factors in reasonably large quantities and under conditions that have allowed reconstitution of an active core and $E\sigma^{55}$ (Halling *et al.*, 1977a). The strategy has included the initial removal of σ^{55} and δ factors from the core, followed by dissociation of the core enzyme and subsequent purification of each of the core subunits.

The σ^{55} and δ factors were separated from the core by dialyzing $E\sigma^{55}$ and $E\delta$ against buffer B (50 m*M* Tris · HCl, 0.1 m*M* EDTA, 10 m*M* dithiothreitol, 1 m*M* PMSF, pH 7.9) containing 10% glycerol and 0.02 *M* KCl at 4°C for 24 hr (Williamson and Doi, 1978). After dialysis the enzyme was passed through a phosphocellulose column. The σ^{55} and δ factors were found in the flow-through fractions, and the core was eluted with buffer B containing 10% glycerol and 0.5 *M* KCl. The low glycerol and KCl concentrations during dialysis must effect some change in conformation in $E\sigma^{55}$ and $E\delta$ that facilitates the release of these factors from the core, since the factors are not readily released from the core without the dialysis step. This idea is supported by the fact that the core released from the phosphocellulose column is less active as measured by reactivation with σ^{55} factor. A two- to threefold increase in this core activity can be obtained by dialyzing the core fraction against buffer A containing 20% glycerol and 0.5 *M* KCl at 4°C and then warming the core preparation at 37°C for 30 min prior to using the core in the σ^{55} factor reactivation assay. However, even after this treatment, full activity of the core is not restored.

The σ^{55} and δ factors obtained in the flow-through fractions of the phosphocellulose column can be readily separated by DEAE–cellulose chromatography (Williamson and Doi, 1978; Tjian *et al.*, 1977). The core obtained from the phosphocellulose column can be used conveniently for the purification of core subunits (Halling *et al.*, 1977a).

The core subunits can be purified quickly and in relatively large quantities by a method involving only two chromatographic columns (Halling *et al.*, 1977a,b). The core is mixed with 9 *M* urea (3 ml of enzyme plus 1 ml of 9 *M* urea) and dialyzed against buffer C (20 m*M* Tris · HCl, 10 m*M* $MgCl_2$, 0.1 m*M* EDTA, 10 m*M* dithiothreitol, 6.5 *M* urea, 10% glycerol, pH 7.9) at 4°C for 4 hr to dissociate the core into its subunits. The dissociated subunits are quickly added to a blue dextran–Sepharose column (Thompson *et al.*, 1975) equilibrated with buffer C and 6.5 *M* urea and washed with the same buffer until protein is no longer eluted; the flow-through fraction contains the α and β subunits. The column is then eluted with buffer C—6.5 *M* urea plus 1 *M* NH_4Cl. The β' subunit, which binds to the column, is eluted by the high-salt buffer as a pure fraction. The α and β subunits in the flow-through fraction are separated by the use of a urea–phosphocellulose column (Ishihama, 1972).

This rapid procedure results in the isolation of pure core subunits and two major polypeptide factors associated with the vegetative cell core. The subunits isolated by this method have been used to identify the core polypeptide bound to the two zinc atoms normally present in the RPase molecule (Halling *et al.*, 1977b) and the core polypeptides responsible for rifampicin (Halling *et al.*, 1977a) and streptolydigin (Halling *et al.*, 1978) sensitivity.

F. Purification of RPase from Sporulating Cells

Essentially the procedures described for the vegetative enzyme have been used to purify RPase from cells at various stages of sporulation. The major precautions in working with sporulating cells have been the use of buffers containing high salt (1 *M* KCl) and serine protease inhibitors for washing cells prior to breakage, and the use of both hemoglobin–Sepharose batch and column methods for the removal of proteases from the crude cell extract (Nakayama *et al.*, 1977). Some other difficulties associated with sporulating cells include obtaining sufficient quantities of cells at a particular state of sporulation, the thicker cell walls which are more difficult to break, the lower content of RPase, the heterogeneity of the RPase, and the presence of only small amounts of sporulation-specific RPase.

For obtaining synchronously sporulating cells in reasonable quantities, the modified Schaeffer medium described by Leighton and Doi (1971) is effective. The use of "stepdown" sporulation media is not recommended, since sporulation synchrony and cell yield are not improved by these procedures.

Reasonable success in isolating sporulation-specific RPase has been achieved by starting with 100 g wet weight of sporulating cells. Starting with less than this amount of cells results in enzyme yields that preclude many useful biochemical analyses, particularly of the minor sporulation-specific forms of the enzyme, which may comprise only 20% of the enzyme. It is important that the stage of sporulation at which cells were harvested be clearly defined, since cells at various stages of sporulation contain different forms of RPase. Usually the cells are defined as being harvested at t_0, t_1, t_2, etc., with the subscripts indicating hours after the end of the log phase of growth. For investigators in this field this is a crucial parameter and should be defined as accurately as possible.

The procedures listed in Table I have been used successfully to obtain reasonably well-defined enzyme preparations from cells at several stages of sporulation. Some of the published methods for purification yield mixtures of forms, particularly if a gradient salt elution is not used at the DNA–cellulose or DNA–agarose column chromatography step. The best way to examine the enzyme eluting from these columns is by SDS–polyacrylamide gel electrophoresis of fractions containing RPase activity and the fractions near the activity peak. This is necessary, since some of the enzyme forms may not show activity with the particular assay method in use but can be detected by their core subunit pattern. A typical elution pattern for RPase from sporulating cells is illustrated in Fig. 4. The subunit pattern of $E\sigma^{29}$ is shown in Fig. 5. Usually the RPase forms are eluted in the following order: $E\delta$, E, $E\sigma^{55}$, $E\sigma^{29}$, EP^{23}, and EP^{34}. The type of form present varies with the sporulation stage; however, the enzyme containing sporulation factors as indicated by the superscripts usually elute later than the vegetative $E\sigma^{55}$ from DNA–cellulose columns (Fukuda and Doi, 1977; Fukuda *et al.,* 1975). The typical salt concentrations at which these forms are eluted from the column are listed in Table II. Several critical points concerning this purification step should be emphasized: (1) It is essential that a linear salt gradient be applied to obtain maximum separation among the various RPase forms; (2) $MgCl_2$ should be added immediately to each fraction to a final concentration of 10 mM to stabilize the enzyme; and (3) the fractions should be assayed for absorption at A_{280}, for enzyme activity, and for their SDS–polyacrylamide gel patterns as soon as possible, since sporulation

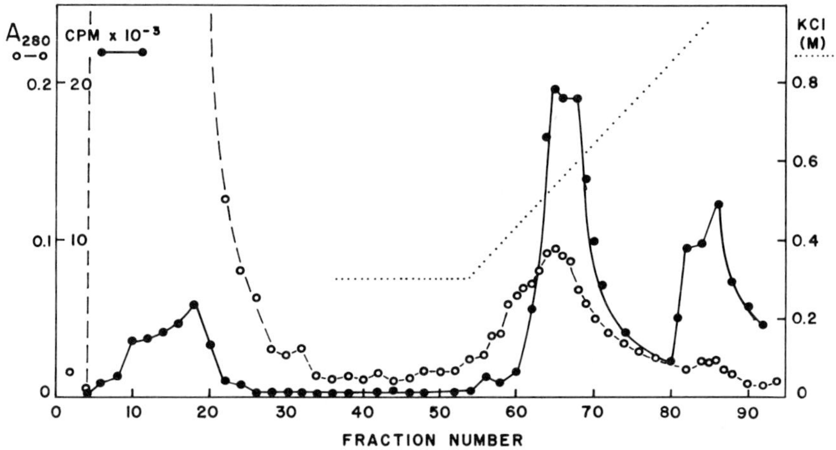

Fig. 4. Elution profile of sporulation cell RPase from a DNA–cellulose column. ○, A_{280}; ●, activity.

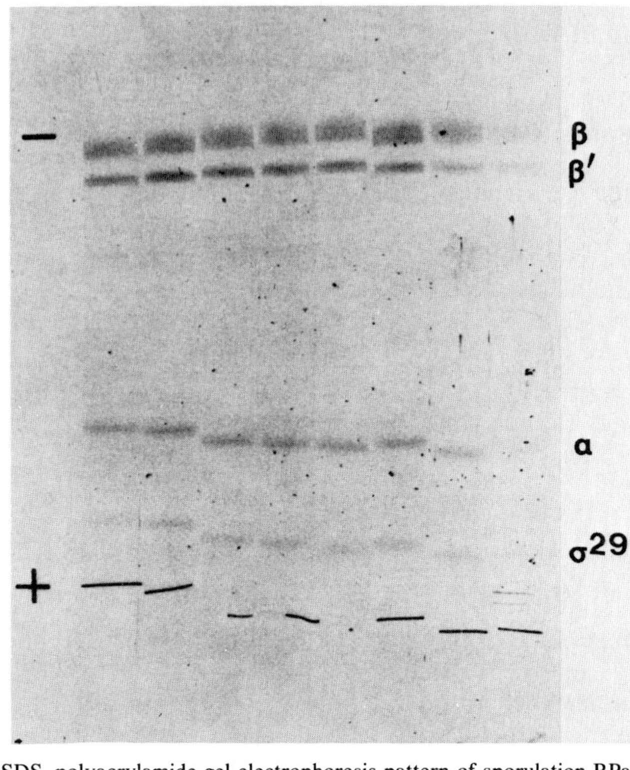

Fig. 5. SDS–polyacrylamide gel electrophoresis pattern of sporulation RPase fractions (Fig. 4, fractions 80–87) from a DNA–cellulose column.

TABLE II RPase Forms Eluted from a DNA–Cellulose Column[a]

Form[b]	Cells	KCl (*M*)[c]
E	Vegetative	0.45
Eδ	Vegetative	0.4
$E\sigma^{55}$	Vegetative	0.55
$E\sigma^{37}$	Vegetative	0.6
$E\sigma^{29}$	Sporulating	0.8
EP^{34}	Sporulating	0.8
EP^{23}	Sporulating	0.8

[a] From Fukuda and Doi (1977), Halling *et al.* (1977a), and Nakayama *et al.* (1978b).
[b] E = $\alpha_2\beta'\beta$; σ = these polypeptides have σ functions; P = these polypeptides stimulate E, but their promoter specificities have not been determined as yet; the superscripts represent the molecular weights of the polypeptides $\times 10^{-3}$.
[c] Approximate concentrations.

RPase forms are unstable at the high salt concentrations at which they are eluted from the column, i.e., about 0.8–0.9 *M* KCl. During initial purification of the enzyme, the sporulation forms were not observed, since the fractions were not assayed immediately and the activity of these fractions had fallen to a negligible level. Fractions are now assayed as they are eluted from the column, pooled, and dialyzed against buffer A containing 20% glycerol to reduce the salt concentration to 0.1 *M* KCl. Buffer A containing 50% glycerol can be used as a storage buffer for the enzyme which is reasonably stable at −20°C.

Suitable conditions for the isolation of pure sporulation-specific σ-like factors have not been achieved as yet; the two points causing difficulty are the low yields of enzyme and the high affinity of some of the sporulation factors for the core. Conditions that readily release vegetative σ^{55} and δ factors from the core have not released the sporulation σ-like factors.

III. Properties of *Bacillus subtilis* RPase

A. Subunit Structure of Vegetative Cell RPase

The RPase core consists of four subunits and has the composition $\alpha_2\beta'\beta$ (Fig. 6). The subunits found in *B. subtilis* core differ in several respects from those present in *E. coli* core (Table III). First, the molecular weights of the subunits are not identical to those of the analogous *E. coli* subunits when analyzed by SDS–polyacrylamide gel electropho-

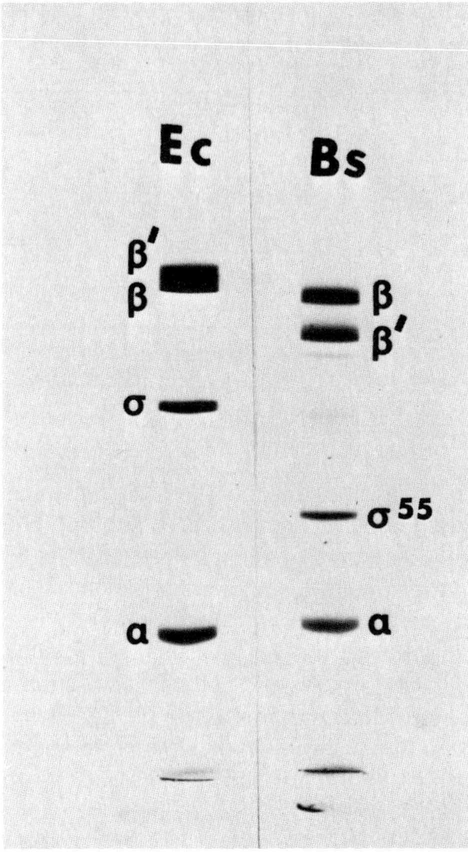

Fig. 6. A comparison of the major subunits of *B. subtilis* and *E. coli* RPase by SDS–polyacrylamide gel electrophoresis.

resis (Halling *et al.*, 1977a). Second, the reconstitution studies with purified core subunits from wild-type and mutant cells have revealed significant differences in the pattern of subunit antibiotic sensitivity. These reconstitution studies showed that the subunit responsible for rifampicin sensitivity in *B. subtilis* was the largest subunit of the core, and this subunit has been designated the β subunit (Halling *et al.*, 1977a) to be consistent with the *E. coli* subunit nomenclature (Rabussay and Zillig, 1969). Furthermore, the second largest *B. subtilis* core subunit was shown to be responsible for streptolydigin sensitivity (Halling *et al.*, 1978), and this subunit also contained the two zinc atoms normally associated with RPase (Halling *et al.*, 1977b); this subunit has been designated β′. In comparison, the second largest core subunit in *E. coli* was shown to be

TABLE III Comparison of *B. subtilis* and *E. coli* RPase Subunits[a]

Subunit or function	*B. subtilis*	*E. coli*
Subunit		
β	140,000 MW	150,000 MW
β′	130,000 MW	160,000 MW
σ	55,000 MW	70,000 MW
α	45,000 MW	42,000 MW
Function		
Rif^r	β subunit	β subunit
Std^r	β′ subunit	β subunit

[a] See text for references. See Table IV for properties of other *B. subtilis* RNA polymerase subunits. From Halling *et al.* (1977a, 1978), Avila *et al.* (1970), Maia *et al.* (1971), Losick *et al.* (1970), Burgess (1976).

responsible for both rifampicin and streptolydigin sensitivity (Iwakura *et al.,* 1973) and had been designated the β subunit (Burgess, 1969). The largest *E. coli* core subunit, designated β′ (Burgess, 1969), has been shown to contain the zinc atoms present in the enzyme (Wu *et al.,* 1977). Thus the relative size of the two largest subunits is reversed in these two organisms, and the streptolydigin sensitivity resides in the β′ subunit of the *B. subtilis* core. Analyses of charge density of the subunits also support these subunit designations (Zillig *et al.,* 1976). Studies utilizing ternary complexes have shown that rifampicin inhibits initiation and streptolydigin inhibits elongation during the transcription process (Kudo and Doi, unpublished observations). The properties of the β and β′ core subunits are summarized in Table III.

Several core-associated polypeptides have been observed for *B. subtilis* RPase. The properties of these polypeptides are summarized in Table IV. The σ factors of *B. subtilis* have much smaller molecular weights than *E. coli* σ factor (Shorenstein and Losick, 1973a,b; Haldenwang and Losick, 1980b; Haldenwang *et al.,* 1981; Wiggs *et al.,* 1981). This smaller size for σ factor has been noted for most gram-positive organisms (Burgess, 1976). In addition to σ factors, there are several other polypeptides associated with the vegetative core in stoichiometric relationships. These include δ factor (Pero *et al.,* 1975; Williamson and Doi, 1978) and ω factors (Duffy and Geiduschek, 1975). Also, there are several other polypeptides with molecular weights between 55,000 and 100,000 that show variable association with the core (Shorenstein and Losick, 1973b; Ishihama, 1972; Halling *et al.,* 1977a; Avila *et al.,* 1971; Maia, 1971). There is less certainty as to whether these function primarily as core-associated polypeptides, whether they associate and regulate core activity only

TABLE IV Molecular Weights of *B. subtilis* RPase Core-Associated Subunits[a]

Subunit	Molecular weight
Vegetative cell RPase	
σ^{55}	55,000
δ	21,000
σ^{37}	37,000
σ^{28}	28,000
ω^1	11,000
ω^2	9,500
Sporulating cell RPase	
σ^{29}	29,000
P^{23}	23,000
P^{34}	34,000
P^{85}	85,000

[a] See Table III for properties of the core subunits. From Shorenstein and Losick (1973a), Pero *et al.* (1975), Duffy and Geiduschek (1975), Fukuda and Doi (1977), Linn *et al.* (1975), Dooley *et al.* (1980), Doi *et al.* (1980), Chamberlin (personal communication).

under certain physiological conditions, whether they can associate with the core in the presence of other factors, or whether they are just contaminating proteins with a high affinity for the core and have no role in regulating its functions.

When the vegetative enzyme is purified by the method of Halling *et al.* (1977a) and eluted from a DNA–cellulose column by a linear salt gradient, two major forms of the core are observed with the compositions $\alpha_2\beta'\beta\delta\omega^1$ (Eδ) and $\alpha_2\beta'\beta\sigma^{55}\omega^1\omega^2$ (Eσ^{55}) (Fig. 1). Eδ has less affinity for the DNA–cellulose column than Eσ^{55} and is eluted earlier. The stoichiometric relationships of the subunits indicate that δ and ω^1 are associated with the same core and that σ^{55} and ω factors are also associated with the same core. Others studies to be discussed indicate that σ^{55} and δ factors do not associate simultaneously with the same core. A minor fraction of the core also is associated with a 37,000-dalton polypeptide and elutes after the major Eσ^{55} peak (Fig. 1 and Table V); this polypeptide has been designated σ^{37}. A third minor holoenzyme has been identified by Wiggs *et al.* (1981) and it consists of core plus a 28,000-dalton sigma factor. This Eσ^{28} form of RPase elutes prior to Eσ^{55} from a DNA–agarose column.

Other methods result in the purification of E and Eσ^{55} with an apparent loss of δ factor from Eδ (Davison *et al.*, 1979) or in a mixture of Eδ and Eσ^{55} (Pero *et al.*, 1975; Spiegelman *et al.*, 1978; Tjian *et al.*, 1977). Since the activities of E, Eδ, and Eσ^{55} are quite different depending on the template and ionic conditions used, it is important that each of these

TABLE V RPase Forms in Vegetative and Sporulating Cells[a]

Vegetative Cells:	E, $E\sigma^{55}$, $E\sigma^{37}$, $E\sigma^{28}$, $E\delta$
Sporulating cells	
Stage 0:	E, $E\sigma^{55}$, $E\sigma^{37}$, $E\delta$
Stage II:	E, $E\sigma^{55}$, $E\sigma^{29}$, EP^{34}
Stage III:	E, $E\sigma^{55}$, $E\sigma^{29}$, EP^{34}, EP^{23}

[a] See text for references. From Fukuda and Doi (1977), Nakayama *et al.* (1978b), Dooley *et al.* (1980), Doi *et al.* (1980), Linn *et al.* (1975), Haldenwang and Losick (1980b), Haldenwang *et al.* (1981), Chamberlin (personal communication). See legend for Table II and text for explanation of different designations of RNA polymerase forms.

forms be purified and that a mixture of forms not be used for critical experiments. In fact the discrepancies in results reported by several groups concerning core activity with the synthetic template poly (dA-dT) can be explained by the presence or absence of δ in the core preparations.

Thus the vegetative RPase is a mixture of at least two major forms of the enzyme and at least two minor forms of holoenzyme. Future purification procedures may result in the identification of additional core-associated polypeptides.

B. Subunit Structure of Sporulating Cell RPase

When RPase is isolated from sporulating cells under conditions in which proteases are inhibited, the core fraction appears to be identical to that found in the vegetative core (Fukuda and Doi, 1977; Linn *et al.*, 1973; Orrego *et al.*, 1973); i.e. the core composition as analyzed by SDS–polyacrylamide gel electrophoresis is $\alpha_2\beta'\beta$. However, since this method of analysis distinguishes only size, it would not detect any minor covalent modifications of the subunits, e.g., phosphorylation, adenylylation, and methylation. Therefore, although the sporulation subunit composition is identical to that of the vegetative core, the data do not exclude the possibility of the presence of covalently modified subunits in the sporulation core.

In addition to $E\sigma^{55}$ several new forms of RPase have been detected in sporulating cells. These sporulation-specific forms consist of the core in association with new polypeptides. Losick and his colleagues (Greenleaf *et al.*, 1973; Greenleaf and Losick, 1973) first reported the presence of a 70,000 to 85,000-dalton polypeptide associated with the core during sporulation. Nishimoto and Takahashi (1974) also demonstrated the presence of a 95,000-dalton polypeptide associated with the core during sporulation. Fukuda *et al.* (1975) first showed the presence of smaller polypeptides (20,000 and 28,000 daltons) associated with the core of sporulating cells and designated them differentiation or δ factors. The presence of one of the sporulation factors has been confirmed by Linn *et al.* (1975) and Haldenwang *et al.* (1981). Doi and his colleagues (Fukuda and Doi, 1977; Fukuda *et al.*, 1975; Nakayama *et al.*, 1978b) have not been able to find the 85,000-dalton polypeptide consistently in their sporulation RPase preparations; the most likely explanation is the difference in purification procedures being used. Furthermore, it has been demonstrated that a mixture of RPase forms exists in sporulating cells and that the population of RPase changes during different stages of sporulation.

Table V summarizes the types of RPase forms that have been observed in vegetative and sporulating cells. The major characteristics of the sporulation forms are (1) the enzymes elute later than the vegetative $E\sigma^{55}$ from the DNA–cellulose column, indicating a higher affinity of these forms for DNA under these ionic conditions; (2) the enzymes contain a low-molecular-weight polypeptide, usually associated with the core in a 1:1 ratio; (3) the enzymes are as active as vegetative $E\sigma^{55}$ with various templates; (4) the sporulation polypeptides are more firmly associated with the core than vegetative σ^{55} and δ factors. They are not removed by phosphocellulose column chromatography (Nakayama *et al.*, 1978b). Vegetative δ factor can displace σ^{55} factor from the core, but not σ^{29} and P^{34} (Williamson and Doi, 1978; Dooley and Doi, unpublished data); and (5) the modified sporulation enzymes are present simultaneously with $E\sigma^{55}$ and comprise only a minor fraction of the total RPase activity.

A comparison of the properties of $E\sigma^{55}$ and $E\sigma^{29}$, one of the sporulation enzymes that has been characterized reasonably well, is presented in Table VI. A most significant observation is the differential effect of netropsin, an antibiotic that binds to AT-rich regions of DNA (Wartell *et al.*, 1974), on transcription by $E\sigma^{55}$ and $E\sigma^{29}$. The activity of $E\sigma^{29}$ was inhibited by a much lower concentration of the antibiotic than the activity of $E\sigma^{55}$ when *B. subtilis* DNA was the template, suggesting that $E\sigma^{29}$ recognized more AT-rich regions of DNA than $E\sigma^{55}$ (Nakayama *et al.*,

TABLE VI Comparison of Vegetative $E\sigma^{55}$ and Eδ with Sporulation E^{29} [a]

Property	$E\sigma^{55}$	Eδ	$E\sigma^{29}$
Elution from DNA–cellulose column	0.6 *M* KCl	0.4 *M* KCl	0.8 *M* KCl
Activity on various templates	High	Very low	High
Factor easily removed from core	Yes	Yes	No
Inhibited by netropsin	High conc.	—	Low conc.

[a] From Fukuda and Doi (1977), Nakayama *et al.* (1978b).

1978b). The drug also inhibited sporulation at concentrations that did not inhibit growth and affected the expression of various genes during sporulation in a selective manner (Beaman *et al.*, 1980; Keilman *et al.*, 1975, 1976).

The obvious question is whether these sporulation forms have different promoter specificities than the vegetative $E\sigma^{55}$. This question will not be resolved until sporulation genes have been isolated, well characterized, and used as templates for these putative sporulation forms of RPase. Losick and his colleagues have isolated a DNA fragment that is transcribed primarily during sporulation (Segall and Losick, 1977) and have isolated an enzyme that transcribes this DNA fragment at a much higher frequency than DNA for vegetative functions (Haldenwang and Losick, 1980a,b). However, this enzyme, which contains a 37,000-dalton polypeptide, is found in log-phase cells (Haldenwang and Losick, 1980a), and its relationship to the sporulation enzymes is still uncertain. However, $E\sigma^{29}$ from sporulating cells does appear to recognize sporulation-specific promoters (Haldenwang *et al.*, 1981).

IV. Functions of RPase

A. Role of σ Factor

The σ^{55} factor of *B. subtilis* was detected by several investigators during early RPase purification studies (Avila *et al.*, 1970; Losick *et al.*, 1970; Maia *et al.*, 1971). The σ^{55} factor was required for the efficient transcription of *B. subtilis* (Avila *et al.*, 1971; Losick *et al.*, 1970) and *E. coli* (Maia *et al.*, 1971) phage templates. The σ^{55} factor of *B. subtilis* appears to have a role similar to that of the *E. coli* σ factor in allowing promoter recognition at a higher efficiency and in causing asymmetric transcription (Inciarte *et al.*, 1976a,b; Kawamura and Ito, 1977; Sogo *et al.*, 1979a,b).

The *B. subtilis* σ^{55} factor can also stimulate the core enzymes of *E. coli* (Shorenstein and Losick, 1973b; Whiteley and Hemphill, 1970) and *Pseudomonas aeruginosa* (Whiteley and Hemphill, 1970), although the templates and ionic conditions used in the reaction mixture markedly affect this stimulatory activity (Whiteley and Hemphill, 1970). In this regard the *B. subtilis* core can also be stimulated by σ factors from *E. coli* (Losick *et al.*, 1970) and *P. aeruginosa* (Whiteley and Hemphill, 1970) under suitable assay conditions. These results with heterologous cores and σ factors are quite remarkable, since the σ factor from *E. coli* is much larger (70,000 daltons) than the *B. subtilis* σ^{55} factor (55,000 daltons) and yet can interact with the *B. subtilis* core in a functional manner. Although the exact degree of promoter specificity and efficiency of these heterologous holoenzymes has not been determined, the results imply that the conformational site of interaction between the core and σ factor has been highly conserved during evolution.

Initial studies with the heterologous mixtures of core and σ factor revealed that the core played a significant role in distinguishing particular phage templates (Shorenstein and Losick, 1973b). The core enzyme and not the source of σ determined the efficiency of transcription of phage DNAs. Subsequent investigations into the nature of RPase–promoter interactions have revealed that the holoenzyme from various organisms can recognize heterospecific promoters with different degrees of efficiency but with a remarkable degree of fidelity (Wiggs *et al.*, 1979). These results suggest that promoters and promoter specificity or RPase in bacteria have been highly conserved.

The role of σ factor during sporulation has been studied extensively. Losick and his colleagues came to the conclusion that, although σ^{55} factor was present in sporulating cells (Tjian and Losick, 1974; Tjian *et al.*, 1974), it did not actively form $E\sigma^{55}$. Part of this conclusion was based on their observation that $E\sigma^{55}$ could not be isolated from sporulating cells and that only a modified RPase could be purified from sporulating cells (Linn *et al.*, 1975). However, Doi and his colleagues (Fukuda and Doi, 1977; Fukuda *et al.*, 1975; Nakayama *et al.*, 1978b) and Szulmajster and his group (Duie *et al.*, 1974; Maia, *et al.*, 1971; Orrego *et al.*, 1973) have consistently found $E\sigma^{55}$ and sporulation-specific forms of RPase in sporulating cells. Thus it appears that the method of purification affects the forms of RPase ultimately isolated and observed. Since early phage genes can be transcribed until late in sporulation (Kawamura and Ito, 1974; Yehle and Doi, 1967), more than 60% of the mRNA of sporulating cells (DiCioccio and Strauss, 1973; Sumida-Yasumoto and Doi, 1974) and dormant spores (Jeng and Doi, 1974) is similar to that found in vegetative cells, and $E\sigma^{55}$ is found in mature spores (Maia *et al.*, 1971), the presence of $E\sigma^{55}$ is not contradictory to the total transcription functions and products observed in the sporulating cell. Therefore it is likely that $E\sigma^{55}$ is the major form of RPase in sporulating cells and that minor forms of sporulation-specific RPase occur simultaneously with $E\sigma^{55}$.

One study has suggested that σ^{55} factor remains associated with the core during transcription (Williamson and Doi, 1979). The interaction of σ^{55} with the core during the elongation phase was analyzed by the nondenaturing gel electrophoresis method of Krakow (1971) who first demonstrated that *Azotobacter vinelandii* σ factor was released from the core after RNA synthesis had been initiated. Using the same method,

Williamson and Doi (1979) confirmed these results with *E. coli* RPase. However, they were not able to show the release of significant amounts of σ^{55} factor during transcription by *B. subtilis* RPase. In another type of approach Spiegelman and Whiteley (1979) analyzed the composition of the *B. subtilis* enzyme during transcription of SP82 DNA and found that most of the σ^{55} factor was released during transcription, in agreement with the results obtained previously with *E. coli* (Williamson and Doi, 1979). It now appears that σ^{55} is in fact released from core during transcription, since the apparent non-release of σ^{55} can be explained by the recent observation that free σ^{55} binds to DNA (Kudo *et al.*, 1981). Thus, σ^{55} released from core binds to DNA. This is the first report of free sigma binding to DNA and this reveals an interesting function for σ factors.

B. Role of δ Factor

The vegetative δ factor was observed first by Pero and her colleagues during their studies on SP01 transcription (Pero *et al.*, 1975). They found that the presence of δ favored the asymmetric transcription of phage middle (Pero *et al.*, 1975) and late (Tjian and Pero, 1976) genes by phage-modified RPases. The δ factor was associated with their holoenzyme fraction from uninfected cells, and therefore δ was not a phage-coded product. This latter observation has been confirmed by several groups (Halling and Doi, 1978; Plevani *et al.*, 1977; Spiegelman *et al.*, 1978).

The use of DNA–cellulose column chromatography has allowed the fractionation of vegetative RPase into at least three fractions including free core (E), core associated with δ factor (Eδ), and core associated with σ^{55} factor ($E\sigma^{55}$) (Halling and Doi, 1978; Plevani *et al.*, 1977). Plevani *et al.* (1977) found a form of vegetative cell enzyme containing the core associated with δ and a 28,000-dalton polypeptide. This form of the enzyme had very low activity relative to $E\sigma^{55}$, but interestingly it could form a rifampicin-resistant initiation complex and was thus more similar to $E\sigma^{55}$ than to core which is incapable of forming a rifampicin-resistant rapid-start complex. This vegetative form of the enzyme is somewhat analogous to the phage-modified RPases capable of transcribing middle and late phage genes and which contain phage-coded small polypeptides associated with the core (Duffy and Geiduschek, 1975; Pero *et al.*, 1975; Spiegelman and Whiteley, 1974b). These observations suggest that δ plays an auxiliary but necessary role in the selection of promoters.

Since the δ factor is found in uninfected vegetative cells, the function of this factor has been studied extensively to determine its role in cellular transcription. The synthetic activity of Eδ itself is very low relative to that of Eσ^{55} on a number of natural templates (Doi *et al.*, 1980; Spiegelman *et al.*, 1978; Tjian *et al.*, 1977; Williamson and Doi, 1978) and varies significantly depending on the template and ionic conditions of the reaction mixture. Under the optimum conditions for RNA synthesis by Eσ^{55}, Eδ had 10-fold lower activity than Eσ^{55}, and its activity varied up to 12-fold depending on the template used (Doi *et al.*, 1980). Generally *Bacillus* phage DNAs were better templates than plasmid or chromosomal DNAs. The activity of Eδ on poly (dA-dT) (Tjian *et al.*, 1977) and various phage templates (Spiegelman *et al.*, 1978) was reduced significantly as the KCl concentration was increased above 0.05 *M*. Although Eδ itself had low synthetic activity, when σ^{55} was added to Eδ in the reaction mixture, the synthetic activity with various phage and plasmid DNAs was stimulated from 3- to 12-fold (Doi *et al.*, 1980); this indicated that σ^{55} was able to interact with E or Eδ in the reaction mixture. (The evidence that σ^{55} and δ compete physically for the core will be discussed later.) However, in spite of the stimulation of Eδ by σ^{55}, the activity observed was still much lower than that seen with Eσ^{55} alone, implying that only a small fraction of E was stimulated by σ^{55} and that δ still exerted a significant inhibitory effect on the enzyme. In the reciprocal experiment, the addition of free δ to Eσ^{55} inhibited its activity from 0 to 87% depending on the template (Doi *et al.*, 1980). Thus δ appears to have an effect on Eσ^{55} interaction with specific promoters. These results *in toto* indicate that δ can have a strong, weak, or no inhibitory effect on transcription, depending on the template or promoter.

Does δ factor have a role in preferential transcription of certain classes of promoters? Since the inhibitory effects of δ are strongest when *B. subtilis* DNA is the template and weakest with phage templates (Doi *et al.*, 1980), several roles for δ are suggested: (1) Since *B. subtilis* DNA contains nicks and single-stranded regions which bind RPase indiscriminately and is in a relatively poor physical state, δ factor may direct the RPase to promoter sites on the DNA and away from nonspecific initiation sites; this would explain the reduced activity on this template; (2) the transcription from early phage promoters, which are very efficient, appears to be affected the least by δ; thus δ factor may have a more significant regulatory effect on weak promoters; (3) δ factor has a significant inhibitory effect on transcription from supercoiled plasmid DNAs; these results suggest that promoters on plasmid DNAs are less efficient than phage early promoters and that the δ factor is actually inhibiting the interaction of RPase with weak promoters; thus again δ appears to favor the interaction of RPase with strong promoters.

In order to understand the mechanism of δ function, the effect of δ on RPase–DNA complex formation, initiation, and elongation processes, and its interaction with core and Eσ, have been investigated. A systematic analysis of the effect of δ on transcription initiation and elongation by Eσ^{55} revealed that the stability of the enzyme–DNA complex was reduced when δ was added to Eσ^{55}, that the rate of chain initiation was not affected, that the presence of δ increased the relative proportion of larger transcripts, and that the translatability of the *in vitro* synthesized transcripts was increased by δ (Spiegelman *et al.*, 1978). It was also shown that E associated with δ and a 28,000-dalton polypeptide obtained from log-phase uninfected cells was able to form an enzyme–DNA complex that was weakly resistant to rifampicin inhibition (Plevani *et al.*, 1977). Since Eδ alone does not form a stable complex with DNA (Spiegelman *et al.*, 1978; Spiegelman and Whiteley, 1979; Williamson and Doi, 1979), it appears that δ in cooperation with the 28,000-dalton polypeptide can function to bind RPase to a promoter site. Thus this particular function is reminiscent of the δ function with phage-modified RPases containing small phage-coded polypeptides. The interpretation of the multiple effects of δ are difficult to reconcile with a simple unifying model. However, these results suggest that one of the major functions of δ is to increase the efficiency of initiation from promoters. In this regard Dickel *et al.* (1980) have demonstrated that δ increases the promoter selectivity of Eσ^{55}. The ionic effect on δ-regulated RNA synthesis suggests that interaction of RPase with strong promoters may be favored by the presence of δ.

Another approach to understanding the function of δ has been to analyze the physical interaction between δ and the core. By using purified preparations of σ^{55}, δ, and core, Williamson and Doi (1978) demonstrated that δ could displace σ^{55} from Eσ^{55} in the absence of DNA to form Eδ and free σ^{55}. The affinity of δ for E was much higher than that of σ^{55} for E, such that an equimolar mixture of E, σ^{55}, and δ always resulted in the formation of Eδ and free σ^{55}. In the presence of DNA, Eδ reacted with DNA to form an E-DNA complex and free δ. This E-DNA complex was able to react with σ^{55} to initiate RNA synthesis. These latter results were also observed by Spiegelman and Whiteley (1978). This series of *in vitro* interactions among Eσ^{55}, Eδ, the factors, and DNA are summarized in Table VII. Two interpretations of these interactions have been proposed. One possibility is that Eδ binds to or near the promoter region of DNA and that this event is followed by the formation of E-DNA and the release of δ; σ^{55} then binds to the E-DNA complex to form an Eσ^{55}-DNA initiation complex; thus δ and σ^{55} would be acting *sequentially* with the core (Williamson and Doi, 1978, 1979). The other possible mode of action is to have δ and σ^{55} factors act *simultaneously* with the core near the promoter region; δ would be released first during this process, and σ^{55} would then initiate RNA synthesis from the promoter

TABLE VII Interaction of σ^{55} and δ Subunits with Core Enzyme[a]

$E\sigma^{55} + \delta \rightarrow E\delta + \sigma^{55}$
$E\delta$ + DNA → E-DNA + δ
E-DNA + $\sigma^{55} \rightarrow E\sigma^{55}$-DNA
$E\sigma^{55}$ + DNA → $E\sigma^{55}$-DNA
$E\sigma^{55} + \delta$ + DNA → $E\sigma^{55}$-DNA + δ
$E\sigma^{55}$-DNA + XTPs → E-DNA-RNA + σ^{55}
E-DNA-RNA → E + DNA + RNA

[a] From Williamson and Doi (1978, 1979), Spiegelman and Whiteley (1979).

site (Spiegelman and Whiteley, 1979). Thus, in one case δ and in the other case δ and σ^{55} would guide the core to a position near the promoter; σ^{55} would initiate RNA synthesis at a specific site in the promoter in both cases. Since δ and σ^{55} cannot form a complex with core simultaneously (Williamson and Doi, 1978), it appears that they must at least act sequentially with the core. However, a transient simultaneous interaction of the factors has not been ruled out.

C. Complex Formation between RPase and Promoters

Chamberlin (1974) has reviewed the various conditions required for the successful formation of complexes between *E. coli* RPase and promoters. This type of analysis is particularly important for the questions currently being considered by investigators interested in (1) transcription mechanisms of *B. subtilis* RPase, which are significantly different in structure and subunit function from those seen with *E. coli* RPase, and (2) transcription of sporulation- and phage-specific genes by polypeptide-modified RPases. Complex formation between RPase and DNA has been used to select restriction fragments containing promoters (Spiegelman and Whiteley, 1978; Talkington and Pero, 1978, 1979; Dickel *et al.*, 1980), to determine whether δ factor has any effect on complex formation (Spiegelman *et al.*, 1978; Dickel *et al.*, 1980), to measure the formation of rapid-start initiation complexes (Duffy and Geiduschek, 1975, 1976; Plevani *et al.*, 1977), and to determine whether polypeptides associated with phage-modified RPases are involved in promoter selection (Spiegelman and Whiteley, 1978; Talkington and Pero, 1978).

The usual analysis of complex formation consists of trapping specific RPase–DNA binary complexes on nitrocellulose filters (Hinkle and Chamberlin, 1972). Previous studies have shown that holoenzymes can form specific and nonspecific complexes with DNA and that holoenzyme–promoter complexes are more stable than holoenzyme–nonpromoter complexes (Hinkle and Chamberlin, 1972). Thus, if a limiting number of RPase molecules are mixed with DNA containing a number of promoter sites, the RPase will bind preferentially to promoters and the interaction with nonspecific sites on the DNA will be reduced.

This technique has been used successfully to identify promoter-containing fragments from restricted DNA (Talkington and Pero, 1979). An increasing amount of RPase was mixed with a fixed amount of restricted DNA; the resulting RPase–DNA complexes were trapped on nitrocellulose filters; the DNA fragments in the complexes were removed from the filters by treatment with SDS and analyzed by polyacrylamide gel electrophoresis. With limiting RPase, fragments containing promoters preferentially formed complexes, whereas with increasing concentrations of RPase, more complexes were formed between RPase and non-promoter-containing fragments. $E\sigma^{55}$ and phage-modified RPase formed tight binding complexes with SP01 DNA in less than 60 sec, and the half-time for dissociation of these complexes was greater than 30 min (Talkington and Pero, 1979).

In some cases heparin, a polyanion that binds to either free RPase or DNA-bound RPase, can be used to distinguish binary complexes formed between RPase and different promoters (Pfeffer *et al.*, 1977; Zillig *et al.*, 1970). Some RPase–promoter complexes are readily destroyed, and others are much more stable to heparin treatment; this fact can be used to enrich these stable complexes. Although this technique works reasonably well with *E. coli* RPase and promoters, it has been found that *B. subtilis* RPase–promoter complexes are generally extremely sensitive to heparin attack (Talkington and Pero, 1979). Some *E. coli* complexes can withstand the presence of 100 μg/ml of heparin for hours (Pfeffer *et al.*, 1977); many *B. subtilis* complexes are destroyed by treatment with 10 μg/ml of heparin for 10 min (Talkington and Pero, 1979). Since some *B. subtilis* complexes can, however, withstand this short treatment with heparin, DNA fragments containing SP01 early promoters have been differentiated by this method into resistant and sensitive classes (Talkington and Pero, 1979).

Complex formation was used to study the function of δ factor, a 21,000-dalton polypeptide found associated with vegetative cell core. Extensive studies by Whiteley and Spiegelman (1978) revealed that the presence of δ factor generally reduced the amount of DNA binding to $E\sigma^{55}$. They felt that the δ factor might be destabilizing nonspecific enzyme–DNA interactions. However, the great excess of RPase over DNA required for their binding studies precluded any definitive conclusions.

The formation of rapid-start complexes (Hinkle *et al.*, 1972; Mangel and Chamberlin, 1974) has been investigated with various *B. subtilis*

RPase forms. In these analyses RPase is initially incubated with DNA to form a binary complex and then challenged with rifampicin and the four ribonucleoside triphosphates. The enzyme bound to DNA can either initiate RNA synthesis or form an inactive RPase–DNA–rifampicin complex. A large fraction of the molecules in a rapid-start complex with the promoter site can initiate RNA synthesis. This criterion has been used to demonstrate that $E\sigma^{55}$ from uninfected and infected cells can make rapid-start complexes with early promoters of phages SPP1 (Plevani *et al.*, 1977), SP01 (Duffy and Geiduschek, 1975, 1976), and ϕ29 (Duffy and Geiduschek, 1976). On the other hand the phage-modified RPase from SP01-infected cells containing the 28,000-dalton polypeptide made rapid-start complexes with only SP01 DNA and not with ϕ29 or ϕ1 DNA (Duffy and Geiduschek, 1976; Petrusek *et al.*, 1976). These and other results suggested strongly that the 28,000-dalton polypeptide directed the core to specific middle gene promoters on SP01 DNA but not to sites on heterologous phage DNAs. Very little rapid-start complex formation has been observed between $E\sigma$ and *B. subtilis* DNA (Duffy and Geiduschek, 1975; Plevani *et al.*, 1977), which is somewhat puzzling and suggests the possibility that most complexes between *B. subtilis* chromosomal promoters and RPase are relatively unstable or extremely rifampicin-sensitive. Since the *B. subtilis* enzyme is structurally quite different from the *E. coli* enzyme (Doi *et al.*, 1980; Halling *et al.*, 1977a, 1978), this is a distinct possibility. Furthermore no rapid-start complexes were formed between T7 DNA and *B. subtilis* RPase, although the enzyme was able to transcribe the early T7 genes (Wiggs *et al.*, 1979). Rice and Meares (1978) showed that rifampicin, when bound to the enzyme, was closely positioned to all four core subunits and to σ.

The problem caused by the unstable nature of binary complexes in the isolation of certain DNA fragments can be overcome by trapping an RPase–DNA–RNA ternary complex at or very near the promoter site (Jones *et al.*, 1977). These ternary complexes are stable to low temperature (0°C) and high ionic strength (0.25 *M* KCl), conditions that dissociate binary complexes. For this purpose RPase, DNA, and three nucleoside triphosphates are incubated to form a ternary complex. This complex, containing a short oligonucleotide, is stable and can be readily isolated by the nitrocellulose filter technique. The enzymatic activity of the ternary complex is resistant to rifampicin but sensitive to streptolydigin inhibition, indicating that initiation has occurred and that streptolydigin inhibits the elongation process (Kudo and Doi, unpublished data). The ability of the RPase–DNA complex to form an initiation complex is confirmation that RPase in fact binds at a promoter site, since some binary complexes that are stable cannot initiate RNA synthesis.

Obviously other reasons may explain the latter observation, e.g., the requirement for accessory positive regulatory factors for initiation from certain promoter sites. But in most cases some adventitious binding occurs with certain fragments having no promoter sites, and these fragments could then be identified by the fact that they could not form stable ternary complexes.

Another variation of this procedure has been to separate the ternary complex by electrophoresis through an agarose gel (Chelm and Geiduschek, 1979). Comigration of the newly synthesized RNA with the template provides a method for identifying the presence of a promoter site on a particular fragment of DNA.

D. *Bacillus subtilis* Promoters and Initiation of Transcription

Three promoter base sequences for *B. subtilis* phage SP01 have been reported by Talkington and Pero (1979). An early gene promoter of SP01 resembles the promoters reported previously for several *E. coli* genes in both the −35 region and the −10 or Pribnow box region (Rosenberg and Court, 1979). The Pribnow box for the early SP01 gene had the sequence CATAAT (where T is hydroxymethyluracil which is present in place of thymine in SP01 DNA) which is similar to the general Pribnow box sequence of TATAAT. The −35 region of this early gene had the sequence TTGACT which is very similar to that found in the −35 region of *E. coli* promoters.

The two base sequences for the middle gene promoters of SP01 were, however, significantly different from that for the early gene (Talkington and Pero, 1979). The Pribnow box region for each of the middle gene promoters had the identical sequence of TTTATTT and was located six bases from the 5′-side of the initiation point of transcription. Each of the middle genes had the sequence AGGAGA in the −35 region, which is quite different from the TTGACT found for the early gene. Since the Pribnow box and −35 region are both involved in binding *E. coli* RPase (Siebenlist and Gilbert, 1980; Simpson, 1979), the differences seen between the early and middle gene promoters indicate that the specificity of the RPases that recognize these sequences must be quite different. In fact it has been shown that a new phage-coded 28,000-dalton polypeptide acts as a σ-like factor guiding the core to the middle gene promoters of SP01 (Duffy and Geiduschek, 1976; Talkington and Pero, 1979) and SP28 (Spiegelman and Whiteley, 1978).

The differences observed between the early and middle gene promoters are extremely significant, since they support the hypothesis that promoters with different specificities exist. Moreover the function of the phage-specified 28,000-dalton polypeptide supports the concept that σ-like polypeptides can regulate the specificity of RPase. This type of regulation may also be involved in the transcription of sporulation-specific genes, since RPase cores associated with small polypeptides have been observed in sporulating cells (Fukuda *et al.*, 1975; Fukuda and Doi, 1977; Linn *et al.*, 1975; Haldenwang *et al.*, 1981).

Another feature of the promoter region of SP01 genes is their high content of AT base pairs. In the region of −1 to −40, 70% of the bases were AT base pairs. In another analysis of a promoter region, a good correlation was obtained between five AT-rich regions and the location of RPase-binding sites on ϕ29 phage DNA (Sogo *et al.*, 1979). Thus one of the features of *B. subtilis* promoter sites appears to be their high AT base pair content.

Initiation of most transcripts *in vitro* by *B. subtilis* RPase occurs with either ATP or GTP (Davison *et al.*, 1979; Jimenez *et al.*, 1974; Kawamura and Ito, 1977). For instance the transcription of early genes of ϕ29 is initiated efficiently with GTP from three sites and with ATP from two sites. Transcription from promoters of middle genes of phages SP01 (Talkington and Pero, 1979) and SP28 (Spiegelman *et al.*, 1978) are also initiated with ATP. Thus the initial bases for transcription by *B. subtilis* RPase resemble those observed with *E. coli* RPase.

In vitro transcription studies have indicated that *B. subtilis* RPase can initiate RNA synthesis accurately and produce transcripts similar to those made *in vivo*. Analysis of ϕ29 transcripts (Davison *et al.*, 1979; Jimenez *et al.*, 1974; Kawamura and Ito, 1977) and T7 transcripts (Wiggs *et al.*, 1979) has shown that the RPase started and terminated transcription at the proper sites. The observations with T7 transcripts are particularly interesting, since the results indicate that the heterologous promoter and terminator sites were both recognized with a high degree of fidelity by the *B. subtilis* RPase.

Since the promoters for *B. subtilis* and *E. coli* genes appear to be quite similar and are recognized and transcribed *in vitro* by the heterologous RPases, no transcription barrier appears to exist between these two species. In support of this idea a number of *B. subtilis* genes have been transformed into *E. coli* by plasmid vectors and have been transcribed and translated into active enzymes (Chi *et al.*, 1978; Ehrlich *et al.*, 1976; Horiuchi *et al.*, 1977; Mahler and Halvorson, 1977; Nagahari and Sakaguchi, 1978). The interesting and curious point is that when *E. coli* genes are transformed into *B. subtilis* cells by plasmid vectors, no phenotypic expression of the *E. coli* genes is observed (Kreft *et al.*, 1978; Mahler and Halvorson, 1977). Since a transcription barrier apparently does not exist between these species, the barrier for expression of *E. coli* genes in *B. subtilis* may reside at the level of the translation mechanism. In this regard, Goldfarb *et al.* (1981) have obtained expression of an *E. coli* gene (chloramphenicol acetyltransferase) in *B. subtilis* by replacing the *E. coli* promoter region with a *B. subtilis* promoter and translation initiation sequence. The *B. subtilis* translation system may be more discriminatory than the *E. coli* system, since the *E. coli* system can translate mRNA not only from *B. subtilis* but even from eukaryotic cells. Since a reasonably large number of translation components exist, it will be of interest to determine the ones that differentiate mRNAs.

E. Transcription Termination

Two types of transcription termination occur in *E. coli,* a ρ-dependent and a ρ-independent termination (Adhya and Gottesman, 1978). The ρ-dependent form is regulated by a protein factor, ρ (Roberts, 1976). This type of factor has also been found recently in *B. subtilis* by Hwang and Doi (1980). The properties of this factor are summarized in Table VIII. The *B. subtilis* factor is smaller than that found in *E. coli* and is a dimeric protein instead of a tetramer. The *B. subtilis* ρ factor has the two basic functions present in the *E. coli* ρ factor, a poly C-, RNA-dependent ATPase activity (Lowery-Goldhammer and Richardson, 1974) and RNA synthesis termination activity (Roberts, 1969). Although this is only the second prokaryotic organism in which ρ factor has been demonstrated, it suggests the possibility of the universal nature of this mechanism for termination in prokaryotes. Two types of termination signals are present in the *E. coli* genome. The ρ-independent termination signals consist usually of a stretch of AT base pairs preceded by a sequence rich in GC pairs (Adhya and Gottesman, 1978). The ρ-dependent termination signals have not revealed any general features as yet. However, for both types of signals there is a region of dyad symmetry that can result in a stem–loop structure in the RNA being synthesized (Adhya and Gottes-

TABLE VIII Comparison of ρ Factor from *B. subtilis* and *E. coli*[a]

Property	*B. subtilis*	*E. Coli*
Molecular weight	94,000	200,000
Subunit molecular weight	47,000	50,000
Number of subunits	2	4
Poly C-dependent ATPase	+	+
Termination activity	+	+

[a] From Huang and Doi (1980) and Roberts (1976).

man, 1978). It will be of interest to see whether the transcription termination signals present in *B. subtilis* will also have similar features.

F. Transcription during Sporulation

During sporulation there is a sequential synthesis of new mRNAs not produced during the log phase of growth (Burgess, 1976; DiCioccio and Strauss, 1973; Sumida-Yasumoto and Doi, 1974; Yamakawa and Doi, 1971). However, the transcription of vegetative genes continues during sporulation, and mRNA from these genes comprise 60–80% of the total mRNA produced in the sporulating cells. Thus it is not surprising that vegetative $E\sigma^{55}$ is found in sporulating cells. The distribution of these mRNAs in the mother cell and forespore is unknown; however, an analysis of dormant spore mRNAs revealed that the relative distribution of vegetative and sporulation-specific mRNAs was about the same as that noted for those of the total sporulating cell (Jeng and Doi, 1974). Thus transcription occurs from both the mother cell and the forespore chromosomes, and sporulation-specific mRNAs may be made from both chromosomes. Nakayama *et al.* (1981) have made the interesting observation that the sporulation specific $E\sigma^{29}$ may exist primarily in the forespore. They and Dooley *et al.* (1980) also have data which suggest that the $E\sigma^{55}$ of sporulating cells is slightly different from that of vegetative cells. The analysis of dormant spore RPase has not revealed any special feature of the enzyme to date; however, the more recent refinements of purification have not been applied, and therefore it is not known whether RPase modifications occur in the developing forespore. It will be of great interest to see whether the modified RPases reported in sporulating cells are concerned with transcription of spore-specific genes. These future studies should reveal whether the promoter specificity of the enzyme has been modified, whether promoters for sporulation-specific genes are different from those for vegetative genes, and whether modified RPases are present in both the mother cell and forespore.

V. Genetics of *Bacillus subtilis* RPase

Rifampicin and streptolydigin resistances have been mapped between the *cysA* and *strA* loci on the *B. subtilis* map (Harford and Sueoka, 1970; Haworth *et al.*, 1973; Sonenshein *et al.*, 1974) (Fig. 7). The gene for streptolydigin resistance (*Std*) always mapped to the right of the gene for rifampicin resistance (*Rfm*), and no interspersion of these markers was ever observed. Also, the mapping distances indicated that Rif and

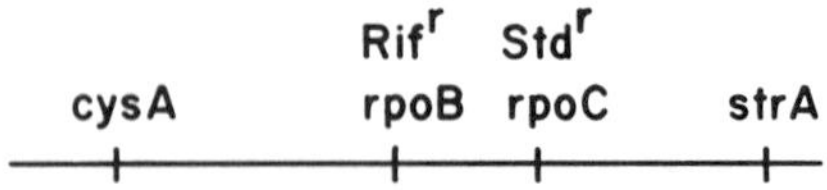

Fig. 7. Genetic map for *B. subtilis* RNA polymerase β and β' subunits. From Haworth and Brown (1973) and Halling *et al.* (1977a, 1978).

Std could either be in the same or contiguous genes. The subunit reconstitution studies for RPase of Halling *et al.* (1977a, 1978) clearly showed that the largest core subunit β was responsible for rifampicin resistance and that the second largest subunit β' was responsible for streptolydigin resistance. Therefore, they were able to show that Rif^r was the *rpo* B locus and Std^R was the *rpo* C locus. In these studies they used the Std mutant OSB406, which mapped the closest to the Rif markers (Haworth *et al.,* 1973). Thus these results showed an interesting difference between the *B. subtilis* and *E. coli* enzymes, since the β subunit in the *E. coli* core is responsible for both Rif and Std sensitivity, whereas these functions are in the β and β' subunits, respectively, in *B. subtilis*. The map distances indicate that the genes for β and β' could be in the same operon.

Sonenshein and his colleagues have demonstrated that one of the core subunits is responsible for lipiarmycin sensitivity (Sonenshein *et al.,* 1977). Two of the Lpm mutants map between the Rif and Std markers, suggesting that either β or β' is responsible for lipiarmycin resistance. *In vitro* the antibiotic blocked formation of the first internucleotide bond and preferentially inhibited Eσ^{55} more than E activity (Sonenshein and Alexander, 1979; Sonenshein *et al.,* 1977). The preferential interaction of lipiarmycin with Eσ^{55} suggests the intriguing possibility of finding σ^{55} mutants of *B. subtilis* with this drug.

Mutations in the β gene causing rifampicin resistance have been analyzed for their effects on sporulation (Doi, 1977b). Several classes of rifampicin-resistant mutants have been observed in terms of their sporulation phenotype (Table IX). In many cases a mutation in the β gene caused no apparent effect on growth or sporulation at any temperature. With other mutants, growth and the degree of sporulation appeared normal; however, morphological aberrations were observed for most of the spores (Doi *et al.,* 1970; Korch and Doi, 1971). Another group grew and sporulated normally at permissive temperatures, and grew normally but did not sporulate at the nonpermissive temperature (Hoganson *et al.,* 1975; Leighton, 1973; Sumida-Yasumoto and Doi, 1977). Thus mutations in the β gene can have a variety of effects on the sporulation process, and the resulting phenotypes indicate that several sites on the β polypeptide are probably involved in binding rifampicin. Depending on the

TABLE IX Classes of Rifampicin-Resistant Mutants[a]

Class	Phenotype
I	Grow and sporulate normally at all temperatures
II	Grow and sporulate at all temperatures, but spores have altered morphology
III	Grow but do not sporulate at any temperature or at reduced frequency (oligosporogenic); spores appear normal
IV	Grow and sporulate normally at permissive temperatures; grow normally but do not sporulate at nonpermissive temperatures; these are Rif^r Spo^{ts}
V	Grow only at permissive temperatures; these are Rif^r *ts*

[a] From Haworth and Brown (1973); Sonenshein *et al.* (1974); Leighton (1973); Sumida-Yasumoto and Doi (1977); Doi *et al.* (1970); Korch and Doi (1971); Santo *et al.* (1973).

site of mutation in the β gene, the activity remained virtually unaffected, resulting in normal growth and sporulation; in other cases growth was normal, but transcription was slightly altered during sporulation, resulting in morphologically altered spores; in the third situation the mutation in the β gene must have affected the enzyme in such a manner that it functioned essentially normally during growth but could not function properly for initiation of sporulation and during sporulation under non-permissive temperatures. The latter results can be interpreted in at least two basic ways: (1) The mutation caused a conformational change in the enzyme, which prevented either a direct interaction with sporulation gene promoters or with factors that directed it to sporulation gene promoters; these factors, for instance, could be highly phosphorylated nucleotides (Nishino *et al.*, 1979; Rhaese and Groscurth, 1976), modifying enzymes such as kinases, or polypeptides that associate with the core to modify its specificity. (2) The mutation altered the enzyme conformation, which prevented normal activity and regulation, resulting in disruption of the normal physiological network system essential for sporulation (Leighton, personal communication); since sporulation occurs under starvation conditions, sporulation would be more sensitive to stress than growth, which would explain the normal growth pattern seen for the Rif^r Spo^{ts} mutants. In fact a combination of these factors could cause the asporogenous state in these mutants.

Mutations in the β subunit gene can cause other pleiotropic effects. Fifteen Rif^r mutants were found to be defective in glutamate synthase activity; among these, 3 were also found to be blocked at stage 0 of sporulation, whereas the other 12 sporulated normally (Ryu, 1978; Ryu and Takayanagi, 1979). These results could be explained by a lack of recognition of the promoters of the glutamate synthase gene and sporulation-specific genes by the mutated RPase; however, another plausible explanation, whenever sporulation is affected, is that the RPase mutation affects the finely tuned physiological network discussed previously by Leighton (personal communication). In another study it was found that a Rif^r mutation could suppress the temperature-sensitive sporulation of an elongation factor G mutant (Hirochika and Kobayashi, 1978). A possible explanation of these results is that some direct or indirect interaction occurs between the transcription and translation machineries that control sporulation. In this regard mutations of the β gene have been observed that partially suppress the Spo0C phenotype and convert it to an Spo0A phenotype, suggesting a possible interaction between RPase β subunit and the Spo0C gene product (Ikeuchi *et al.,* 1979). The complex nature of the Rif^r mutation is further exemplified by the observation that infection of Rif^r Spo^{ts} cells by phage PMB12 converts these cells to Spo$^+$ (Bramucci *et al.,* 1977). The suppression of the Spo^{ts} phenotype can be interpreted in a number of ways, including the interaction of a phage-coded polypeptide with the host RPase, which corrects its function, or substitution of a phage-coded RPase for the mutated host RPase in the transcription of sporulation-specific genes. Although firm biochemical explanations are not available for these phenomena, the results suggest strongly that the RPase interacts with a number of other cellular components during sporulation and that a mutation in the β subunit affects this interaction, resulting either in the lack of expression of certain sporulation genes or in derangement of the highly balanced physiological state necessary for sporulation.

The genes for the β and σ^{55} subunits have been conserved highly during the evolution of bacilli. It has been shown that the antibodies for these two subunits from *B. subtilis* are able to cross-react with the subunits from a variety of *Bacillus* species including *B. stearothermophilus* (52% G + C), *B. licheniformis* (45% G + C), *B. amyloliquefaciens* (44% G + C), *B. pumilus* (41% G + C), *B. megaterium* (37% G + C), and *B. cereus* (32% G + C) (Doi *et al.*, 1980); also, cross-reaction has been noted with the RPase from several other gram-positive organisms including *Lactobacillus casei* (46% G + C), *Streptococcus lactis* (38% G + C), and *Micrococcus luteus* (72% G + C). Since the antibody cross-reacted with the σ and β subunits from such a wide range of organisms, it indicated that these genes may have evolved early and maintained their basic properties in a highly conservative manner. On the other hand, cross-reaction studies with gram-negative organisms including *E. coli* did not provide unequivocal results in one study (Doi *et al.*, 1980), although cross-reaction has been reported by others (Fukuda *et al.*, 1977). Thus it is possible that these genes have been conserved highly among both gram-positive and gram-negative organisms. The fact that heterologous combinations of core and σ factor are active (Shorenstein and Losick, 1973b; Whiteley and Hemphill, 1970) supports this possibility.

VI. RPase of Phage PBS2

The infection of *B. subtilis* by PBS2 can occur in the presence of rifampicin which ordinarily is a potent inhibitor of host RPase. An analysis of this phenomenon has revealed the probable presence of a rifampicin-resistant RPase in the phage particle, which is involved in the early transcription of the phage genome (Price and Frabotta, 1972; Price *et al.*, 1974). Furthermore, for late transcription, a rifampicin-resistant RPase encoded by the phage has been identified (Clark, 1978; Clark *et al.*, 1974). The complex enzyme contains four subunits with molecular weights of 80,000, 76,000, 58,000, and 48,000. The enzyme is resistant to both rifampicin and streptolydigin and anti-*B. subtilis* RPase antibody. Therefore it is probable that the PBS2 promoters may be quite distinct from the *B. subtilis* chromosomal promoters and require two new enzymes for transcription of the early and late genes of this phage. These observations again suggest that nonoverlapping spectra of promoters exist in *B. subtilis* cells and their phages. Further analysis of the PBS2 system should clarify this point.

VII. Conclusions

The structural complexity of the RPase in vegetative and sporulating cells of *B. subtilis* indicates that a major mechanism for regulating its activity and specificity may involve a family of σ-like polypeptides that guide the enzyme to different promoter sequences. This is in stark contrast to the *E. coli* transcription machinery in which only a single holoenzyme has been observed. Furthermore, the basic structural and functional differences between *B. subtilis* and *E. coli* RPases and their RPase–DNA binary complexes emphasize a separate evolutionary pathway for these organisms and perhaps for all gram-positive and gram-negative organisms.

The need for a high degree of transcription specificity, particularly during sporulation, is evident from the fact that sporulation occurs when cells are under extreme stress and failure of sporulation would result in lysis and death of the cell. The polarity of events during sporulation resembles the polarity of sequential events during replication of phage, which also have evolved a highly specific transcription apparatus to ensure against an abortive replication cycle that would result in death. Thus during differentiation regulation of the transcription machinery does not have to be reversible, and the conversion of part of the enzyme to a highly specific form may have evolved to accommodate this need.

The implication of this type of regulatory mechanism is that spectra of promoter sequences are present and, although the evidence is quite meager to date, the data do suggest that promoters for different forms of RPase have different recognition sequences. Future studies on promoter sequences in *B. subtilis* should be very enlightening not only in terms of differential transcription in bacilli but in terms of the comparative properties of promoters in various prokaryotes.

The successful development of the methods for RPase purification and characterization and the future utilization of DNA recombinant technology for the isolation of specific promoter-containing DNA fragments should allow rapid progress in our understanding of gene expression in *B. subtilis*.

Acknowledgments

The author thanks his former colleagues Shirley M. Halling, Kenneth C. Burtis, Valerie Williamson, Margaret M. Dooley, Tatsuo Nakayama, Ryuji Fukuda, Chikako Sumida-Yasumoto, Toshio Yamakawa, Javier Sanchez-Anzaldo, Terrance Leighton, Phillip K. Freese, Gerald Keilman, Leatrice Santo, Yoshito Sadaie, Cynthia Dickel, and Lyle Brown for their contributions to the study of RNA polymerase in his laboratory. The author also expresses his appreciation to Peter Geiduschek, Richard Losick, A. L. Sonenshein, and Margarita Salas for sending him reprints pertinent to this review. The author's reported research was supported in part by grants from the National Science Foundation and the National Institute of General Medical Sciences.

References

Adhya, S., and Gottesman, M. (1978). *Ann. Rev. Biochem.* **47,** 967–996.

Avila, J., Hermoso, J. M., Vinuela, E., and Salas, M. (1970). *Nature (London)* **226,** 1244–1245.

Avila, J., Hermoso, J. M., Vinuela, E., and Salas, M. (1971). *Eur. J. Biochem.* **21,** 526–535.

Babinet, C. (1967). *Biochem. Biophys. Res. Commun.* **26,** 639–644.

Beaman, B. L., Burtis, K. C., Doi, R. H., Yeggy, J. P., and Stahly, D. P. (1980). *Can. J. Microbiol.* **26,** 420–426.

Bramucci, M. G., Keggins, K. M., and Lovett, P. S. (1977). *J. Virol.* **24,** 194–200.

Burgess, R. R. (1969). *J. Biol. Chem.* **244,** 6160–6167.

Burgess, R. R. (1976). *In* "RNA Polymerase" (R. Losick and M. Chamberlin, eds.), pp. 69–100. Cold Spring Harbor Laboratory, Cold Spring Harbor, New York.

Chamberlin, M. (1974). *Ann. Rev. Biochem.* **43,** 721–775.

Chelm, B. K., and Geiduschek, E. P. (1979). *Nucleic Acids Res.* **7,** 1851–1867.

Chi, N.-Y. W., Ehrlich, S. D., and Lederberg, J. (1978). *J. Bacteriol* **133,** 816–821.

Clark, S. (1978). *J. Virol.* **25,** 224–237.

Clark, S., Losick, R., and Pero, J. (1974). *Nature (London)* **252,** 21–24.

Davison, B. L., Leighton, T., and Rabinowitz, J. C. (1979). *J. Biol. Chem.* **254,** 9220–9226.

DiCioccio, R. A., and Strauss, N. (1973). *J. Mol. Biol.* **77,** 325–336.

Dickel, C. D., Burtis, K. C., and Doi, R. H. (1980). *Biochem. Biophys. Res. Commun.* **95,** 1789–1795.

Doi, R. H. (1977a). *Bacteriol. Rev.* **41,** 568–594.

Doi, R. H. (1977b). *Ann. Rev. Genet.* **11,** 29–48.

Doi, R. H., and Igarashi, R. T. (1964). *Proc. Nat. Acad. Sci. USA* **52,** 755–762.

Doi, R. H., Brown, L. R., Rodgers, G., and Hsu, Y.-P. (1970). *Proc. Nat. Acad. Sci. USA* **66,** 404–410.

Doi, R. H., Halling, S. M., Williamson, V. M., and Burtis, K. C. (1980). *In* "Genetics and Evolution of RNA Polymerase, tRNA, and Ribosomes" (S. Osawa, H. Ozeki, H. Uchida and T. Yura, eds.), pp. 117–136, Univ. of Tokyo Press, Tokyo.

Dooley, M. M., Halling, S. M. and Doi, R. H. (1980). *Biochim. Biophys. Acta* **610,** 158–166.

Duffy, J. J., and Geiduschek, E. P. (1975). *J. Biol. Chem.* **250,** 4530–4541.

Duffy, J. J., and Geiduschek, E. P. (1976). *Cell* **8,** 595–604.

Duffy, J. J., Petrusek, R. L., and Geiduschek, E. P. (1975). *Proc. Nat. Acad. Sci. USA* **72,** 2366–2370.

Duie, P., Kaminski, M., and Szulmajster, J. (1974). *FEBS Lett.* **48,** 214–217.

Ehrlich, S. D., Bursztyn-Pettegrew, H., Stroynowski, I., and Lederberg, J. (1976). *Proc. Nat. Acad. Sci. USA* **73,** 4145–4149.

Fukuda, R., and Doi, R. H. (1977). *J. Bacteriol.* **129,** 422–432.

Fukuda, R., Keilman, G., McVey, E., and Doi, R. H. (1975). *In* "RNA polymerase pattern of sporulating *Bacillus subtilis*" (P. Gerhardt, R. N. Costilow, and H. L. Sadoff, eds.), pp. 213–220. American Society for Microbiology, Washington, D.C.

Fukuda, R., Ishihama, A., Saitoh, T., and Taketo, M. (1977). *Mol. Gen. Genet.* **154,** 135–144.

Goldfarb, D. S., Doi, R. H., and Rodriguez, R. L. (1981). *Nature (London),* **293,** 309–311.

Gonzalez, N., Wiggs, J., and Chamberlin, M. J. (1977). *Arch. Biochem. Biophys.* **182,** 404–408.

Greenleaf, A. L., and Losick, R. (1973). *J. Bacteriol.* **116,** 290–294.

Greenleaf, A. L., Linn, T. G., and Losick, R. (1973). *Proc. Nat. Acad. Sci. USA* **70,** 490–494.

Haldenwang, W. G., and Losick, R. (1980a). *Nature (London)* **282,** 256–260.

Haldenwang, W. G., and Losick, R. (1980b). *Proc. Nat. Acad. Sci. USA* **77,** 7000–7004.

Haldenwang, W. G., Lang, N., and Losick, R. (1981). *Cell* **23,** 615–624.

Halling, S. M., and Doi, R. H. (1978). *Biochem. Biophys. Res. Commun.* **80,** 349–354.

Halling, S. M., Burtis, K. C., and Doi, R. H. (1977a). *J. Biol. Chem.* **252,** 9024–9031.

Halling, S. M., Sanchez-Anzaldo, F. J., Fukuda, R., Doi, R. H., and Meares, C. F. (1977b). *Biochemistry* **16,** 2880–2884.

Halling, S. M., Burtis, K. C., and Doi, R. H. (1978). *Nature (London)* **272,** 837–839.

Harford, N., and Sueoka, N. (1970). *J. Mol. Biol.* **51,** 267–286.

Haworth, S. R., and Brown, L. R. (1973). *J. Bacteriol.* **114,** 103–113.

Hinkle, D. C., and Chamberlin, M. J. (1972). *J. Mol. Biol.* **70,** 157–185.

Hinkle, D. C., Mangel, W. F., and Chamberlin, M. J. (1972). *J. Mol. Biol.* **70,** 209–220.
Hirochika, H., and Kobayashi, Y. (1978). *J. Bacteriol.* **136,** 983–993.
Hoganson, D. A., Irgens, R. L., Doi, R. H., and Stahly, D. P. (1975). *J. Bacteriol.* **124,** 1628–1629.
Horiuchi, S., Uozumi, T., Hoshino, T., Ozaki, A., Nakajima, S., Beppu, T., and Arima, K. (1977). *Mol. Gen. Genet.* **157,** 175–182.
Hwang, J.-Y., and Doi, R. H. (1980). *Eur. J. Biochem.* **104,** 313–320.
Ikeuchi, T., Babasaki, K., and Kurahashi, K. (1979). *J. Bacteriol.* **139,** 327–332.
Inciarte, M. R., Lazaro, J. M., Salas, M., and Vinuela, E. (1976a). *Virology* **74,** 314–323.
Inciarte, M. R., Vinuela, E., and Salas, M. (1976b). *Eur. J. Biochem.* **71,** 77–83.
Ishihama, A. (1972). *Biochemistry* **11,** 1250–1258.
Iwakura, Y., Ishihama, A., and Yura, T. (1973). *Mol. Gen. Genet.* **121,** 181–196.
Jaehning, J. A., Wiggs, J. L., and Chamberlin, M. J. (1979). *Proc. Nat. Acad. Sci. USA* **76,** 5470–5474.
Jeng, Y.-H., and Doi, R. H. (1974). *J. Bacteriol.* **119,** 514–521.
Jiminez, F., Avila, J., Vinuela, E., and Salas, M. (1974). *Biochim. Biophys. Acta* **349,** 320–327.
Jones, B. B., Chan, H., Rothstein, S., Wells, R. D., and Reznikoff, W. S. (1977). *Proc. Nat. Acad. Sci. USA* **74,** 4914–4918.
Kawamura, F., and Ito, J. (1974). *Virology* **62,** 414–425.
Kawamura, F., and Ito, J. (1977). *J. Virol.* **23,** 562–577.
Keilman, G. R., Tanimoto, B., and Doi, R. H. (1975). *Biochem. Biophys. Res. Commun.* **67,** 414–420.
Keilman, G. R., Burtis, K., Tanimoto, B., and Doi, R. H. (1976). *J. Bacteriol.* **128,** 80–85.
Korch, C. T., and Doi, R. H. (1971). *J. Bacteriol.* **105,** 1110–1118.
Krakow, J. S. (1971). *Methods Enzymol.* **22,** 520–528.
Kreft, J., Bernard, K., and Goebel, W. (1978). *Mol. Gen. Genet.* **162,** 59–67.
Kudo, T., Jaffe, D., and Doi, R. H. (1981). *Mol. Gen. Genet.* **181,** 63–68.
Leighton, T. J. (1973). *Proc. Nat. Acad. Sci. USA* **70,** 1179–1183.
Leighton, T. J. (1977). *J. Biol. Chem.* **252,** 268–272.
Leighton, T. J., and Doi, R. H. (1971). *J. Biol. Chem.* **246,** 3189–3195.
Leighton, T. J., Doi, R. H., Warren, R. A. J., and Kelln, R. A. (1973). *J. Mol. Biol.* **76,** 103–122.
Lill, U. I., Kniep-Behrendt, E. M., Bock, L., and Hartmann, G. R. (1977). *Hoppe-Seylers's Z. Physiol. Chem.* **358,** 1591–1603.
Linn, T. G., Greenleaf, A. L., Shorenstein, R. G., and Losick, R. (1973). *Proc. Nat. Acad. Sci. USA* **70,** 1865–1869.
Linn, T., Greenleaf, A. L., and Losick, R. (1975). *J. Biol. Chem.* **250,** 9256–9261.
Litman, R. M. (1968). *J. Biol. Chem.* **243,** 6222–6233.
Losick, R. (1972). *Ann. Rev. Biochem.* **41,** 409–446.
Losick, R., and Pero, J. (1976a). *Adv. Enzymol.* **44,** 165–185.
Losick, R., and Pero, J. (1976b). *In* "RNA Polymerase" (R. Losick and M. Chamberlin, eds.), pp. 227–246. Cold Spring Harbor Laboratory, Cold Spring Harbor, New York.
Losick, R., Shorenstein, R. G., and Sonenshein, A. L. (1970). *Nature (London)* **227,** 910–913.
Lowery-Goldhammer, C., and Richardson, J. P. (1974). *Proc. Nat. Acad. Sci. USA* **71,** 2003–2007.
Mahler, I., and Halvorson, H. O. (1977). *J. Bacteriol.* **131,** 374–377.
Maia, J. C. C., Kerjan, P., and Szulmajster, J. (1971). *FEBS Lett.* **13,** 269–274.
Mangel, W. F., and Chamberlin, M. J. (1974). *J. Biol. Chem.* **249,** 3007–3013.

Nagahari, K., and Sakaguchi, K. (1978). *Mol. Gen. Genet.* **158,** 263–270.

Nakayama, T., Munoz, L., and Doi, R. H. (1977). *Anal. Biochem.* **78,** 165–170.

Nakayama, T., Munoz, L. E., Sadaie, Y., and Doi, R. H. (1978a). *J. Bacteriol.* **135,** 952–960.

Nakayama, T., Williamson, V., Burtis, K., and Doi, R. H. (1978b). *Eur. J. Biochem.* **88,** 155–164.

Nakayama, T., Irikura, M., Kurogi, Y., and Matsuo, H. (1981). *J. Biochem. (Tokyo)* **89,** 1681–1691.

Nishimoto, H., and Takahashi, I. (1974). *Can. J. Biochem.* **52,** 966–973.

Nishino, T., Gallant, J., Shalit, P., Palmer, L., and Wehr, T. (1979). *J. Bacteriol.* **140,** 671–679.

Orrego, C., Kerjan, P., Manca de Nadra, M. C., and Szulmajster, J. (1973). *J. Bacteriol.* **116,** 636–647.

Pero, J., Nelson, J., and Fox, T. D. (1975). *Proc. Nat. Acad. Sci. USA* **72,** 1589–1593.

Petrusek, R., Duffy, J. J., and Geiduschek, E. P. (1976). *In* "RNA Polymerase" (R. Losick and M. J. Chamberlin, eds.), pp. 567–585. Cold Spring Harbor Laboratory, Cold Spring Harbor, New York.

Pfeffer, S. R., Stahl, S. J., and Chamberlin, M. J. (1977). *J. Biol. Chem.* **252,** 5403–5407.

Plevani, P., Albertini, A. M., Galizzi, A., Adamoli, A., Mastromei, G., Riva, S., and Cassani, G. (1977). *Nucleic Acids Res.* **4,** 603–623.

Price, A. R., and Frabotta, M. (1972). *Biochem. Biophys. Res. Commun.* **48,** 1578–1585.

Price, A. R., Hitzeman, R., Frato, J., and Lombardi, K. (1974). *Nucleic Acids Res.* **1,** 1497–1502.

Rabussay, D., and Zillig, W. (1969). *FEBS Lett.* **5,** 104–106.

Rhaese, H. J., and Groscurth, R. (1976). *Proc. Nat. Acad. Sci. USA* **73,** 331–335.

Rice, L. S., and Meares, C. F. (1978). *Biochem. Biophys. Res. Commun.* **80,** 26–32.

Roberts, J. W. (1969). *Nature (London)* **224,** 1168–1174.

Roberts, J. W. (1976). *In* "RNA Polymerase" (R. Losick and M. Chamberlin, eds.), pp. 247–271. Cold Spring Harbor Laboratory, Cold Spring Harbor, New York.

Rosenberg, M., and Court, D. (1979). *Ann. Rev. Genet.* **13,** 319–353.

Ryu, J. (1978). *J. Bacteriol.* **135,** 408–414.

Ryu, J.-I., and Takayanagi, S. (1979). *J. Bacteriol.* **139,** 652–656.

Santo, L., Leighton, T. J., and Doi, R. H. (1973). *J. Bacteriol.* **115,** 703–706.

Schaller, H., Nusslein, C., Bonhoeffer, F. J., Kurz, C., and Nietzschmann, I. (1972). *Eur. J. Biochem.* **26,** 474–481.

Segall, J., and Losick, R. (1977). *Cell* **11,** 751–761.

Shorenstein, R. G., and Losick, R. (1973a). *J. Biol. Chem.* **248,** 6163–6169.

Shorenstein, R. G., and Losick, R. (1973b). *J. Biol. Chem.* **248,** 6170–6173.

Siebenlist, U., and Gilbert, W. (1980). *Proc. Nat. Acad. Sci. USA* **77,** 122–126.

Simpson, R. B. (1979). *Proc. Nat. Acad. Sci. USA* **76,** 3233–3237.

Sogo, J. M., Inciarte, M. R., Corral, J., Vinuela, E., and Salas, M. (1979a). *J. Mol. Biol.* **127,** 411–436.

Sogo, J. M., Rodeno, P., Koller, T., Vinuela, E., and Salas, M. (1979b). *Nucleic Acids Res.* **7,** 107–120.

Sonenshein, A. L., and Alexander, H. B. (1979). *J. Mol. Biol.* **127,** 55–72.

Sonenshein, A. L., Cami, B., Brevet, J., and Cote, R. (1974). *J. Bacteriol.* **120,** 253–265.

Sonenshein, A. L., Alexander, H. B., Rothstein, D. M., and Fisher, S. H. (1977). *J. Bacteriol.* **132,** 73–79.

Spiegelman, G. B., and Whiteley, H. R. (1974a). *J. Biol. Chem.* **249,** 1476–1482.

Spiegelman, G. B., and Whiteley, H. R. (1974b). *J. Biol. Chem.* **249,** 1483–1489.

Spiegelman, G. B., and Whiteley, H. R. (1978). *Biochem. Biophys. Res. Commun.* **81,** 1058–1065.
Spiegelman, G. R., and Whiteley, H. R. (1979). *Biochem. Biophys. Res. Commun.* **87,** 811–817.
Spiegelman, G. B., Hiatt, W. R., and Whiteley, H. R. (1978). *J. Biol. Chem.* **253,** 1756–1765.
Sternbach, H., Engelhardt, R., and Lezius, A. G. (1975). *Eur. J. Biochem.* **60,** 51–55.
Stetter, K. O. (1977). *Hoppe-Seyler's Z. Physiol. Chem.* **358,** 1093–1104.
Stetter, K. O., and Zillig, W. (1974). *Eur. J. Biochem.* **48,** 527–540.
Sumida-Yasumoto, C., and Doi, R. H. (1974). *J. Bacteriol.* **117,** 775–782.
Sumida-Yasumoto, C., and Doi, R. H. (1977). *J. Bacteriol.* **129,** 433–444.
Talkington, C., and Pero, J. (1978). *Proc. Nat. Acad. Sci. USA* **75,** 1185–1189.
Talkington, C., and Pero, J. (1979). *Proc. Nat. Acad. Sci. USA* **76,** 5465–5469.
Thompson, S. T., Cass, K. H., and Stellwagen, E. (1975). *Proc. Nat. Acad. Sci. USA* **72,** 669–672.
Tjian, R., and Losick, R. (1974). *Proc. Nat. Acad. Sci. USA* **71,** 2872–2876.
Tjian, R., and Pero, J. (1976). *Nature (London)* **262,** 753–757.
Tjian, R., Stinchcomb, D., and Losick, R. (1974). *J. Biol. Chem.* **250,** 8824–8828.
Tjian, R., Losick, R., Pero, J., and Hinnebush, A. (1977). *Eur. J. Biochem.* **74,** 149–154.
Wartell, R. M., Larson, J. E., and Wells, R. D. (1974). *J. Biol. Chem.* **249,** 6719–6731.
Whiteley, H. R., and Hemphill, H. E. (1970). *Biochem. Biophys. Res. Commun.* **41,** 647–654.
Wiggs, J. L., Bush, J. W., and Chamberlin, M. J. (1979). *Cell* **16,** 97–109.
Wiggs, J. L., Gilman, M. Z., and Chamberlin, M. J. (1981). *Proc. Natl. Acad. Sci. USA* **78,** 2762–2766.
Williamson, V. W., and Doi, R. H. (1978). *Mol. Gen. Genet.* **161,** 135–141.
Williamson, V. M., and Doi, R. H. (1979). *Mol. Gen. Genet.* **174,** 47–52.
Wu, C.-W., Wu, F.Y.-H., and Speckhard, D. C. (1977). *Biochemistry* **16,** 5449–5454.
Yamakawa, T., and Doi, R. H. (1971). *J. Bacteriol.* **106,** 305–310.
Yehle, C. O., and Doi, R. H. (1967). *Virology* **1,** 935–947.
Yura, T., and Ishihama, A. (1979). *Ann. Rev. Genet.* **13,** 59–97.
Zillig, W. *et al.* (1970). *Cold Spring Harbor Symp. Quant. Biol.* **35,** 47–58.
Zillig, W., Palm, P., and Heil, A. (1976). *In* "RNA Polymerase" (R. Losick and M. Chamberlin, eds.), pp. 101–125. Cold Spring Harbor Laboratory, Cold Spring Harbor, New York.
Zillig, W., Stetter, K. O., Tobien, M. (1978). *Eur. J. Biochem.* **91,** 193–199.
Zillig, W., Stetter, K., and Janekovic, D. (1979). *Eur. J. Biochem.* **96,** 597–604.

4

The Translational Apparatus of *Bacillus subtilis*

ISSAR SMITH

I. Introduction

The genetic determinants for the translational apparatus must be regulated in such a manner that the synthesis of the multiple components of the protein synthetic machinery and their interactions can rapidly respond to changes in the external environment. In this way instantaneous changes in the rate of growth and synthesis of new enzymes enable the prokaryotic cell to adjust to the altered milieu. Also, bacilli sporulate under certain environmental conditions. New proteins are formed during this developmental cycle.

The Molecular Biology
of the Bacilli

ISBN 0-12-222701-8

The basic mechanisms of protein synthesis have been elucidated [these are reviewed thoroughly in Weissbach and Pestka (1977)], and much is known about the genetics of the translation apparatus, mainly from studies on *Escherichia coli* (Nomura, 1976). Much remains to be done in the area of regulation, however. How is the synthesis of the multiple components of the protein synthetic machinery regulated to respond quickly to environmental changes? This, among other questions, will be discussed in this chapter, with specific reference to bacilli. Since most work in this area has been performed on *E. coli,* for reasons cited below, there will be an attempt to compare salient features of the translational apparatus in both *E. coli* and bacilli, mostly *Bacillus subtilis.*

This study is not intended to be an exhaustive review but tries to examine, critically, the state of our knowledge of the physiology and the genetics of translation in bacilli with an emphasis on problems and future directions.

II. The Translation Mechanism in *Escherichia coli* and Bacilli

The mechanism of protein synthesis is an exceedingly complex one. Over 200 components are involved in the process whereby the information encoded in a linear RNA molecule, formed by the transcription of a DNA sequence, is translated into a finished polypeptide. Since most of the components of these systems can be interchanged between *E. coli* and bacilli, an exhaustive discussion of each of the steps of protein synthesis would be redundant in light of the excellent reviews that have appeared in the last few years (Weissbach and Pestka, 1977). Examples of the interchangeability of the various components are illustrated by the fact that *Bacillus stearothermophilus* 16 S rRNA can be used to reconstitute functional 30 S ribosomes with *E. coli* 30 S ribosomal proteins, and vice versa. *Escherichia coli* 5 and 23 S RNA likewise can be used to reconstitute 50 S ribosomes in the presence of *B. stearothermophilus* 50 S ribosomal proteins (Nomura, 1973). Initiation factors from *B. stearothermophilus* will work with *E. coli* ribosomes, and vice versa (Lodish, 1969). *Escherichia coli* elongation factors EF-G and EF-Tu can complement *B. subtilis* protein synthetic systems that have temperature-sensitive lesions in these enzymes (Dubnau *et al.,* 1976). These factors in *B. stearothermophilus* and *B. subtilis* will also work with *E. coli* ribosomes (Waterson *et al.,* 1970; Sala *et al.,* 1974).

A good illustration of the molecular basis of the interchangeability of some components is observed when the N-terminal amino acid sequences of all the 30 S ribosomal proteins and several of the 50 S proteins from

E. coli, B. stearothermophilus, and *B. subtilis* are compared (Yaguchi *et al.*, 1978; Higo *et al.*, 1980). In the sequences compared there was 84% correlation between the two bacilli and 50% between *B. stearothermophilus* and *E. coli*. This high degree of similarity has allowed a functional correlation of ribosomal proteins in the three organisms and a unified nomenclature for 30 S ribosomal proteins in *B. subtilis* (Tables I and II), as has already been published for *E. coli–B. stearothermophilus* (Yaguchi *et al.*, 1978; Higo *et al.*, 1980).

There are some minor differences in the requirements for protein synthesis *in vitro;* i.e., potassium ions inhibit *B. subtilis* elongation reactions, unlike the case in *E. coli* (Sala *et al.*, 1974), and *B. subtilis* ribosomes may be more unstable than their counterparts in *E. coli* (Takeda and Lipmann, 1966). However, the most important difference is the interspecific expression barrier found between bacilli (and other gram-positive bacteria) and *E. coli*. This is manifested by an inability of *E. coli* genes to express in *B. subtilis* and *in vitro* by the discrimination exhibited by

TABLE I Revised Nomenclature for *B. subtilis* 30 S Ribosomal Proteins

Old nomenclature[a]	New nomenclature[b]
S1	S2
S2	S3
S3	—
S4	S4
S5	S5
S6	—
S7	S7
S8	S8
S9	S6
S10	S9
S11	S11
S12	S12
S13	S10
S14	S13
S15	—
S16	S17
S17	S16
S18	S15
S19	S19
S20	S20
S21	S18
—	S14
—	S21

[a] According to Geisser *et al.* (1973).
[b] Higo *et al.* (1980).

TABLE II Correlation between Some *E. coli* and *B. subtilis* 50 S Ribosomal Proteins[a]

B. subtilis 50 S ribosomal protein	*E. coli* counterpart
B-L2	E-L2
B-L7	E-L3
B-L8	E-L6
B-L9	E-L12
B-L11	E-L11
B-L15	E-L17
B-L17	E-L22
B-L21	E-L20
B-L23	E-L24

[a] Data from Higo *et al.* (1980) and Smith *et al.* (1980a).

bacillus ribosomes toward *E. coli* mRNAs, preventing translation of these messages. The barriers are much less restrictive in the other direction, so that *E. coli* ribosomes can translate gram-positive mRNAs *in vivo* and *in vitro*. This one-way barrier has been reported for *E. coli* plasmid genes determining resistance to various antibiotics cloned together with *B. subtilis* plasmids determining resistance to other drugs. These chimeras, so-called bridge plasmids, can replicate in either host. Such chimeras express both types of antibiotic resistance in *E. coli,* but only the *B. subtilis* type expresses in *B. subtilis* (Ehrlich, 1978; Kreft *et al.,* 1978). However, it has been recently reported that the plasmid-encoded Tet^r gene from *E. coli* cloned into a bacillus plasmid will express in both organisms (Goebel *et al.,* 1979). This has not been confirmed in other laboratories as yet. Restriction of expression has also been observed with *E. coli* chromosomal genes cloned in *B. subtilis* (Grandi *et al.,* 1981).

Ribosomes from gram-positive bacteria do not translate *E. coli* mRNA or RNA isolated from RNA bacteriophages (Stallcup *et al.,* 1976) except for certain cases that will be discussed below. Although the actual *in vivo* mechanism(s) for the restriction of expression of *E. coli* genes in *B. subtilis* is not known with certainty, *in vitro* experiments indicate that the block is at the level of translation. The fact that coliphage T7 DNA fragments cloned into *B. subtilis* are transcribed is in agreement with this idea (Scherzinger *et al.,* 1980).

The most thoroughly studied *in vitro* example of translational restriction of protein synthesis is the *B. stearothermophilus* ribosome–coliphage RNA system. *Escherichia coli* ribosomes translate R17 RNA and produce three major proteins: the coat protein, the replicase, and the A or maturation protein. Recently a fourth protein, which is coded for by an overlapping gene in the related RNA phages, MS2 and f2, has been described (Atkins *et al.*, 1979; Beremand and Blumenthal, 1979). *Bacillus stearothermophilus* ribosomes translate only the A protein (Lodish, 1969). This limitation was observed to be a function of the 30 S ribosomes (Lodish, 1969, 1970); i.e., when *E. coli* 30 S particles were reassociated with *B. stearothermophilus* 50 S subunits, all three proteins were read. Subsequent to this it was shown that the 30 S proteins were responsible for the effect (Goldberg and Steitz, 1974), although other experiments demonstrated an effect of both 16 S RNA and protein S12 (Held *et al.*, 1974).

The latter results are at variance with those of other workers who showed that *E. coli* S1, when added to intact *B. stearothermophilus* ribosomes, allowed replicase and coat protein synthesis (Isono and Isono, 1975). It is difficult to compare these results, because in one case single *B. stearothermophilus* components were added to an otherwise complete *E. coli* 30 S reconstitution system to see which element was responsible for the decreased reading of R17 DNA relative to poly U incorporation. In the other experiments, *E. coli* S1 was added to intact 30 S *B. stearothermophilus* ribosomes and an *increased* reading of the two usually nontranslated peptides was observed. Interestingly, no protein corresponding to S1 has been observed in *B. stearothermophilus,* either by functional or immunological assays (Isono and Isono, 1976).

S1 is involved in the binding of mRNA to the ribosome, presumably by stabilizing the interaction between the purine-rich Shine–Delgarno sequence on the mRNA and the pyrimidine-rich 3′-end of 16 S rRNA. S1 is required for good binding between the Shine–Delgarno sequence of the coat protein and replicase messages and 16 S rRNA. The A-protein Shine–Delgarno sequence has better complementarity to the 3′-end of the *E. coli* 16 S rRNA and does not require S1 (Steitz, 1979). This would explain why *B. stearothermophilus* ribosomes can translate the A protein in the absence of *E. coli* S1. Examination of the 3′-end of 16 S RNA for *B. stearothermophilus, B. subtilis,* and *E. coli* shows high conservation of sequence (Shine and Delgarno, 1975):

Escherichia coli	G A U C A C C U C C U U A
Bacillus stearothermophilus	G A U C A C C U C C U U U C U (A)
Bacillus subtilis	G Z Z Z Z Z Z Z Y C U U U C U

where Z is any nucleotide but G and Y is pyrimidine. Unless the 3′-terminal UCU(A) present in bacilli 16 S rRNA is the determining factor in the rRNA–mRNA interaction, it appears that associated protein factors may be important in the correct registration and/or translation of the mRNA. It has been reported that *E. coli* mRNAs bind, although poorly, to *B. subtilis* ribosomes (Legault-Démare and Chambliss, 1975; Sharrock *et al.*, 1979), and it has been postulated that translation in bacilli may require another step, after initiation and before the beginning of elongation, that is not present in gram-negative bacteria (Sharrock *et al.*, 1979). It is suggested that some structural feature of gram-positive mRNA, not found in gram-negative messages, is involved in this transitional step (Sharrock *et al.*, 1979). This idea is an intriguing one but requires the sequencing data obtained from the 5′-end of several gram-positive mRNAs. This sequence information could then be compared with equivalent regions of *E. coli* mRNA species. With the development of cloning techniques in *B. subtilis* (Chapter 10, this volume), it would be feasible to clone an *E. coli* structural gene into a *B. subtilis* plasmid downstream from the leader sequence of a known gene. In this way the structural components, present in this mRNA sequence, necessary for *in vivo* translation of the *E. coli* gene in *B. subtilis* could be elucidated.

In vitro protein synthetic systems have been developed in *B. subtilis* that are dependent on RNA and DNA (Legault-Démare and Chambliss, 1974; Leventhal and Chambliss, 1979). The major technical difficulties encountered in *B. subtilis in vitro* systems have been high endogenous protease levels. This obstacle has been successfully eliminated by the use of extensive high-salt washing of cells and agents such as the protease inhibitor phenylmethylsulfonyl chloride and hemoglobin–sepharose, which quantitatively removes protease activity from cellular extracts (Nakayama *et al.*, 1978). These *in vitro* systems should be useful in translating mRNAs from the proposed composite *E. coli–B. subtilis* plasmids. This will allow identification of factors necessary for the mRNA recognition and initiation steps of protein synthesis in *B. subtilis*.

III. Chromosomal Organization of Genetic Determinants for Protein Synthesis

A. Methodology

A requirement for the analysis of genetics of a particular system has traditionally been the existence of mutant alleles that cause a loss of function. The presence of the mutated or wild-type phenotype is then used in genetic selection and consequent mapping in relation to other genes. This classic method is invaluable for studying genes for supplementable growth requirements such as amino acids or vitamins, etc., because the mutated cells can be kept alive. Nonfunctional mutations in the translational apparatus are lethal, and other methods have been developed for the isolation and mapping of such lesions.

1. *Conditional Lethal Mutations*

Mutations that cause nonsynthesis or malfunction of a protein or nucleic acid under certain environmental conditions, but not others, have been successfully employed in studying mutational changes in many components of the transcriptional and translational machinery. These methods have the advantage that primary selection can be employed. Temperature-sensitive mutations are the most frequently studied. These prevent growth at the restricting temperature.

Suppressible mutations, i.e., termination or missense mutations, which are lethal to the cell in the absence of informational suppressors, can be isolated in the presence of suppressor mutations. These, in turn, have been studied and mapped in *E. coli* and are normally mutations in genes for minor tRNA species. Temperature sensitivity has been effectively used to isolate mutations in the translational apparatus of both *E. coli* and *B. subtilis*. The enormous body of information concerning the genetics and physiology of suppressors in *E. coli* has allowed the isolation of several nonsense mutations in genes for the translational apparatus (Bollen *et al.,* 1980). Suppressor mutations have also been isolated in *B. subtilis,* but very little is known about their genetics or physiology of action. They have not been used to isolate conditional mutants in the protein synthesis machinery as yet.

2. *Resistance to Antibiotics and Other Growth Inhibitors*

In some cases, positive selection for mutations that confer resistance to bacteriostatic or bacteriocidal agents has been employed to isolate mutations in the translational apparatus. Resistance to amino acid analogs can result from mutations in genes coding for aminoacyl-tRNA synthetases, as the mutated enzyme displays an elevated K_i for the analog. Similarly, many antibiotics, which act at the level of the ribosome or protein synthesis elongation, can be used to select resistant mutations that cause lowered binding of the drug, and these can be mapped by conventional genetic techniques. This method has been widely employed to map aminoacyl-tRNA synthetase, protein synthesis elongation factor, and ribosomal protein genes in *E. coli* and *B. subtilis*. It has been especially valuable in the latter organism, since the absence of the lipopolysaccharide layer in the cell cortex of gram-positive organisms renders these bacteria sensitive to the action of many antibiotics, especially those affecting 50 S ribosomes. *Escherichia coli* and other gram-negatives are naturally impermeable and thus resistant to these drugs.

3. *Strain and Species Differences*

Frequently there are differences in the physical or biochemical properties of homologous components of the translational apparatus when two or more genetically compatible strains or species are compared. The phenotypic difference, i.e., migration of a protein through an acrylamide gel resulting from a charge or size difference or a differing nucleotide or primary amino acid sequence which is resolved by conventional sequence analysis, is then used as an unselected marker in crosses between such strains. The genetic determinants for ribosomal proteins, elongation factors, and rRNA have been mapped in *E. coli* and *B. subtilis* with these methods (Smith, 1977).

4. *Synchrotransfer Methods*

Synchronized cells have been used to study the synthesis of rRNA and tRNA in *E. coli* using the increase in chromosome gene copy number throughout the cell cycle and the resulting augmentation in transcription of stable RNA species (Rudner *et al.*, 1964; Cutler and Evans, 1967).

An elegant variation of this technique was employed by Sueoka and co-workers to map *B. subtilis* chromosomal genes (Yoshikawa and Sueoka, 1963; Oishi *et al.*, 1964). *Bacillus subtilis* spores have completed chromosomes. When spores germinate, synchronous DNA replication ensues. If density-labeled spores are allowed to germinate in a normal medium, or vice versa, DNA isolated from different stages of germination can be fractionated by isopycnic centrifugation. Heavy, hybrid, and light bands can then be assayed for transforming activity for selected markers and RNA–DNA hybridization. Markers that replicate earlier undergo density shifts at earlier times. Unfortunately, the initial assumption in these experiments was that the *B. subtilis* chromosome was linear and replicated in a linear fashion. However, it has been shown more recently that the *B. subtilis* chromosome is circular and that replication is bidirectional (Henner and Hoch, 1980; and Chapter 1). Thus, all studies in which genes were located on the *B. subtilis* chromosome solely on the basis of density transfer studies must be reevaluated.

5. *Cloning of Translational Genes*

The explosion in the amount of information presently available concerning the mapping and regulation of the translational apparatus has come about through the pioneering efforts of Nomura and other workers with *E. coli* (Nomura, 1976). Originally, several F′ plasmids with translational genes were isolated. Then, specialized transducing bacteriophages $\phi 80$ and λ were constructed by growth on *E. coli* mutants having the normal phage attachment sites on the *E. coli* chromosome deleted.

Insertion of the lamboid phages then can occur throughout the chromosome, and transducing phages containing different regions of the chromosome have been isolated. Some of these insertion events occurred in the major ribosomal regions found at 72′ and 88′ (see below). "Third-generation" clones have been developed using DNA isolated from specialized transducing phages and as vectors, high-copy plasmids, usually Col E1 derivatives or modified λ phages. The cloned DNA fragments have been used to study the map position of various translational and transcriptional genes, to localize promoters for these determinants, and to study their regulation.

Bacillus subtilis and *B. thuringiensis* rRNA genes have been cloned in *E. coli* with Col E1 derivatives. Unfortunately, F factors do not exist in *B. subtilis* and, although specialized transduction does occur in this organism (Chapter 9), the system is still too primitive to be used for cloning. For these reasons, i.e., the availability of powerful cloning techniques that start with biologically purified genes (on F′ or bacteriophage vectors), the study of the genetics of the translational mechanism in *E. coli* is ahead of similar work in bacilli.

B. Mapping of Genes for the Translational Apparatus

1. *Aminoacyl-tRNA Synthetases and tRNA Genes*

In *E. coli,* 19 genes coding for aminoacyl synthetases and 34 tRNA cistrons have been mapped (Bachman and Low, 1980). Only two genes affecting these enzymes have been mapped in *B. subtilis*. In one case the mutations causing a temperature-sensitive fluorotryptophan-resistant tryptophanyl-tRNA synthetase was located at 100°, between *arcC* and *metA* (Steinberg and Anagnostopoulos, 1971). A lysyl-tRNA synthetase mutation, causing a temperature-sensitive phenotype was mapped in another region of the chromosome, 5°, between *sul* and *purA* (Racine and Steinberg, 1974).

Even fewer genetic data are available for tRNA in *B. subtilis*. By means of the synchrotransfer method (see above) two clusters of tRNA complementary sequences were observed, one near the origin of replication and major ribosomal protein and rRNA gene cluster (see below), and the second near the chromosome terminus (Dubnau *et al.,* 1965a; Oishi *et al.,* 1966; Smith *et al.,* 1968). The close proximity of tRNA genes to rRNA cistrons is consistent with the situation in *E. coli* (see below).

Several nonsense suppressor mutations, presumably affecting the structural genes for minor tRNA species, have been isolated in *B. subtilis* (Okubo and Yanagida, 1968; Georgopoulos, 1969; Mellado *et al.,* 1976).

The *sup-3* mutation has been mapped between *aroI* and *mtlB,* 30′ on the *B. subtilis* chromosome (Henner and Steinberg, 1979a). The evidence that tRNAs have been mutated in any of the *sup* mutations is, at this moment, largely circumstantial. It is known that nonsense (shorter) peptides can be suppressed *in vivo* to give longer proteins (Shub, 1975; Mellado *et al.,* 1976). However, the biochemical nature of the suppression, i.e., the amino acid inserted to reverse termination, the termination codon (UAA, UGA, or UAG), and the tRNA altered, remain unknown. It is perhaps premature to consider any *B. subtilis sup* mutations as alterations in tRNA genes until further experimentation is performed.

2. *rRNA Genes*

There are seven operons coding for 16, 23, and 5 S rRNA in *E. coli* (Nomura, 1976). The tandem promoters for these cistrons are at the 16 S end of the cistrons (deBoer *et al.,* 1979; Gilbert *et al.,* 1979; Glaser and Cashel, 1979; Young and Steitz, 1979). Most if not all rRNA transcription units have tRNA in the spacers between the 16 and 23 S RNA genes, and some of the rRNA cistrons have tRNA genes at their 3′-ends (Morgan *et al.,* 1978).

A similar number of rRNA genes are found in *B. subtilis* in two clusters (Dubnau *et al.,* 1965a; Oishi and Sueoka, 1965; Smith *et al.,* 1968). The major cluster is found very close to the attachment site for the lysogenic bacteriophage, *spo2* (Chow and Davidson, 1973), which has been mapped between *spcA* and *lin* (Smith and Smith, 1973). The second cluster of rRNA genes was localized, by density transfer experiments, near the terminus of the chromosome (Smith *et al.,* 1968).

Gene banks of *B. thuringiensis* (Klier *et al.,* 1979) and *B. subtilis* (Moran and Bott, 1979a, b; Rapoport *et al.,* 1979) have been cloned in *E. coli* using Col E1 derivatives, and the structures of the rRNA cistrons have been analyzed. The order of the rRNA genes is 16, 23, 5 S, as in *E. coli.* Interestingly, no spacers have been observed between 16 and 23 S RNA genes in *Bacillus* rRNA transcription units, indicating that tRNA genes are not found between the rRNA determinants as in *E. coli.* This seems to be inconsistent with the idea that a function of the intragenic tRNA spacers was to prevent unequal recombination between different rRNA cistrons. All the rRNA cistrons are not necessary for viability, as it has been shown that *rrnE* in *E. coli* can be deleted with no apparent adverse effect on cell growth (Ellwood and Nomura, 1980).

3. *Ribosomal Protein Genes*

Twenty of the twenty-one 30 S ribosomal protein genes, and 27 of the 34 50 S genes have been mapped in *E. coli* (Isono, 1980). Twenty-seven of the ribosomal protein genetic determinants map in the 72′ region of the chromosome, but there are several other clusters, notably at 88′. In *B. subtilis,* 23 ribosomal protein genes have been mapped using antibiotic resistance and strain and species differences, and all have so far been found in the *strA* region (Osawa *et al.,* 1978; Smith *et al.,* 1980a; Fig. 1). A new system of nomenclature has been established for *B. subtilis* ribosomal proteins relating these proteins to the *E. coli* homologs. Use has been made of N-terminal amino acid sequence and of functional homology in comparing the ribosomal proteins from the two species (Higo *et al.,* 1980; Tables I and II). When the map position of genes for homologous ribosomal proteins are compared in *E. coli* and *B. subtilis* (Table III), all the *E. coli* genes, save the determinants for S20 (Osawa, 1976) and L11 (Smith and Smith, 1973), in which direct comparison can be made, show map positions similar to those of their *B. subtilis* counterparts. The actual order of the genes in the *strA* (72′) region is also quite similar when the two organisms are compared (Fig. 2). The case of the L11 is extremely interesting in this regard. The *B. subtilis* l11 has a blocked N-terminal amino acid, and primary sequence data are not available to allow comparisons between *E. coli* and *B. subtilis* l11 proteins (Higo *et al.,* 1980). However, resistance to the antibiotic thiostrepton is associated with the loss of protein L11 from the ribosomes of *B. subtilis* (Wienen *et al.,* 1979) and *Bacillus megaterium* (Cundliffe *et al.,* 1979), so they cannot bind thiostrepton. The L11 from bacilli shows an immunological cross-reaction with *E. coli* L11 (Cundliffe *et al.,* 1979; Wienen *et al.,* 1979). Sporangiomycin, a related antibiotic, presumably acts in the same manner as thiostrepton, and sporangiomycin-resistant mutants map close to *cysA,* as does thiostrepton resistance (Smith and Smith, 1973; Bazzicalupo *et al.,* 1975). The *tsp* mutations also cause a *relC* phenotype in *B. subtilis* (Smith *et al.,* 1978, 1980b) in a manner analogous to *relC* mutations in *E. coli,* which are found in *rplK,* the genetic determinant for L11 (Parker *et al.,* 1976). Similar studies with *B. megaterium tsp* mutations have confirmed the *B. subtilis* results (Stark and Cundliffe, 1979).

The reason for an absence of L11 in the thiostrepton-resistant mutants is presently not known. The L11 protein, and the ability of *E. coli* and *B. subtilis* 50 S ribosomes to bind thiostrepton and related antibiotics is removed from ribosomes by relatively gentle washing in high salt (Bazzicalupo *et al.,* 1975; Highland *et al.,* 1975; Parker *et al.,* 1976; Thompson *et al.,* 1979). It is possible that the *tsp* mutation alters the binding of L11 to the ribosome either by changing the L11 primary sequence or, less likely, by altering another protein that is part of the thiostrepton-binding site. However, it has not been possible to detect

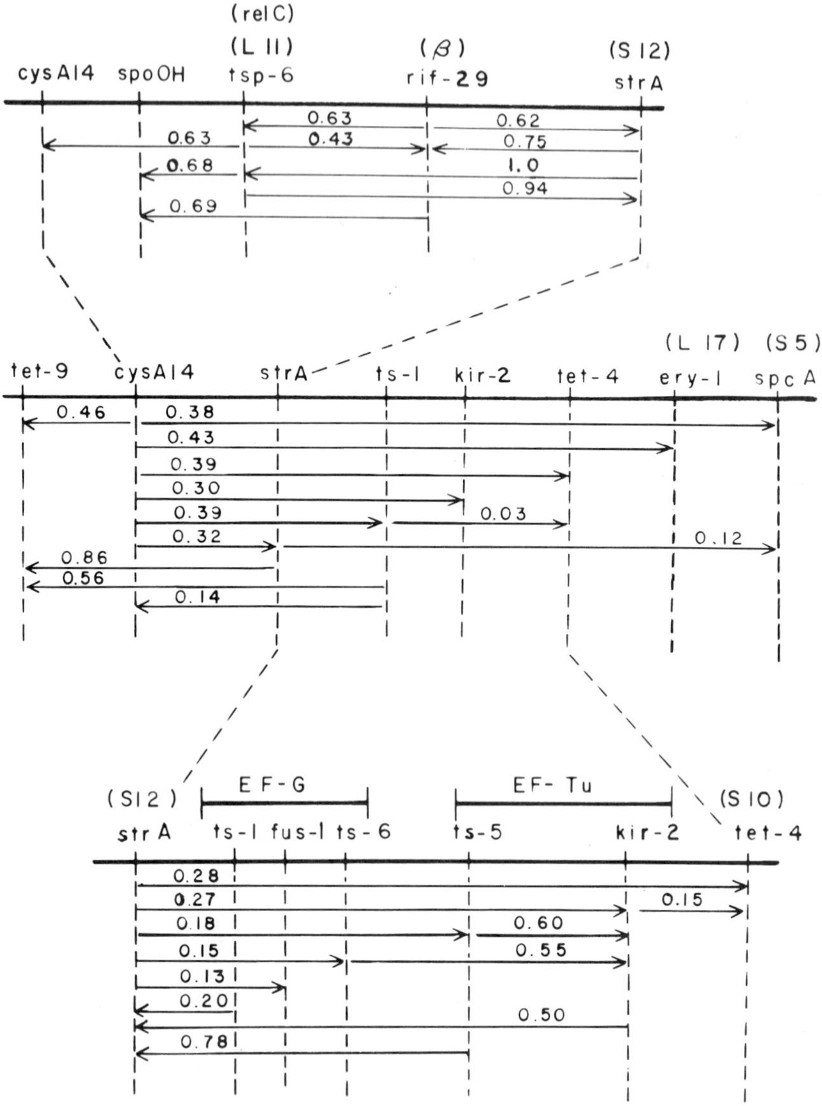

Fig. 1. Map of the *strA* chromosomal region of *B. subtilis*. The middle section of the map represents the order and distances of the indicated genetic markers obtained from three-factor transduction crosses. The upper and lower maps were compiled from three-factor transformation crosses. The head of each arrow points to the recipient marker of the recombinant class. Numbers represent the fraction of recombination between markers. Where known, the gene product is placed over the respective genetic determinant. The data are from published experiments from the author's laboratory.

TABLE III Chromosomal Location of *B. subtilis* and *E. coli* Ribosomal Protein Genes[a]

B. subtilis protein	Gene map position	Counterpart *E. coli* protein	Gene map position (min)
S3	*str* region	S3	72
S5	*str* region	S5	72
S8	*str* region	S8	72
S10	*str* region	S10	72
S12	*str* region	S12	72
S17	*str* region	S17	72
S19	*str* region	S19	72
S20	*str* region	S20	0
L1	*str* region	—	—
L2	*str* region	L2	72
L4	*str* region	—	—
L5	*str* region	—	—
L6	*str* region	—	—
L8	*str* region	L6	72
L11	*str* region	L11	88
L12	*str* region	—	—
L14	*str* region	—	—
L15	*str* region	L17	72
L16	*str* region	—	—
L17	*str* region	L22	72
L22	*str* region	—	—
L23	*str* region	L24	72
L25	*str* region	—	—

[a] Data from Higo *et al.* (1980), Isono (1980), and Smith *et al.* (1980a).

L11 in whole-cell extracts of *tsp* mutants even in the presence of protease inhibitors (Smith *et al.*, 1978). It is also possible that the *tsp* mutation is of the nonsense variety, but it is not suppressed by known *B. subtilis* suppressors (Smith *et al.*, 1978). In any case, all the *tsp* mutations we have analyzed, 20 independent mutations isolated spontaneously or after EMS and nitrosoguandine mutagenesis, lack L11 (Smith *et al.*, 1980a). Perhaps selecting for resistance to sporangiomycin or siomycin, another antibiotic related to thiostrepton, would demonstrate an altered, not missing, L11. Another approach might be to mutagenize locally in the *rif spo0H* region of *B. subtilis* (Fig. 1), selecting for *rif* and scoring for *relC* as an unselected marker (Smith *et al.*, 1980b). In this way, many of the problems associated with thiostrepton resistance, i.e., poor growth, secondary mutations, etc., might be avoided (Smith *et al.*, 1980b).

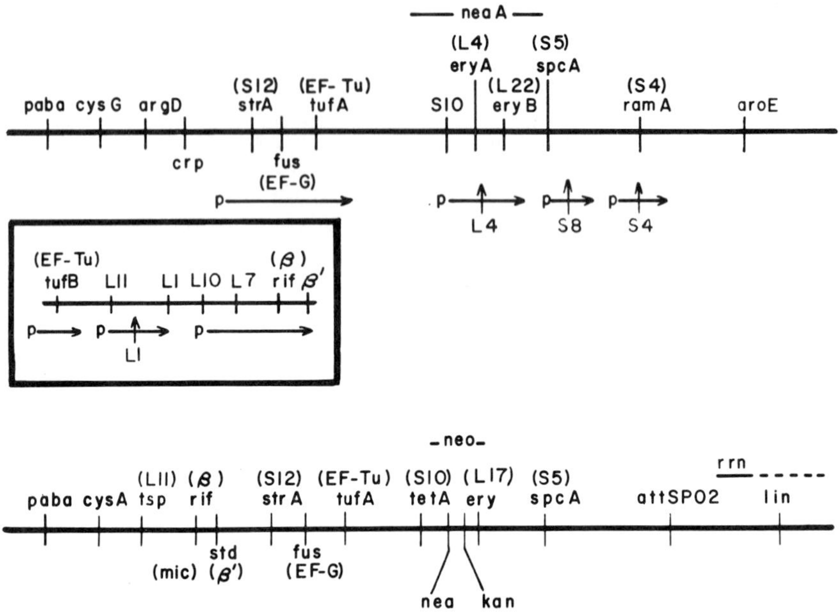

Fig. 2. Comparison of the *E. coli* and *B. subtilis* major ribosomal protein gene clusters. The upper map represents some of the genes in the 72′ segment of the *E. coli* chromosome, and the lower one depicts the *B. subtilis strA* region. The inset illustrates the 88′ segment of *E. coli*. Some genes have been eliminated to facilitate the comparison. The arrows intersecting some of the *E. coli* promoters indicate autogeneous repression of translation of these operons. The maps, compiled from references cited in the text, have not been drawn to scale.

The relationship between the *E. coli rplK* and the *B. subtilis tsp* gene is of more than passing interest, since both genes are close to the RNA polymerase subunits β and β′ (*rpoB* and *rpoC*) loci in the respective organisms. In *E. coli,* however, the *rplK, rpoB, rpoC* cluster is at 88′ (Fig. 2), whereas this region is closely linked to *strA* in *B. subtilis*. This point is more fully discussed below.

Thus, all the mapped *B. subtilis* ribosomal protein genes are found in the *strA* region. However, fewer than half the total number of these cistrons have been localized. All the *E. coli* ribosomal proteins that have been correlated with their genetically mapped *B. subtilis* counterparts, save S20 and L11, also are found in the *E. coli strA* (72′) region (Table III). It is thus possible that many of the unmapped *B. subtilis* ribosomal protein genes may be found in other parts of the genome.

In a study designed to isolate ribosomal protein genes outside the *B. subtilis strA* region, 150 mutants, showing temperature— sensitive protein synthesis, were isolated after ultraviolet (UV) and ethyl methanesulfonate (EMS) mutagenesis (S. Osawa and I. Smith, unpublished results). They were mapped by transduction into three classes, all in relation to the *strA* gene: greater than 90% linked, less than 5% linked, and lastly, all intermediately linked genes. It was found that 130 mutations were highly linked to *strA*, 9 showed less than 5%, and 11 showed intermediate linkage. The 50 and 30 S ribosomal proteins of mutants from the latter two classes were analyzed by two-dimensional gel electrophoresis, but no alterations were observed.

It is of interest to compare these preliminary data with an earlier and much more thorough study on *E. coli* (Isono *et al.*, 1976, 1978). Out of a total of 2000 *ts* mutants, isolated after heavy nitrosoguanidine mutagenesis, 150 primary alterations in 39 ribosomal proteins were observed by means of two-dimensional electrophoresis. Forty-four mutations were in 16 proteins whose genetic determinants were distant from the 72′ region. The remaining 106 mutations were in 23 proteins mapping at 72′.

Obviously, more *ts* mutants will have to be analyzed before we can say, with certainty, whether or not all ribosomal protein genes are clustered in the *B. subtilis strA* region. The striking preponderance of protein synthesis *ts* mutations closely linked to *strA* (86%) in our studies implies much greater clustering for translational genes than is observed in *E. coli*.

4. *Elongation Factors*

Genes for protein synthesis elongation factors EF-Tu and EF-G map adjacent to *rpsL*, the genetic determinant for S12 in both *B. subtilis* and *E. coli* (Nomura *et al.*, 1977; Smith *et al.*, 1980b; Figs. 1 and 2), and it has been demonstrated in *E. coli* that *tufA*, *ef-g*, and *rpsL* are in one operon (Nomura, 1976). Two genes coding for EF-Tu are found in *E. coli*. *tufB*, coding for an almost identical protein (Furano, 1977), maps in the 88′ region (Yamamoto and Nomura, 1979).

It appears that only one functional gene for EF-Tu is found in *B. subtilis*. A *ts* mutation, *ts-5*, affecting EF-Tu *in vitro*, has been isolated in *B. subtilis* (Dubnau *et al.*, 1976). If there were two functional *tuf* genes, one would expect a loss of approximately 50% in EF-Tu activity, *in vitro*, after heating, instead of the more than 90% loss observed. In addition, the mapping of several spontaneous kirromycin (an EF-Tu-specific antibiotic) resistance mutations very close to *ts-5* also indicates that this region codes for EF-Tu (Smith and Paress, 1978; Fig. 1). All the EF-Tu protein from each of several *kir* strains showed altered migration in isoelectric-focusing gels, again suggesting that only one gene is expressed *in vivo*. If two functional genes were present in *B. subtilis*, as in *E. coli,* it would be impossible to move the *kir* marker into sensitive recipients because kirromycin sensitivity is dominant both *in vivo* and *in vitro* (Fischer *et al.,* 1977; van de Klundert *et al.,* 1977). It has been possible to transform *kir* into sensitive *B. subtilis* recipients either by primary selection or as an unselected marker.

These results do not rule out the existence of two closely linked (not separated by transformation) genes for EF-Tu. However, the spontaneous origin of the *kir* mutations at a relatively high frequency, 10^{-7}, would require both genes to mutate to kirromycin resistance at a frequency greater than 10^{-4}, an unlikely possibility.

Experiments in which nick-translated *E. coli tuf* genes have been used as a hybridization probe have also demonstrated that all gram-positive bacteria have one *tuf* gene, whereas all gram-negatives have two (Furano, 1978; D. Filer and A. V. Furano, personal communication).

The gene for elongation factor EF-Ts, which has been localized at 4′ in *E. coli* (Nomura, 1976), has not been mapped in *B. subtilis* as yet.

5. *Evolution of Prokaryotic Translation Genes*

Examination of the chromosomal maps of many prokaryotes shows that the rifampicin resistance (*rif*) locus coding for the β subunit of RNA polymerase is closely linked to the *strA* (*rpsL*) locus. This is true for *Neisseria, Streptomyces,* and *Pseudomonas,* as well as *B. subtilis* (Smith *et al.,* 1980a). On the other hand, these loci are unlinked in the enteric *E. coli* and *Salmonella* (Fig. 2). In a recent study, Filer and Furano investigated the number and location of *tuf* genes coding for elongation factor EF-Tu in various prokaryotes (D. Filer and A. V. Furano, personal communication). It was observed that, in all enterics studied, two unlinked *tuf* genes were found. In five other nonenteric gram-negative bacteria, including *Pseudomonas,* two closely linked *tuf* genes were observed. The two gram-positives, *B. subtilis* and *Micrococcus luteus,* had one *tuf* gene. These workers postulate that some time early in the evolution of gram-negative bacteria, after divergence from the gram-positive branch, postulated to have occurred $1–2 \times 10^9$ yr ago (Hori and Osawa, 1979), a duplication of the *tuf* gene appeared. Subsequent to this event, a translocation of the chromosome region containing the duplicated *tuf* gene as well as *rpoB, rpoC,* and *rplK,* occurred at some point after divergence of the enterics from other gram-negative bacteria. This suggests that the *B. subtilis*-type chromosomal organization is more primitive. It is possible that the apparently higher clustering of most translational genes in *B. subtilis* may also reflect an earlier type of chromosomal organization as well.

IV. Regulation of the Translational Apparatus

One can envisage several types of control that will affect the ability of a cell to synthesize proteins. The most obvious is to change the quantity and quality of mRNA molecules so that a constant number of ribosomes and other components of the translational apparatus can alter the amounts and types of protein being synthesized. This would not necessarily affect the total capacity of the cell to make proteins. This kind of regulation involves the repression and derepression of various enzymes and includes the involvement of various regulatory elements such as the cyclic AMP–CAP system of most prokaryotes.

Another type of control would be to adjust the levels of the protein synthetic machinery to reflect growth rate. When cells grow rapidly, more proteins must be synthesized. Since the rate of amino acid and nucleotide polymerization to form proteins and nucleic acids is essentially constant (Maaløe, 1979), increases in the rates of translation and transcription can only come about by increasing the numbers of functioning RNA polymerase molecules, ribosomes, and other components of the translational apparatus. This could be done by having an excess of unused translational–transcriptional components that can be mobilized to meet the needs of the new growth rate, or by increased numbers of these components through *de novo* synthesis.

These general features of control of translation exist in gram-positive as well as gram-negative bacteria. Another general form of regulation of translation exists in bacilli, however, Under appropriate environmental conditions, i.e., starvation for carbon, nitrogen, and phosphate, bacilli sporulate. New kinds of RNA and proteins appear during the course of the developmental stages (Yamakawa and Doi, 1971; DiCioccio and Strauss, 1973; Petit-Glatron and Rapoport, 1975; Linn and Losick, 1976; and Segall and Losick, 1977). It is of great importance to understand how these changes in transcription and translation are regulated. The reversal of sporulation in which dormant spores germinate, outgrow, and resume vegetative growth also shows an ordered sequence of transcription and translation (Hansen *et al.*, 1970; Yeh and Steinberg, 1978).

Because of limitations of space, this chapter will not attempt to discuss the first type of regulation except to stress the fact that bacilli reportedly do not contain cyclic AMP or adenylate cyclase (reviewed in Hanson, 1976). Catabolite repression is mediated through levels of cyclic AMP in other prokaryotes, and catabolite repression is an important factor in growth and differentiation in bacilli (Chapter 6). It is this extremely important that the absence of cyclic AMP in bacilli be reevaluated using the more sensitive detection methods now available.

Bacilli also possess an elaborate system of regulation for the production and export of extracellular enzymes. Since these are discussed in detail in Chapter 11, this topic will also not be dealt with here.

A. Coordinate Control of the Translational Apparatus

The coordinate control of the ribosomal components, elongation factors, tRNA, and RNA polymerase subunits can be explained to a large extent by clustering of the genetic determinants for these factors in polycistronic operons. This would require overall transcriptional control, by means of active promoters and/or positive and negative transcriptional regulators, acting at the level of RNA polymerase–DNA interactions. The 16, 23, and 5 S genes and many tRNA genes have the same promoter (Nomura *et al.*, 1977; Morgan *et al.*, 1978). Genes for elongation factors and RNA polymerase subunits are closely linked to ribosomal protein genes and are usually found in common transcription units (Nomura, 1976) (Fig. 2). The controls acting at the levels of these polycistronic units are extremely complex. For example, *rpoB*, and *rpoC* are in a common transcription unit with *rplJ* and *rplL* in *E. coli* (Yamamoto and Nomura, 1978; and Fig. 2). Yet the L10, L7/12, and the β and β′ subunits of RNA polymerase are present in different molar amounts (Lindahl *et al.*, 1977) and their genetic determinants are transcribed in different amounts (Dennis, 1977a). Nucleotide sequencing of the intergenic space between *rplJl* and *rpoBC* shows attenuator-like sequences (Post *et al.*, 1979). *In vivo* attenuation and processing of mRNA from this region have been found to occur approximately 200 nucleotides past the end of the *rplL* gene (Barry *et al.*, 1980).

Posttranscriptional control is also observed when ribosomal protein genes are cloned onto high-copy plasmids (Dennis and Fiil, 1979; Lindahl and Zengel, 1979; Dean and Nomura, 1980; Zengel *et al.*, 1980) or λ derivatives (Fallon *et al.*, 1979a; Holowachuk *et al.*, 1980). Transcription of these clones reflect copy number but not translation. A type of autogenous control is exerted by certain ribosomal proteins, so that the translation of polycistronic mRNAs coding for a given ribosomal protein and others on the same transcription unit is inhibited. This translational inhibition is also observed *in vitro* Brot *et al.*, 1980; Yates and Nomura, 1980; Yates *et al.*, 1980). Ribosomal proteins S4, S8, L1, and L4 seem to be specific autogenous translational repressors for their respective polycistronic mRNAs (Fig. 2). Autogenous control of RNA polymerase subunits β and β′ at the level of translation has also been reported (Fukuda *et al.*, 1980). Although it is not clear whether such autogenous regulation occurs normally in cells not harboring many copies of the ribosomal genes, it is a way to keep the numbers of ribosomal proteins coordinated with rRNA. Excess ribosomal proteins would decrease their rates of synthesis, and the presence of adequate levels of rRNA would trap free ribosomal proteins in the process of ribosome assembly.

While posttranscriptional regulation of translational genes has not yet been observed in bacilli, it has been discussed at some length here because similar controls, if extant in this genus, would have tremendous significance for strategies for cloning these genes in bacilli (Section V).

B. The *rel* System

Levels of the regulatory nucleotides (p)ppGpp also have a role in adjusting the levels of the protein synthetic machinery, as the *in vivo* synthesis of rRNA and ribosomal proteins is regulated by ppGpp (Dennis and Nomura, 1974, 1975; Dennis, 1977b; Nierlich, 1978; Maaløe, 1979). Bacterial promoters that respond to stringent control (ppGpp) show high conservation of a heptanucleotide sequence spanning nucleotide positions −5 to +2 at the origin of transcription (Travers, 1980). This may be the site at which ppGpp exerts its inhibiting effect on the transcription of these genes.

There are still ambiguities in our understanding of the role of (p)ppGpp in the *in vivo* regulation of rRNA, ribosomal protein, tRNA, and aminoacyl-tRNA synthetases (Reeh *et al.*, 1976), and there has been a suggestion that other guanosine derivatives such as ppGp may be the actual regulatory agents (Pao and Gallant, 1979). However, the role of the system in amino acid starvation is the most clearly understood mechanism for regulation of the translational apparatus.

Starvation for an amino acid or lack of tRNA charging in *E. coli* causes an immediate cessation of the synthesis of components of the transcription–translation apparatus (Dennis and Nomura, 1975). The guanosine nucleotides ppGpp (guanosine 3′-diphosphate 5′-diphosphate) and pppGpp (guanosine 3′-diphosphate 5′-triphosphate) are produced *in vivo* upon commencement of amino acid starvation in *E. coli* (Cashel and Gallant, 1969) and in *B. subtilis* (Gallant and Margason, 1972).

In *E. coli* the 50 S ribosome modulates the synthesis of (p)ppGpp depending on the binding of aminoacyl-tRNA or uncharged tRNA to the ribosomal A site in the presence of the correct triplet codon. In the presence of aminoacyl-tRNA, peptide bond formation occurs by transfer of the peptidyl group from peptidyl-tRNA at the P site to aminoacyl-tRNA. In the presence of uncharged tRNA, the product of the *relA* gene, ATP:GTP 3′-pyrophosphotransferase (stringent factor), catalyzes the synthesis of the regulatory nucleotides ppGpp and pppGpp in association with the ribosome–mRNA–uncharged tRNA complex (Haseltine and Block, 1973; Pedersen *et al.*, 1973; Sy and Lipmann, 1973). These compounds then act at the transcriptional level to regulate the synthesis of RNA in an operon-specific manner: Synthesis of rRNA, ribosomal proteins, and other components of the translational mechanism is decreased by these nucleotides, but the syntheses of some mRNAs are stimulated and others are blocked (Yang *et al.*, 1974; Reiness *et al.*, 1975). Friesen and co-workers characterized another *rel* locus in *E. coli* (*relC*) and showed that this gene codes for 50 S ribosomal protein L11 (*rplK*) (Friesen *et al.*, 1974). Mutations in this gene caused a relaxed phenotype *in vivo* (production of RNA after amino acid starvation) and *in vitro* (inability of *relC* ribosomes to interact with $relA^+$ stringent factor to produce the regulatory guanosine polyphosphates). Several other genes involved in the production of (p)ppGpp have also been described (Table IV). It seems that the regulatory nucleotides (p)ppGpp are not indispensable for growth. A recent report has described a new *E. coli* mutation, *relS*, which in a *relA* background does not produce detectable (p)ppGpp under any conditions (Engel *et al.*, 1979).

In bacilli, the *tsp* mutations described above, which are associated with a missing L11, also cause a relaxed phenotype (Smith *et al.*, 1978, 1980b; Stark and Cundliffe, 1979). *In vitro* assays have shown that ribosomes cannot synthesize (p)ppGpp from *tsp* cells, whereas wild-type ribosomes can. Ribosomes isolated from revertants to thiostrepton sensitivity also regain the ability to synthesize the regulatory guanosine nucleotides *in vitro* (Smith *et al.*, 1978). Thus, the *tsp* mutation alters 50 S ribosome protein L11 so that it cannot function, *in vivo* or *in vitro*,

TABLE IV Genes Involved in (p)ppGpp Formation and Regulation

	Map position		
Gene	*E. coli* (min)[a]	*B. subtilis* (deg)[b]	Phenotype or component affected
relA	59	230–250	ATP:GTP 3-pyrophosphotransferase
relB	34	—	Glucose starvation·sensitivity
relC (*rplK*)	89	10	Ribosomal protein L11
relS	12–14[c]	—	Complete absence of ppGpp in *relA* mutants[c]
relX	59	—	ppGpp regulation in carbon downshift
spoT	81	—	(p)ppGpp pyrophosphatase

[a] Data from Bachmann and Low (1980).
[b] Smith *et al.* (1980b). Map coordinates from Henner and Hoch (1980).
[c] Engel *et al.* (1979).

as part of the (p)ppGpp-synthesizing system. By analogy with *E. coli,* this mutation is called *relC.*

A mutant of *B. subtilis* that showed a relaxed response to amino acid starvation was isolated by Swanton and Edlin (1972). This mutant synthesized RNA when starved for required amino acids and did not synthesize the regulatory nucleotides ppGpp and pppGpp under these conditions. In the presence of wild-type stringent factor, ribosomes from this strain synthesized (p)ppGpp *in vitro,* and it was postulated that this mutation was of the *relA* type (Smith *et al.,* 1978). Preliminary mapping data also indicated that the Swanton and Edlin mutation was unlinked to *strA* by transduction, whereas *tsp* (*relC*) was over 95% linked.

Stringent factor, the product of the *relA* gene in *B. subtilis,* has now been purified (Smith *et al.,* 1980b). When stringent factor was isolated in similar fashion from *rel* strains and their isogenic wild-type parents, it was found that the Swanton and Edlin mutant possessed less than 3% of the stringent factor activity (expressed either as specific activity or total activity) of an isogenic parent strain. As expected, *relC* strains had high levels of stringent factor activity (Smith *et al.,* 1980b).

There are reports in the literature indicating that *B. subtilis* and *E. coli* may differ in the stimuli required for *in vivo* synthesis of the regulatory quanosine nucleotides. Rhaese and co-workers have suggested that *B. subtilis* inhibition of the binding of charged or uncharged tRNA to ribosomes activates the *rel* system to form (p)ppGpp (Rhaese *et al.,* 1975). However, extensive *in vitro* work in *E. coli,* as discussed above, has demonstrated that only binding of uncharged tRNA to the ribosomal A site, with its appropriate codon, is the stimulus for (p)ppGpp synthesis. Ribosomes, even if exhaustively high-salt-washed, contain sufficient tRNA and mRNA so that there is no requirement for these macromolecules in the synthesis of (p)ppGpp in *E. coli* (Pedersen *et al.,* 1973) or *B. subtilis* (Fortnagel and Bergmann, 1974; Rhaese and Groscurth, 1974; Smith *et al.,* 1978). When *E. coli* ribosomes are dissociated by dialysis against $10-^{-4}$ *M* Mg^{2+} and 30 and 50 S ribosomal subunits are separated, the formation of (p)ppGpp is completely dependent on 50 and 30 S subunits, stringent factor, uncharged tRNA, and the proper codon (Haseltine and Block, 1973; Pedersen *et al.,* 1973).

It has now been shown that, when *B. subtilis* 50 and 30 S subunits are separated and then mixed with purified stringent factor, there is an absolute dependence on the presence of an uncharged tRNA and its specific codon in the synthesis of (p)ppGpp. In addition, tetracycline and thiostrepton completely inhibit (p)ppGpp formation *in vitro,* as is the case in *E. coli* (Smith *et al.,* 1980b). Tetracycline, of course, prevents tRNA binding to ribosomes (Pestka, 1977), and thiostrepton binds to 50

S ribosomal protein L11 (Highland *et al.*, 1975). These data indicate that the *rel* system of *B. subtilis* is identical to that of *E. coli* in terms of direct stimulus required.

The *B. subtilis relA* marker has been mapped by transduction, utilizing the nine reference recipients of Dedonder (Dedonder *et al.*, 1977). These strains, carrying markers that overlap the entire *B. subtilis* genome, allow rapid mapping of *B. subtilis* genes. In this way it has been shown that the *relA* mutation is found between *aroD* and *leu*, a region very distant from the major ribosomal cluster of *B. subtilis* (Smith *et al.*, 1980b; and Table IV).

Several other nucleotides seem to be related to the *rel* response in *B. subtilis*. Wild-type cells (rel^+) produce several nucleotides with the properties of pppApp, ppApp, pGpp, and ppGp, as well as (p)ppGpp, after amino acid starvation (Nishino *et al.*, 1979). The Swanton and Edlin *relA* mutant did not form these compounds after starvation, indicating the involvement of stringent factor. (The unusual nucleotides of *B. subtilis* and their involvement in sporulation are discussed more fully below.)

Bacillus brevis has two distinct (p)ppGpp synthetic enzymes (Sy and Akers, 1976; Sy, 1979). One of these enzymes is associated with ribosomes, requires a ribosome–mRNA–uncharged tRNA complex for activity, and is inhibited by thiostrepton and tetracycline. It is thus homologous to the *E. coli* and *B. subtilis* stringent factor. The other enzyme is soluble, does not require the ribosome complex for activity, and has a much lower (Mg^{2+}) optimum than the ribosomal stringent factor. The two enzyme activities are roughly equivalent during vegetative growth. In stationary cultures, much less stringent factor activity was observed (see below).

Bacillus stearothermophilus also possesses a ribosome-independent (p)ppGpp synthetase (Fehr *et al.*, 1979). The function of these nonribosomal enzymes is not known at present, and their relationship to other nonribosomal (p)ppGpp synthetases (Zabos *et al.*, 1976) is not clear. It is possible that basal levels of (p)ppGpp, not affected by *relA* function, i.e., amino acid starvation, are controlled by these enzymes. If this is true, *relS* would be a good candidate for the genetic determinant of the nonribosomal (p)ppGpp synthetases.

C. Sporulation and Germination

1. *Transcription Controls*

Most current hypotheses dealing with the control of sporulation revolve around the idea that sporulation-specific genes are recognized by modified RNA polymerases (Chapters 3 and 6), since many new types of mRNA and proteins make an ordered appearance during the stages of sporulation (Yamakawa and Doi, 1971; DiCioccio and Strauss, 1973; Petit-Glatron and Rapoport, 1975; Linn and Losick, 1976; Segall and Losick, 1977). Specific evidence for transcriptional control is the following: (1) Modified RNA polymerases appear during the course of sporulation in *B. thuringiensis* (Klier and Lecadet, 1974) and *B. subtilis* (Chapters 3 and 6). Some early sporulation negative mutants do not show these modifications. (2) Mutations in RNA polymerase genes sometimes result in sporulation-deficient genotypes (Leighton, 1977; Chapters 3 and 6, this volume). (3) A cloned *B. subtilis* sporulation gene is not read by the predominant vegetative RNA polymerase and is read by a modified enzyme present in spores (Haldenwang and Losick, 1979; and Chapter 6).

2. *Translational Effects*

While the idea of spore regulation at the level of transcription is the most appealing theory at present, there are many observations that complicate this relatively straightforward hypothesis. Several antibiotic resistance mutations in *B. subtilis* that affect the translation apparatus also are associated with a sporulation-negative phenotype. These mutations have been found in genes that affect 30 and 50 S ribosomes and the elongation factor EF-G (Table V). The mutations in streptomycin resistance (*str*) that affect sporulation presumably alter protein S12, whereas S5 is altered in mutation in spectinomycin resistance (*spc*) which lead to a sporulation-negative phenotype (Cannon and Bott, 1979). Ri-

TABLE V Mutations in the Translational Apparatus That Affect Sporulation

Mutation	Phenotype or component affected	Sporulation phenotype	Reference
str	30 S ribosome (protein S12?)	Temperature-sensitive sporulation	Leighton (1974)
str	30 S ribosome (protein S12?)	Sporulation-negative	Campbell and Chambliss (1977)
spc	Ribosomal protein S5	Sporulation-negative in presence of spectinomycin	Cannon and Bott (1979)
ery	Ribosomal protein L17	Temperature-sensitive sporulation	Tipper *et al.* (1977)
ery	50 S ribosome (L17?)	Sporulation-negative in presence of erythromycin	Domoto *et al.* (1975).
ksg	Ribosomes	Sporulation-negative in presence of kasugamycin	Tominaga and Kobayashi (1978)
fus	Elongation factor EF-G	Temperature-sensitive sporulation	Kobayashi *et al.* (1977)

bosomal S12 is involved in the initiation step of protein synthesis (Held *et al.*, 1974) and tRNA recognition (Gorini, 1974; Yates, 1979). Mutations in its genetic determinant, *rpsL,* determine resistance to streptomycin, which affects protein synthesis by causing misreading (Gorini, 1974). Mutations in the gene *rpsE,* coding for ribosomal protein S5, not only cause spectinomycin resistance but also give rise to the streptomycin-independent phenotype in *E. coli* (Hasenbank *et al.*, 1973; Itoh and Wittmann, 1973), *B. stearothermophilus* (Isono, 1974), and *B. subtilis* (Itoh *et al.*, 1975). Certain lesions in *rpsE* also cause ribosomal ambiguity (*ram*) phenotypes (Piepersberg *et al.*, 1975) which are characterized by their ability to counteract the effects of the *strA* mutation on translational fidelity (Rosset and Gorini, 1969). Thus, the S5 and S12 proteins must interact in the ribosome,and mutations in these proteins may affect the ribosome's ability to interact with certain sporulation-specific mRNAs or to read them correctly.

On the other hand, the 50 S ribosome is presumably not involved in the initiation of protein synthesis or in the mRNA recognition step. Yet, certain mutations to erythromycin resistance (*ery*) that alter protein L17 and mutations to fusidic acid resistance (*fus*) that affect elongation factor EF-G cause conditional sporulation. It has been observed that the temperature-sensitive sporulation caused by a specific *fus* mutation, affecting EF-G, is suppressed by a class of *rif* mutants that alter RNA polymerase (Hirochika and Kobayashi, 1978). This suggests a specific interaction between translation and transcription during sporulation.

Interactions between RNA polymerase and ribosome mutations have been reported in *E. coli.* A temperature-sensitive lethal mutant of *E. coli* was constructed by combining a temperature-insensitive *rif* mutation and a temperature-insensitive *strA* mutation (Chakrabarti and Gorini, 1977). Another interaction between *str* and *rif* mutations was observed in the T7 infection of permissive male strains (Chakrabarti and Gorini, 1975). Certain *strA* mutations allowed growth of T7 in male strains. This permissiveness was associated with enhanced T7 transcription. Introduction of a nonpermissive *rif* gene abolished this enhancement of transcription.

In vitro coupling between transcription and translation has also been observed in *E. coli* (Jacobs *et al.*, 1978). The formation of *lac* mRNA *in vitro* was increased severalfold by the addition of ribosomes. Chloramphenicol or erythromycin blocked the ribosomal stimulation of *lac* transcription. It is conceivable that certain mutations in the translational apparatus may slow down the movement of ribosomes along newly synthesized mRNAs, thus preventing or slowing down transcription (and/or exposing the nascent mRNA to nucleolytic degradation). A similar mechanism could occur during sporulation.

Other reports have suggested that protein synthesis initiation factors from sporulating cells have specificity different from that of those isolated from vegetative cells (Chambliss and Legault-Demare, 1977), but these reports have not been confirmed by other workers (Guha and Szulmajster, 1977). Similarly, it has been reported that certain ribosomal proteins are altered during sporulation (Fortnagel and Bergmann, 1973; Kobayashi, 1973). These changes have not been found in other laboratories (Guha *et al.*, 1975). Changes in tRNA posttranscriptional modification occurring within the anticodon loop during sporulation have been reported (Singhal and Vold, 1976; Vold, 1978). In general, though, all resolvable tRNA species are present in equivalent amounts throughout all growth stages (germination, vegetative growth, and sporulation) (Henner and Steinberg, 1979b).

It is known that different mRNAs are produced during sporulation. Some of these new messages have longer half-lives than their vegetative counterparts, as measured by stability in the presence of rifampicin (Petit-Glatron and Rapoport, 1976; Segall and Losick, 1977). It has also been reported that mRNAs isolated from sporulating *B. subtilis* have poly-A tails of 160–180 nucleotides at their 3′-ends. Poly-A tails were not found in vegetative cells nor in mRNA protein isolated from an asporogenous mutant (Kerjan and Szulmajster, 1980). These extremely interesting results await confirmation and further study.

3. *Unusual Nucleotides*

A significant theory has been proposed that relates the appearance of several highly phosphorylated nucleotides (HPN) to the sporulation process (Rhaese and Groscurth, 1976). It has been reported that some of these nucleotides, pppApp and ppApp, are synthesized on ribosomes isolated from sporulating but not vegetative cells (Rhaese and Groscurth, 1974, 1978). Ribosomes isolated from an early sporulation mutant (*spo0F*) did not produce these HPN (Rhaese *et al.*, 1977). Conversely, sporulating ribosomes and soluble factors did not synthesize (p)ppGpp. These studies indicate that more than basal levels of (p)ppGpp are not needed for sporulation, a result that is in agreement with the observation that *relA* and *relC* mutants can sporulate (Rhaese *et al.*, 1976; Nishino *et al.*, 1979; Smith *et al.*, 1980b). However, other results of the Rhaese group have not been confirmed in other laboratories. As discussed above, some of the HPN nucleotides, i.e., pppApp and ppApp, are produced in wild-type vegetative cells after amino acid starvation but not in *relA* mutants (Nishino *et al.*, 1979). This indicates their synthesis is under stringent, not sporulation, control. According to Rhaese hypotheses, the nucleotide, pppApp, is a key regulator in the onset of sporulation (Rhaese *et al.*, 1977). Nishino *et al.*, could find no trace of this compound during vegetative growth or sporulation. Another major discrepancy deals with the ability of sporulating cells to make (p)ppGpp. Other *in vivo* experiments (Fortnagel and Bergmann, 1974) and *in vitro* experiments (Sy, 1979; Smith *et al.*, 1980b) have shown that, whereas there are losses in the ability of cells or extracts to synthesize (p)ppGpp during sporulation, significant activity remains up to several hours after the onset of this developmental process.

Since the pivotal role of HPN in sporulation remains to be confirmed by other laboratories, it is premature at this point to speculate on how this compound might affect the transcription or translation of spore-specific mRNAs.

4. *Germination*

Germinating spores in bacilli have been extensively studied with regard to the synthesis of cellular macromolecules (Hansen *et al.*, 1970). This process can be thought of as a type of shift-up in transitional growth. Dormant cells are placed to a suitable medium, and RNA synthesis immediately starts, followed by protein and finally DNA synthesis. It has been reported that RNA synthesis is biphasic during the early stages of germination and outgrowth (Armstrong and Sueoka, 1968). For 0 to 10 min after the initial suspension of spores in a germination medium, it was found that only rRNA was synthesized. Starting at 15 min, mRNA and protein, as well as rRNA and tRNA, were synthesized. Similar results were obtained by Woese and Bleyman (1969). Other experiments, however, in *B. subtilis* (Balassa and Contesse, 1965) and *Bacillus cereus* (Spiegelman *et al.*, 1969; Torriani *et al.*, (1969) indicated that all classes of RNA were synthesized early after the initiation of germination. More recent studies in *B. subtilis* using RNA–DNA hybridization have shown that mRNA as well as rRNA is synthesized as early as 3 min after germination begins and that the relative content of rRNA synthesized during the germination process ranges from 30 to 60% of the total RNA transcripts (Margulies *et al.*, 1978; Sloma and Smith, 1979). These experiments did not deal with the transcription of individual mRNAs that turn off and on during the transcription process, and there are several reports indicating that this occurs (Hansen *et al.*, 1970; Setoguchi *et al.*, 1978; Yeh and Steinberg, 1978).

Another type of control may be exerted at the level of translation during germination. Dormant spores of *B. cereus* do not have functional mRNA, and protein synthesis during germination is dependent on the synthesis of new mRNA (Kobayashi *et al.*, 1965). In addition, ribosomes of dormant spores cannot be programmed with synthetic mRNAs. Soon after germination begins, ribosomes regain the ability to respond to mRNAs *in vitro* (Kobayashi *et al.*, 1965; Idriss and Halvorson, 1969). The physiological significance is not clear at the moment.

D. Induced Macrolide–Lincosamide–Streptogramin B Resistance

It is fitting at this time to finish this discussion of the regulation of the translational apparatus with a description of the control of induced macrolide–lincosamide–streptogramin B (MLS) resistance in *B. subtilis*. This system, which has the features of autogenous, posttranscriptional regulation of a specific protein, also requires ribosome modification for expression of resistance. It is, therefore, an interesting system in its own right, as well as being a model for other types of regulation in bacilli, e.g., sporulation.

Inducible MLS resistance in *S. aureus* is plasmid-determined and results from specific N^6-dimethylation of 23 S rRNA (Lai *et al.*, 1973). A *S. aureus* plasmid determining resistance to erythromycin, pE194, has been introduced into *B. subtilis,* and a similar type of resistance mechanism, i.e., 23 S rRNA methylation, has been found in this organism (Weisblum *et al.*, 1979). Study of the *B. subtilis*–pE194 system has now provided much information on the regulation of induced MLS resistance. Erythromycin induction seems to work by posttranscriptional modification rather than by increased transcription of the rRNA methylase gene (Shivakumar *et al.*, 1980). Increased synthesis of a 29K protein (E3) is observed after erythromycin addition. Erythromycin-sensitive deletion mutants produced truncated E3 proteins, thus indicating that this protein determines resistance to the antibiotic and is most likely the 23 S rRNA methylase (Shivakumar *et al.*, 1979, 1980). The E3 protein may be regulated autogenously, since its level in induced wild-type plasmid-carrying strains is strongly controlled and the deletion mutants, although deficient in methylase production, overproduce truncated E3 protein.

Plasmid-bearing *B. subtilis* strains prossessing the *ole-1* mutation (Goldthwaite and Smith, 1972) cannot be induced for MLS resistance. This mutation affects ribosomal protein L-17 (Tipper *et al.*, 1977; Smith *et al.*, 1980a), and ribosomes from *ole-1* strains bind erythromycin with lower affinity than wild-type ribosomes (Shivakumar *et al.*, 1980). The ribosomal binding of erythromycin is therefore an important requirement for the induction of methylase. The exact role of the erythromycin–ribosome complex is not known yet.

From an analysis of the leader sequence of 29K protein mRNA, Dubnau and co-workers have suggested that the specificity of the 29K protein induction is found in the structure of the mRNA (Shivakumar *et al.*, 1980; D. Dubnau, personal communication, 1980). It is hoped that a study of this complex control system will be performed soon *in vitro*. This should allow a more detailed description of the factors involved in the transcription and translation of the MLS resistance gene.

V. Future Directions

During the course of this chapter, the author has attempted to discuss various areas of research on the nature of the translational system of bacilli with a view toward which questions are important and what approaches might be feasible. Some of these problems were (1) the specificity of mRNA recognition in bacilli; (2) the clustering of r-protein genes in *B. subtilis;* (3) transcriptional and/or translational control of sporulation; (4) the regulation of extracellular enzyme synthesis, although not discussed in this chapter; and (5) the nature of catabolite repression in bacilli.

These problems are diverse, and each will require different approaches and methodologies. However, it is clear, to this reviewer at least, that applications of recombinant DNA technology will be an important part of the solution in each case. For example, sporulation is a complex process involving 30–40 genes whose expression is controlled in a temporal sequence (Piggot and Coote, 1976). It is consequently very difficult to analyze the controls for this process in a meaningful way. The cloning of sporulation genes has already provided much information on the control of two spore loci (Chapter 6). In the near future, hopefully, several more spore-specific genetic determinants will be cloned. These genes can then be used to program *in vitro* transcription and translation, and the role of modified RNA polymerases and altered ribosomes, as well as other cell components, can be tested.

It is worthwhile to analyze some of the different recombinant DNA strategies and the problems that may arise, with specific reference to cloning bacillus translational genes. As a case in point we can discuss the cloning of ribosomal protein genes in bacilli. Direct shotgun cloning in bacilli with *S. aureus*-derived plasmids using presently available techniques is dependent on host rec^{+} function, and this requires insert DNA that is nonhomologous (Chapters 5 and 10). This precludes the use of DNA from *Bacillus licheniformis* or other bacilli, since there is significant DNA homology between genes for the translational apparatus in bacilli (Dubnau *et al.*, 1965b). In attempted cloning experiments, using DNA from *B. licheniformis* and *B. subtilis* rec^{+} hosts, all transformants for translational genes are chromosomal integrants (E. Dubnau and I. Smith, unpublished results).

Thus, we must use a completely nonhomologous system for cloning. *Bacillus subtilis* and *B. thuringiensis* rRNA genes have been cloned in *E. coli* using high-copy Col E1 derivative plasmids. However, it is doubtful that such vectors will be useful for cloning ribosomal protein genes. High-copy plasmids with ribosomal protein genes in *E. coli* show inhibition of translation (see above) and are frequently unstable (Fallon *et al.*, 1979b). There is a high probability that clones formed between bacillus ribosomal protein genes and high-copy *E. coli* plasmids such as pBR322 would be lethal for the host *E. coli* cell and would not be adequately represented in the gene banks formed by the ligation of bacillus chromosomal DNA and pBR322 or related plasmids (Chapter 10). A feasible approach would be to prepare a bacillus gene bank in λ derivatives like the Charon phages (Blattner *et al.*, 1977). Recombinant molecules can be packaged *in vitro,* giving high titers of infectious, lytic phage. These phage will lyse *E. coli* after several rounds of replication. Amplified phage DNA can be prepared and used to transform genetically marked *B. subtilis* recipients. The advantage of this system is that viability of the clone and its *E. coli* host are not a requirement for selection. Since pieces of DNA of up to 20 kb can be cloned in the Charon system, many fewer recombinant clones must be analyzed to obtain a good representation of the bacillus genome than in the pBR322–bacilli gene banks. To this end, a Charon 4–*B. subtilis* DNA gene library has been constructed recently (E. Ferrari, D. J. Henner, and J. A. Hoch, personal communication).

It is expected that in the near future we will know much more about the control of translation in bacilli. When a new volume in this series is published, perhaps a chapter on the translational system of the bacilli will present fewer problems and more solutions, less theories and more hard facts.

Acknowledgments

The author would like to thank his many colleagues who furnished reprints and preprints of their unpublished work. He is especially grateful to Dr. S. Osawa and the members of his group, Dr. K. Isono, Dr. M. Nomura, and Dr. A. Matheson for providing information which the author has freely used for this review. Discussions with Eugenie Dubnau, David Dubnau, Thomas Gryczan, and Leonard Mindich were extremely helpful and, without the expert secretarial assistance of A. Howard, this chapter would still be in the process of translation. Research from the author's laboratory discussed in this chapter was supported by USPHS National Institutes of Health grants GM19693 and GM28407.

References

Armstrong, R. L., and Sueoka, N. (1968). *Proc. Nat. Acad. Sci. USA* **59,** 153–160.
Atkins, J. F., Steitz, J. A., Anderson, C. W., and Model, P. (1979). *Cell* **18,** 247–256.
Bachmann, B. J., and Low, K. B. (1980). *Microbiol. Rev.* **44,** 1–56.
Balassa, G., and Contesse, G. (1965). *Ann. Inst. Pasteur* **109,** 684–705.
Barry, G., Squires, C., and Squires, C. L. (1980). *Proc. Nat. Acad. Sci. USA* **77,** 3331–3335.
Bazzicalupo, M., Parisi, B., Pirali, G., Polsinelli, M., and Sala, F. (1975). *Antimicrob. Agents Chemother.* **8,** 651–656.
Beremand, M. N., and Blumenthal, T. (1979). *Cell* **18,** 257–266.
Blattner, F. R. *et al.* (1977). *Science* **196,** 161–169.
Bollen, A., Cabezon, T., Delcuve, G., Herzog, A., Denicourt, D., and Corchuelo, M. C. (1980). *In* "RNA Polymerase, tRNA and Ribosomes: Their Genetics and Evolution" (S. Osawa, H. Ozeki, H. Uchida, and T. Yura, eds.). Univ. of Tokyo Press, Tokyo.
Brot, N., Caldwell, P., and Weissbach, H. (1980). *Proc. Nat. Acad. Sci. USA* **77,** 2592–2595.
Campbell, K. M., and Chambliss, G. H. (1977). *Mol. Gen. Genet.* **158,** 193–200.
Cannon, J. G., and Bott, K. F. (1979). *Mol. Gen. Genet.* **174,** 149–162.
Cashel, M., and Gallant, J. (1969). *Nature (London)* **221,** 838–841.
Chakrabarti, S. L., and Gorini, L. (1975). *Proc. Nat. Acad. Sci. USA* **72,** 2084–2087.
Chakrabarti, S. L., and Gorini, L. (1977). *Proc. Nat. Acad. Sci. USA* **74,** 1157–1161.
Chambliss, G. H., and Legault-Démare, L. (1977). *J. Bacteriol.* **132,** 13–22.
Chow, L. T., and Davidson, N. (1973). *J. Mol. Biol.* **75,** 265–279.
Cundliffe, E. *et al.* (1979). *J. Mol. Biol.* **132,** 235–252.
Cutler, R. G., and Evans, J. E. (1967). *J. Mol. Biol.* **26,** 91–105.
Dean, D., and Nomura, M. (1980). *Proc. Nat. Acad. Sci. USA* **77,** 3590–3594.
DeBoer, H. A., Gilbert, S. F., and Nomura, M. (1979). *Cell* **17,** 201–209.
Dedonder, R. A., Lepesant, J.-A., Lepesant-Kejzlarová, J., Billault, A., Steinmetz, M., and Kunst, F. (1977). *Appl. Environ. Microbiol.* **33,** 989–993.
Dennis, P. P. (1977a). *J. Mol. Biol.* **115,** 603–625.
Dennis, P. P. (1977b). *J. Bacteriol.* **129,** 580–538.
Dennis, P. P., and Fiil, N. P. (1979). *J. Biol. Chem.* **254,** 7540–7547.
Dennis, P. P., and Nomura, M. (1974). *Proc. Nat. Acad. Sci. USA* **71,** 3819–3823.
Dennis, P. P., and Nomura, M. (1975). *Nature (London)* **255,** 460–465.
DiCioccio, R. A., and Strauss, N. (1973). *J. Mol. Biol.* **77,** 325–336.
Domoto, T., Kobayashi, K., and Kobayashi, Y. (1975). *In* "Spores VI" (P. Gerhardt, R. N. Costilow, and H. L. Sadoff, eds.), pp. 307–313. American Society for Microbiology, Washington, D.C.
Dubnau, D., Smith, I., and Marmur, J. (1965a). *Proc. Nat. Acad. Sci. USA* **54,** 724–730.
Dubnau, D., Smith, I., Morell, P., and Marmur, J. (1965b). *Proc. Nat. Acad. Sci. USA* **54,** 491–498.
Dubnau, E., Pifko, S., Sloma, A., Cabane, K., and Smith, I. (1976). *Mol. Gen. Genet.* **147,** 1–12.
Ehrlich, S. D. (1978). *Proc. Nat. Acad. Sci. USA* **75,** 1433–1436.
Ellwood, M., and Nomura, M. (1980). *J. Bacteriol.* **143,** 1077–1080.
Engel, J. A., Sylvester, J., and Cashel, M. (1979). *In* "Regulation of Macromolecular Synthesis by Low Molecular Weight Mediators" (G. Koch and D. Richter, eds.), pp. 25–38. Academic Press, New York.
Fallon, A. M., Jinks, C. S., Strycharz, G. D., and Nomura, M. (1979a). *Proc. Nat. Acad. Sci. USA* **76,** 3411–3415.

Fallon, A. M., Jinks, C. S., Yamamoto, M., and Nomura, M. (1979b). *J. Bacteriol.* **138,** 383–396.
Fehr, S., Godt, F., Isono, K., and Richter, D. (1979). *FEBS Lett.* **97,** 91–95.
Fischer, E., Wolf, H., Hantke, K., and Parmeggiani, A. (1977). *Proc. Nat. Acad. Sci. USA* **74,** 4341–4345.
Fortnagel, P., and Bergmann, R. (1973). *Biochim. Biophys. Acta* **299,** 136–141.
Fortnagel, P., and Bergmann, R. (1974). *Biochem. Biophys. Res. Commun.* **56,** 264–272.
Friesen, J. D., Fiil, N. P., Parker, J. M., and Haseltine, W. A. (1974). *Proc. Nat. Acad. Sci. USA* **71,** 3465–3469.
Fukuda, R., Kajitani, M., Muto, A., and Ishihama, A. (1980). *In* "RNA Polymerase, tRNA and Ribosomes: Their Genetics and Evolution" (S. Osawa, H. Ozeki, H. Uchida, and T. Yura, eds.). Univ. Tokyo Press, Tokyo.
Furano, A. V. (1977). *J. Biol. Chem.* **252,** 2154–2157.
Furano, A. V. (1978). *Proc. Nat. Acad. Sci. USA* **75,** 3104–3108.
Gallant, J., and Margason, G. (1972). *J. Biol. Chem.* **247,** 2289–2294.
Geisser, M., Tischendorf, G. W., and Stöffler, G. (1973). *Mol. Gen. Genet.* **127,** 129–145.
Georgopoulos, C. P. (1969). *J. Bacteriol.* **97,** 1397–1402.
Gilbert, S. F., deBoer, H. A., and Nomura, M. (1979). *Cell* **17,** 211–224.
Glaser, G., and Cashel, M. (1979). *Cell* **16,** 111–121.
Goebel, W., Kreft, J., and Burger, K. J. (1979). *In* "Plasmids of Medical, Environmental and Commercial Importance" (K. N. Timmis and A. Puhler, eds.), pp. 471–480. Elsevier/North-Holland Publ., Amsterdam.
Goldberg, M. L., and Steitz, J. A. (1974). *Biochemistry* **13,** 2123–2129.
Goldthwaite, C., and Smith, I. (1972). *Mol. Gen. Genet.* **114,** 190–204.
Gorini, L. (1974). *In* "The Ribosome" (M. Nomura, A. Tissières, and P. Lengyel, eds.), pp. 791–803. Cold Spring Harbor Laboratory, Cold Spring Harbor, New York.
Grandi, G., Mattes, M., and Sgaramella, V. (1981). *Plasmid* **6,** 99–111.
Guha, S., and Szulmajster, J. (1977). *J. Bacteriol.* **131,** 866–871.
Guha, S., Roth, H. E., and Nierhaus, K. (1975). *Mol. Gen. Genet.* **138,** 299–307.
Haldenwang, W. G., and Losick, R. (1979). *Nature (London)* **282,** 256–260.
Hansen, J. N., Spiegelman, G., and Halvorson, H. O. (1970). *Science* **168,** 1291–1298.
Hanson, R. S. (1976). *In* "Spores VI" (P. Gerhardt, H. L. Sadoff, and R. N. Costilow, eds.), pp. 318–326. American Society for Microbiology, Washington, D.C.
Haseltine, W. A., and Block, R. (1973). *Proc. Nat. Acad. Sci. USA* **70,** 1564–1568.
Hasenbank, R., Guthrie, C., Stöffler, G., Wittmann, H. G., Rosen, L., and Apirion, D. (1973). *Mol. Gen. Genet.* **127,** 1–18.
Held, W. A., Gette, W. R., and Nomura, M. (1974). *Biochemistry* **13,** 2115–2121.
Henner, D. J., and Hoch, J. A. (1980). *Microbiol. Rev.* **44,** 57–82.
Henner, D. J., and Steinberg, W. (1979a). *J. Bacteriol.* **139,** 668–670.
Henner, D. J., and Steinberg, W. (1979b). *J. Bacteriol.* **140,** 555–566.
Highland, J. H., Howard, G. A., Ochsner, E., Stöffler, G., Hasenbank, R., and Gordon, J. (1975). *J. Biol. Chem.* **250,** 1141–1146.
Higo, K.-I., Itoh, T., Kumazaki, T., and Osawa, S. (1980). *In* "RNA Polymerase, tRNA and Ribosomes: Their Genetics and Evolution" (S. Osawa, H. Ozeki, H. Uchida, and T. Yura, eds.), pp. 655–666. Univ. Tokyo Press, Tokyo.
Hirochika, H., and Kobayashi, Y. (1978). *J. Bacteriol.* **136,** 983–993.
Holowachuk, E. W., Friesen, J. D., and Fiil, N. P. (1980). *Proc. Nat. Acad. Sci. USA* **77,** 2124–2128.
Hori, H., and Osawa, S. (1979). *Proc. Nat. Acad. Sci. USA* **76,** 381–385.
Idriss, J. M., and Halvorson, H. O. (1969). *Arch. Biochem. Biophys.* **133,** 442–453.

Isono, K. (1974). *Mol. Gen. Genet.* **133,** 77–86.
Isono, K. (1980). *In* "RNA Polymerase, tRNA and Ribosomes: Their Genetics and Evolution" (S. Osawa, H. Ozeki, H. Uchida, and T. Yura, eds.). Univ. Tokyo Press, Tokyo.
Isono, K., and Isono, S. (1976). *Proc. Nat. Acad. Sci. USA* **73,** 767–774.
Isono, K., Krauss, J., and Hirota, Y. (1976). *Mol. Gen. Genet.* **149,** 297–302.
Isono, S., and Isono, K. (1975). *Eur. J. Biochem.* **56,** 15–22.
Isono, S., Isono, K., and Hirota, Y. (1978). *Mol. Gen. Genet.* **165,** 15–20.
Itoh, T., and Wittmann, H. G. (1973). *Mol. Gen. Genet.* **127,** 19–32.
Itoh, T., Kosugi, H., Higo, K., and Osawa, S. (1975). *Mol. Gen Genet.* **139,** 293–301.
Jacobs, K. A., Shen, V., and Schlessinger, D. (1978). *Proc. Nat. Acad. Sci. USA* **75,** 158–161.
Kerjan, P., and Szulmajster, J. (1980). *Biochem. Biophys. Res. Commun.* **93,** 201–208.
Klier, A., and Lecadet, M.-M. (1974). *Eur. J. Biochem.* **47,** 111–119.
Klier, A. F., Kunst, F., and Rapoport, G. (1979). *Nucleic Acids Res.* **7,** 997–1010.
Kobayashi, Y. (1973). *In* "Régulation de la Sporulation Microbienne," pp. 107–108. Colloq. Int. C.N.R.S. No. 227, Paris.
Kobayashi, Y., Steinberg, W., Higa, A., Halvorson, H. O., and Levinthal, C. (1965). *In* "Spores III" (L. Campbell and H. O. Halvorson, eds.), pp. 200–212. American Society for Microbiology, Washington, D.C.
Kobayashi, H., Kobayashi, K., and Kobayashi, Y. (1977). *J. Bacteriol.* **132,** 262–269.
Kreft, J., Bernhard, K., and Goebel, W. (1978). *Mol. Gen. Genet.* **162,** 59–67.
Lai, C.-J., Dahlberg, J. E., and Weisblum, B. (1973). *Biochemistry* **12,** 457–460.
Legault-Démare, L., and Chambliss, G. H. (1974). *J. Bacteriol.* **120,** 1300–1307.
Legault-Démare, L., and Chambliss, G. H. (1975). *Mol. Gen. Genet.* **142,** 277–287.
Leighton, T. (1974). *J. Mol. Biol.* **86,** 855–863.
Leighton, T. (1977). *J. Biol. Chem.* **252,** 268–272.
Leventhal, J. M., and Chambliss, G. H. (1979). *Biochim. Biophys. Acta* **564,** 162–171.
Lindahl, L., and Zengel, J. M. (1979). *Proc. Nat. Acad. Sci. USA* **76,** 6542–6546.
Lindahl, L., Yamamoto, M., Nomura, M., Kirschbaum, J. B., Allet, B., and Rochaix, J.-D. (1977). *J. Mol. Biol.* **109,** 23–47.
Linn, T., and Losick, R. (1976). *Cell* **8,** 103–114.
Lodish, H. L. (1969). *Nature (London)* **244,** 867–870.
Lodish, H. L. (1970). *Nature (London)* **226,** 705–707.
Maaløe, O. (1979). *In* "Biological Regulation and Development" (R. F. Goldberger, ed.), Vol. 1, pp. 487–542. Plenum Press, New York.
Margulies, L., Setoguchi, Y., and Rudner, R. (1978). *Biochim. Biophys. Acta* **521,** 708–718.
Mellado, R. P., Viñuela, E., and Salas, M. (1976). *Eur. J. Biochem.* **65,** 213–223.
Moran, C. P., Jr., and Bott, K. F. (1979a). *J. Bacteriol.* **140,** 99–105.
Moran, C. P., Jr., and Bott, K. F. (1979b). *J. Bacteriol.* **140,** 742–744.
Morgan, E. A., Ikemura, T., Lindahl, L., Fallon, A. M., and Nomura, M. (1978). *Cell* **13,** 335–344.
Nakayama, T., Munoz, L. E., Sadaie, Y., and Doi, R. H. (1978). *J. Bacteriol.* **135,** 952–960.
Nierlich, D. P. (1978). *Ann. Rev. Microbiol.* **32,** 393–432.
Nishino, T., Gallant, J., Shalit, P., Palmer, L., and Wehr, T. (1979). *J. Bacteriol.* **140,** 671–679.
Nomura, M. (1973). *Science* **179,** 864–873.
Nomura, M. (1976). *Cell* **9,** 633–644.
Nomura, M., Morgan, E. A., and Jaskunas, S. R. (1977). *Ann. Rev. Genet.* **11,** 297–347.
Oishi, M., and Sueoka, N. (1965). *Proc. Nat. Acad. Sci. USA* **54,** 483–491.

Oishi, M., Yoshikawa, H., and Sueoka, N. (1964). *Nature (London)* **204,** 1069–1073.
Oishi, M., Oishi, A., and Sueoka, N. (1966). *Proc. Nat. Acad. Sci. USA* **55,** 1095–1103.
Okubo, S., and Yanagida, T. (1968). *J. Bacteriol.* **95,** 1187–1188.
Osawa, S. (1976). *Mol. Gen. Genet.* **144,** 49–51.
Osawa, S., Tukui, A., and Saito, H. (1978). *Mol. Gen. Genet.* **164,** 113–129.
Pao, C. C., and Gallant, J. (1979). *J. Biol. Chem.* **254,** 688–692.
Parker, J., Watson, R. J., and Friesen, J. D. (1976). *Mol. Gen. Genet.* **144,** 111–114.
Pedersen, F. S., Lund, E., and Kjeldgaard, N. O. (1973). *Nature (London) New Biol.* **243,** 13–15.
Pestka, S. (1977). *In* "Molecular Mechanisms of Protein Biosynthesis" (H. Weissbach and S. Pestka, eds.), pp. 467–553. Academic Press, New York.
Petit-Glatron, M.-F., and Rapoport, G. (1975). *In* "Spores VI" (P. Gerhardt, R. N. Costilow, and H. L. Sadoff, eds.), pp. 255–264. American Society for Microbiology, Washington, D.C.
Petit-Glatron, M.-F., and Rapoport, G. (1976). *Biochimie* **58,** 119–129.
Piepersberg, W., Böck, A., and Wittmann, H. G. (1975). *Mol. Gen. Genet.* **140,** 91–100.
Piggot, P. J., and Coote, J. G. (1976). *Bacteriol. Rev.* **40,** 908–962.
Post, L. E., Strycharz, G. D., Nomura, M., Lewis, H., and Dennis, P. P. (1979). *Proc. Nat. Acad. Sci. USA* **76,** 1697–1701.
Racine, F. M., and Steinberg, W. (1974). *J. Bacteriol.* **120,** 384–389.
Rapoport, G., Klier, A., Billault, A., Fargette, F., and Dedonder, R. (1979). *Mol. Gen. Genet.* **176,** 239–245.
Reeh, S., Pedersen, S., Friesen, J. D. (1976). *Mol. Gen. Genet.* **149,** 279–289.
Reiness, G., Yang, H.-L., Zubay, G., and Cashel, M. (1975). *Proc. Nat. Acad. Sci. USA* **72,** 2881–2885.
Rhaese, H.-J., and Groscurth, R. (1974). *FEBS Lett.* **44,** 87–93.
Rhaese, H.-J., and Groscurth, R. (1976). *Proc. Nat. Acad. Sci. USA* **73,** 331–335.
Rhaese, H.-J., and Groscurth, R. (1978). *Eur. J. Biochem.* **85,** 517–528.
Rhaese, H.-J., Dichtelmüller, H., and Grade, R. (1975). *Eur. J. Biochem.* **56,** 385–392.
Rhaese, H.-J., Grade, R., and Dichtelmüller, H. (1976). *Eur. J. Biochem.* **64,** 205–213.
Rhaese, H.-J., Hoch, J. A., and Groscurth, R. (1977). *Proc. Nat. Acad. Sci. USA* **74,** 1125–1129.
Rosset, R., and Gorini, L. (1969). *J. Mol. Biol.* **39,** 95–112.
Rudner, R., Prokop-Schneider, B., and Chargaff, E. (1964). *Nature (London)* **203,** 479–483.
Sala, F., Bazzicalupo, M., and Parisi, B. (1974). *J. Bacteriol.* **119,** 821–829.
Scherzinger, E., Grieder-Lauppe, H., Voll, N., and Wanke, M. (1980). *Nucleic Acids Res.* **8,** 1287–1305.
Segall, J., and Losick, R. (1977). *Cell* **11,** 751–761.
Setoguchi, Y., Margulies, L., and Rudner, R. (1978). *Biochim. Biophys. Acta.* **521,** 719–725.
Sharrock, W. J., Gold, B. M., and Rabinowitz, J. C. (1979). *J. Mol. Biol.* **135,** 627–638.
Shine, J., and Delgarno, L. (1975). *Nature (London)* **254,** 34–38.
Shivakumar, A. G., Hahn, J., and Dubnau, D. (1979). *Plasmid* **2,** 279–289.
Shivakumar, A. G., Hahn, J., Grandi, G., Kozlov, Y., and Dubnau, D. (1980). *Proc. Nat. Acad. Sci. USA* **77,** 3903–3907.
Shub, D. A. (1975). *J. Bacteriol.* **122,** 788–790.
Singhal, R. P., and Vold, B. (1976). *Nucleic Acids Res.* **3,** 1249–1261.
Sloma, A., and Smith, I. (1979). *Mol. Gen. Genet.* **175,** 113–120.
Smith, I. (1977). *In* "Molecular Mechanisms of Protein Biosynthesis" (H. Weissbach and S. Pestka, eds.), pp. 627–700. Academic Press, New York.
Smith, I., and Paress, P. (1978). *J. Bacteriol.* **135,** 1107–1117.

Smith, I., and Smith, H. (1973). *J. Bacteriol.* **114,** 1138–1142.
Smith, I., Dubnau, D., Morell, P., and Marmur, J. (1968). *J. Mol. Biol.* **33,** 123–140.
Smith, I., Paress, P., and Pestka, S. (1978). *Proc. Nat. Acad. Sci. USA* **75,** 5993–5997.
Smith, I., Dubnau, E., Williams, G., Cabane, K., and Paress, P. (1980a). *In* "RNA Polymerase, tRNA and Ribosomes: Their Genetics and Evolution" (S. Osawa, H. Ozeki, H. Uchida, and T. Yura, eds.), pp. 341–361. Univ. Tokyo Press, Tokyo.
Smith, I., Paress, P., Cabane, K., and Dubnau, E. (1980b). *Mol. Gen. Genet.* **178,** 271–279.
Spiegelman, G., Dickinson, E., Idriss, J., Steinberg, W., Rodenberg, S., and Halvorson, H. O. (1969). *In* "Spores IV" (L. Campbell, ed.), pp. 235–246. American Society for Microbiology, Washington, D.C.
Stallcup, M. R., Sharrock, W. J., and Rabinowitz, J. C. (1976). *J. Biol. Chem.* **251,** 2499–2510.
Stark, M. J. R., and Cundliffe, E. (1979). *Eur. J. Biochem.* **102,** 101–105.
Steinberg, W., and Anagnostopoulos, C. (1971). *J. Bacteriol.* **105,** 6–19.
Steitz, J. A. (1979). *In* "Biological Regulation and Development" (R. F. Goldberger, ed.), Vol. I, pp. 349–399. Plenum Press, New York.
Swanton, M., and Edlin, G. (1972). *Biochem. Biophys. Res. Commun.* **46,** 583–588.
Sy, J. (1979). *In* "Regulation of Macromolecular Synthesis by Low Molecular Weight Mediators" (G. Koch and D. Richter, eds.), pp. 95–106. Academic Press, New York.
Sy, J., and Akers, H. (1976). *Biochemistry* **15,** 4399–4403.
Sy, J., and Lipmann, F. (1973). *Proc. Nat. Acad. Sci. USA* **70,** 306–309.
Takeda, M., and Lipmann, F. (1966). *Proc. Nat. Acad. Sci. USA* **56,** 1875–1882.
Thompson, J., Cundliffe, E., and Stark, M. (1979). *Eur. J. Biochem.* **98,** 261–265.
Tipper, D. J., Johnson, C. W., Ginther, C. L., Leighton, T., and Wittmann, H. G. (1977). *Mol. Gen. Genet.* **150,** 147–159.
Tominaga, A., and Kobayashi, Y. (1978). *J. Bacteriol.* **135,** 1149–1150.
Torriani, A., Garrick, L., and Silberstein, Z. (1969). *In* "Spores IV" (L. Campbell, ed.), pp. 247–261. American Society for Microbiology, Washington, D.C.
Travers, A. A. (1980). *J. Bacteriol.* **141,** 973–976.
van de Klundert, J. A. M., Den Turk, E., Borman, A. H., Van der Meide, P. H., and Bosch, L. (1977). *FEBS Lett.* **81,** 303–307.
Vold, B. S. (1978). *J. Bacteriol.* **135,** 124–132.
Waterson, J., Beaud, G., and Lengyel, P. (1970). *Nature (London)* **227,** 34–38.
Weisblum, B., Graham, M. Y., Gryczan, T., and Dubnau, D. (1979). *J. Bacteriol.* **137,** 635–643.
Weissbach, H., and Pestka, S. (1977). "Molecular Mechanisms of Protein Biosynthesis." Academic Press, New York.
Wienen, B. *et al.* (1979). *J. Biol. Chem.* **254,** 8031–8041.
Woese, C. R., and Bleyman, M. (1969). *In* "Spores IV" (L. Campbell, ed.), pp. 223–234. American Society for Microbiology, Washington, D.C.
Yaguchi, M., Visentin, L. P., Nazar, R. N., and Matheson, A. T. (1978). *In* "Biochemistry of Thermophily" (S. M. Friedman, ed.), pp. 169–191. Academic Press, New York.
Yamakawa, T., and R. H. Doi (1971). *J. Bacteriol.* **106,** 305–310.
Yamamoto, M., and Nomura, M. (1978). *Proc. Nat. Acad. Sci. USA* **75,** 3891–3895.
Yamamoto, M., and Nomura, M. (1979). *J. Bacteriol.* **137,** 584–594.
Yang, H.-L., Zubay, G., Urm, E., Reiness, G., and Cashel, M. (1974). *Proc. Nat. Acad. Sci. USA* **71,** 63–67.
Yates, J. L. (1979). *J. Biol. Chem.* **254,** 11550–11554.
Yates, J. L., and Nomura, M. (1980). *Cell* **21,** 517–522.

Yates, J. L., Arfsten, A. E., and Nomura, M. (1980). *Proc. Nat. Acad. Sci. USA* **77,** 1837–1841.
Yeh, E. C., and Steinberg, W. (1978). *Mol. Gen. Genet.* **158,** 287–296.
Yoshikawa, H., and Sueoka, N. (1963). *Proc. Nat. Acad. Sci. USA* **49,** 559–566.
Young, R. A., and Steitz, J. A. (1979). *Cell* **17,** 225–234.
Zabos, P., Bauer, P., Schlotthauer, J., and Horvath, I. (1976). *FEBS Lett.* **64,** 107–110.
Zengel, J. M., Muecki, D., and Lindahl, L. (1980). *Cell* **21,** 523–535.

5

Genetic Transformation in *Bacillus subtilis*

DAVID DUBNAU

The Molecular Biology
of the Bacilli

ISBN 0-12-222701-8

I. Introduction

A. Importance of Genetic Transformation

Genetic transformation in *Bacillus subtilis* was first reported by Spizizen (1958). Since then, the ability to transfer genetic information via DNA molecules in solution has proven to be a powerful tool in the genetic analysis of this organism. More recently, transformation using plasmid DNA has been central to the development of molecular cloning technology in *B. subtilis* (Chapter 10). In addition to these useful aspects, the process of transformation is of intrinsic interest. Understanding the mechanism of DNA uptake is relevant to the broader question of the interaction of the cell with its environment. The nature and development of the competent state presents interesting problems in regulation and in the study of prokaryotic sexuality. Finally, the integration of transforming DNA presents an opportunity, in some ways unique, to examine genetic recombination at the molecular level.

B. Scope

Although all these applications and aspects of transformation will be touched on in this chapter, I will not attempt an exhaustive compilation of references but will instead focus on certain aspects of the transformation process. It is hoped that the reader will be left with a clearer understanding of the molecular mechanisms involved in the uptake, processing, and integration of transforming DNA. In addition, this chapter will attempt to present a more-or-less unified view of transformation using plasmid, chromosomal, and bacteriophage DNA. Recent reviews dealing with bacterial transformation and transfection have been prepared by Trautner and Spatz (1973), Notani and Setlow (1974), Dubnau (1976), Lacks (1977), Fox (1978), Low and Porter (1978), Venema (1979), Prozorov (1980), and Smith *et al.* (1981).

C. Transformation and the Study of Genetic Recombination

Genetic transformation in *B. subtilis* offers several advantages as a system for the study of recombination on the molecular level. Integration is efficient; after uptake the integration probability for a marker carried on a given DNA fragment is 0.5–1.0 (Bodmer, 1966; Somma and Polsinelli, 1970; Dubnau and Cirigliano, 1972c). Transformation proceeds by the administration of DNA in solution. This fact permits extensive manipulation of one of the molecular partners in the recombinational event. In a sense then, transformation is a semi *in vitro* process. In *B.*

subtilis, nucleic acid synthesis proceeds at a very low rate in the competent state (Dooley *et al.*, 1971), and DNA replication is not required for integration (Dubnau and Cirigliano, 1973b; Levin and Landman, 1973). As a result, the integration process may be studied in the absence of complexities due to DNA synthesis. For instance, the primary product of the recombinational event can be readily isolated and examined. Thus, "population" problems inherent in other systems, e.g., phage recombination, need not be faced. There are, however, two major technical problems and, one hopes, a temporary limitation. The classic transforming DNA preparation consists of a randomly sheared collection of chromosomal DNA fragments which are polydisperse with respect to length and heterogeneous with respect to sequence. This complicates analyses of the uptake, processing, and integration events. Further, the recipient partner in integration consists of about 2.5×10^9 daltons and possesses a fragile (unknown) tertiary structure. It is thus difficult, it not impossible, to study changes in the resident chromosome accompanying transformation. The final limitation stems from the paucity of information concerning *B. subtilis* deoxyribonucleases, topoisomerases, and other DNA-binding proteins. It is hoped that this state of ignorance will be short-lived.

II. The Competent State

The competent state may be defined as that which permits the uptake and integration of DNA from solution. Competence may be measured by the binding or uptake of radioisotopically labeled DNA or by the yield of transformant colonies (using chromosomal or plasmid DNA) or of infectious centers (using phage DNA). The use of chromosomal DNA and, as we shall see, phage DNA, depends on the presence of a functioning recombination (*rec*) system. The *rec* system may be viewed as at least partly independent of competence, since transduction, which is *rec*-dependent, occurs using noncompetent cultures.

Several growth regimens have been described that produce competent cultures (for instance, Anagnostopoulos and Spizizen, 1961; Wilson and Bott, 1968; Dubnau and Davidoff-Abelson, 1971). These empirically derived and varying procedures all demonstrate that competence develops after most of the growth of the culture has ceased. In this, competence is reminiscent of other phenomena such as sporulation, antibiotic production, protease, and amylase synthesis. Rather than postulate that competence is somehow related to sporulation or is a presporulation event, as some authors have done, it seems more fruitful to suggest that common regulatory mechanisms are reflected in these phenomena, perhaps triggered by depletion of a nutrient.

A complicating and unexplained aspect of the *B. subtilis* system is the population heterogeneity of competence. Analysis of cotransformation frequencies of "unlinked" markers suggested that only a minority of cells achieved competence (Nester and Stocker, 1963). This was demonstrated directly both by autoradiography (Somma and Polsinelli, 1970) and when methods were developed permitting the separation of competent and noncompetent cell fractions (Singh and Pitale, 1967; Cahn and Fox, 1968; Hadden and Nester, 1968). These methods are based on the fact that competent cells have a lower buoyant density in renografin and are somewhat smaller than noncompetent cells. It is not known why only 10–20% of the cells in a given culture become competent, but this heterogeneity implies that competence does not develop simply in response to a diffusible factor.

Competence is accompanied by a decrease in the rate of nucleic acid synthesis (McCarthy and Nester, 1967; Dooley *et al.*, 1971) and by a latency in cell multiplication (Nester and Stocker, 1963). The extent and nature of the blockage of DNA synthesis are not clear. In one study, degradation products of [^{3}H]thymidine-labeled DNA, which were released into the medium during uptake as 5′-TMP, thymidine, and thymine, were not measurably reincorporated into DNA by competent cells (Dubnau and Cirigliano, 1972b). In another report, no [^{3}H]thymidine incorporation was observed into the competent subfraction after separation on a renografin gradient (Dubnau and Cirigliano, 1973b). On the other hand, Piechowska *et al.* (1975) and Loveday (1978) have reported that degradation of radiolabeled *Escherichia coli* and of transfecting $\phi\varepsilon$ DNA by competent cells leads to reincorporation of the products. This incorporation is at least partly sensitive to 6-(*p*-hydroxyphenylazo)uracil (HPU), a specific inhibitor of DNA polymerase III. These reports are difficult to compare meaningfully. Different strains and competence regimens were used, and the specific activities of the radioactive precursors are not known in the case of the degradation–reincorporation experiments. Loveday (1978) has suggested that the blockage is in nucleotide metabolism rather than in replication per se. Whatever the case, semiconservative DNA replication is not required for transformation, since HPU has no effect on the yield of transformant colonies or on the physical integration of donor DNA segments (Dubnau and Cirigliano, 1973b; Levin and Landman, 1973).

In general, extreme caution must be employed in relating physiological and biochemical changes that *accompany* competence to the development of competence itself. Since the competent state emerges during transition to the stationary phase of growth, it is reasonable to expect profound changes to occur, many of which may have little or nothing to do with transformation. It is essential to define molecular components directly involved in the transformation process and to study their synthesis and organization during competence development. It is also necessary to use a genetic approach—to analyze competence mutants and their gene products.

Very little work has been done that describes *specific* biochemical alterations that can be related to the competent state. Three such reports, however, deserve mention. Eisenstadt *et al.* (1975) have described a single-strand DNA-binding protein that was synthesized at an increased rate during the development of competence. This protein was not detected in a noncompetent mutant (which was also asporogenous). Joenje *et al.* (1972) have described a factor that can accelerate the induction of competence in precompetent cultures. This factor has a molecular weight of about 700 and is resistant to heat and proteolytic enzymes (Venema, 1979). Finally, Joenje and Venema (1975) have described a very interesting exonuclease activity located on the cell surface, which develops at the same time as competence and is found only in the competent subfraction of the culture.

III. Binding

A. Definitions

The first stable attachment of transforming DNA to the cell surface has been referred to as binding. It is distinct from uptake which is operationally distinguished by the development of DNase resistance and carries the implication of intracellularity.

B. Measurement and Nature of Binding

Binding can be measured in several ways: by competition experiments between radioisotopically labeled and cold DNA, by detecting rapidly sedimenting complexes of radioactive DNA with cells, and by filter retention of [^{3}H]DNA–cell complexes using polycarbonate filters. It may also be visualized by electron microscopy (J. Poindexter and D. Dubnau, unpublished). These studies have led to several general conclusions which will now be described.

The binding of duplex DNA by competent cells is not species-specific. However, glucosylated DNA, double-strand RNA, and various synthetic polymers bind poorly, if at all (Ciferri *et al.*, 1970; Soltyk *et al.*, 1975, 1976). This is in marked contrast to the *Hemophilus* system in which base sequence-specific regonition occurs, presumably by some element on the cell surface (Sisco and Smith, 1979; Deich and Smith, 1980). DNA is probably bound noncovalently to *B. subtilis* competent cells, since it can be quantitatively removed by treatment with detergents and by phenol, although it has not been shown that the DNA so released is not covalently modified (Dubnau and Cirigliano, 1974; Dubnau, 1976).

Binding appears to be an all-or-nothing event, since the mass of DNA bound is proportional to the molecular weight (Dubnau, 1976), indicating that the entire molecule remains attached to the cell. The bound DNA molecule is retained on the cell surface in such a way that the DNA molecule is physically extended, since it can be removed by shear and by exogenous DNase (Dubnau and Cirigliano, 1972c). This can be visualized by electron microscopy, as shown in Fig. 1 (J. Poindexter and D. Dubnau, unpublished). These studies were carried out using T7 DNA. After brief (2- to 5-min) exposures of competent cultures to the DNA, the cell–DNA complexes were isolated free of unbound DNA by sedimentation through a column of sucrose. The DNA–cell complexes were prepared for microscopy by formamide spreading, using cytochrome *c* (Kleinschmidt, 1968). Figure 1 shows some typical examples selected from many complexes examined. In general, these images reveal DNA molecules attached to the cell surface, often at more than one point, with most of the mass of a given molecule extended freely into the liquid phase. The extended portions range downward in size from T7 length to small loops or segments with free ends.

Based on measurements of the extent of binding of radioactive DNA, the number of binding sites per cell appears to be 20–50 (Dubnau and Cirigliano, 1972c; Singh, 1972). This is similar to the number (30–80) measured in *Streptococcus pneumoniae* (Fox and Hotchkiss, 1957) and much larger than the number (about 2) determined for *Hemophilus* (Stuy and Stern, 1964).

The initial binding interaction probably need not occur at a molecular terminus, since circular covalently continuous (CCC) plasmid DNA binds to competent cells and competes with linear duplex DNA for surface sites (Contente and Dubnau, 1979a; Lacks, 1979; and S. Contente and D. Dubnau, unpublished). The requirements for binding are somewhat unclear. It has been reported that EDTA permits binding to occur but interferes with subsequent steps (Morrison, 1971). A more recent report concludes that, although binding is Mg^{2+}-independent, EDTA irreversibly damages the DNA receptor sites (Garcia *et al.*, 1978).

The precise site of binding on the cell surface is also unclear; it is not known whether the receptors are located on the cell wall or on the membrane. Low levels of lysozyme stimulate transformation (Prozorov, 1965; Wilson and Bott, 1970), whereas spheroplasts of competent cells are not normally transformable (Tichy and Landman, 1969). These data can perhaps best be interpreted as indicating a role for the cell wall in initial binding. Mild lysozyme treatment, which weakens this structure without disrupting it totally, may then facilitate permeation of bound molecules. Vermuelen and Venema (1974a,b) have suggested, based on electron microscope autoradiography, that uptake, and therefore possibly binding, tend to occur in association with mesosomal membrane structures.

IV. Cleavage and Kinetic Analysis

Much of the bound DNA recovered from the cell surface following lysis (Dubnau and Cirigliano, 1972b) or after removal by detergent–phenol treatment (Dubnau and Cirigliano, 1974; Dubnau, 1976) is found to be cleaved but still double-stranded. However, shortly after binding, intact donor-sized molecules are also recovered. The kinetics of the processes implied by these observations have been studied. Transformation may be interrupted by a variety of treatments: competition with heterologous DNA (which prevents further binding), vortex mixing (which removes longer and extended bound molecules), and the addition of DNase (which removes all surface-bound DNA). In this way the kinetics of binding, fragmentation, and uptake can be studied. The binding process was shown to occur with no discernible lag at 37°C, as measured by the appearance of transformants when the process was interrupted by the addition of competing DNA (Dubnau and Cirigliano, 1972c). This was also shown using [^{3}H]DNA (Dubnau, 1976). Shear resistance appeared with a lag of about 0.5–1 min, and DNase resistance with a lag of 1.5–2.0

Fig. 1. Complexes of competent *B. subtilis* with T7 DNA (J. Poindexter and D. Dubnau, unpublished). The DNA (2 μg/ml) was added to competent cultures and incubated for 2–5 min at 37°C. The samples were chilled and then sedimented through columns of 15% sucrose, 0.1 *M* EDTA, and 0.01 *M* NaCl, pH 6.9. The pellets were resuspended gently and prepared for electron microscopy using the formamide-spreading procedure (Kleinschmidt, 1968). The figure is a composite of complexes recorded in four separate experiments. Control experiments using competent cells to which DNA was added immediately before sedimentation through sucrose did not reveal any extracellular DNA sedimenting with the cells.

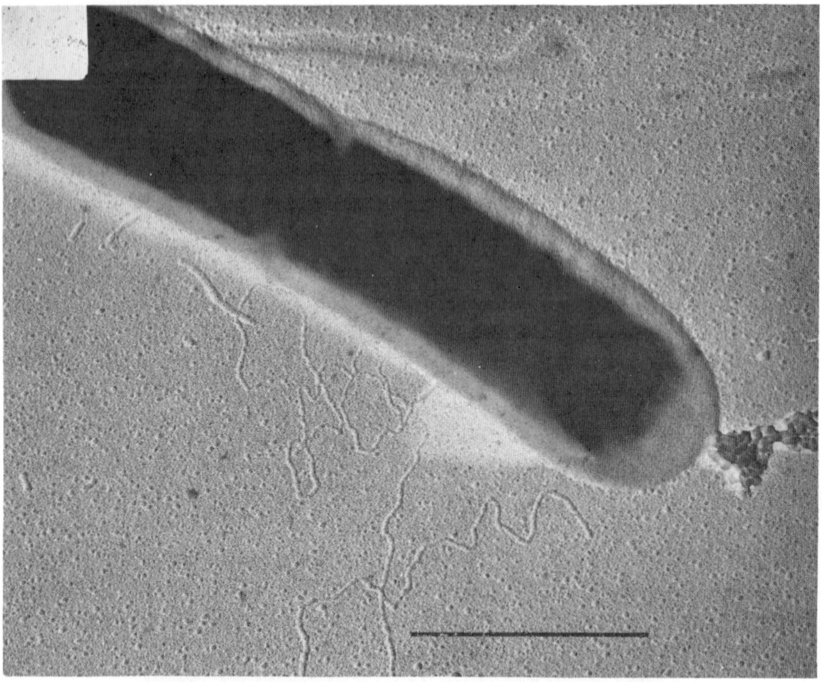
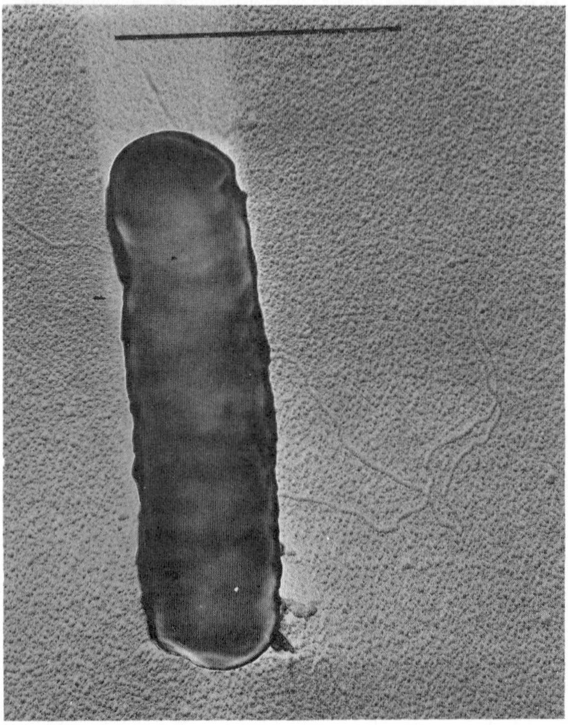

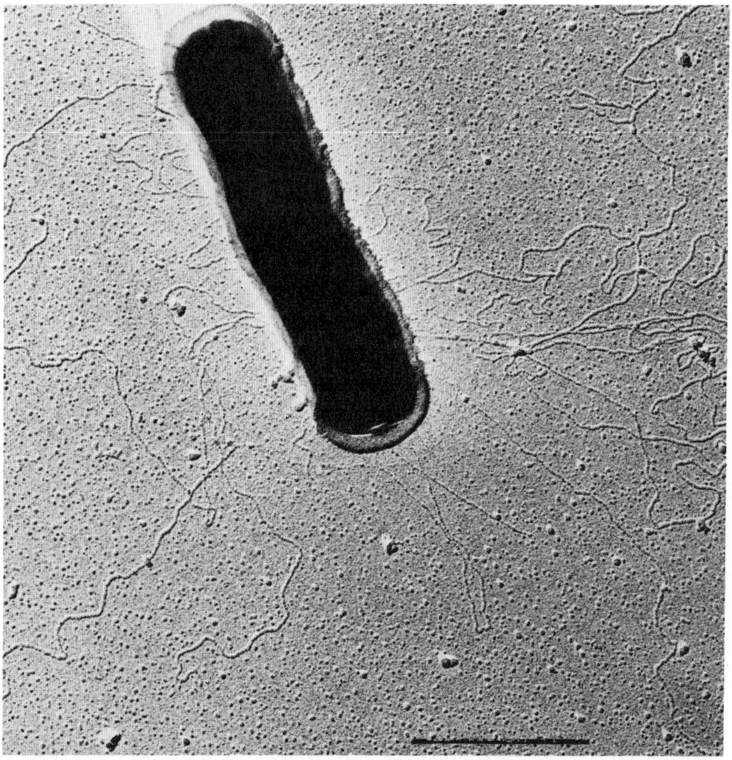
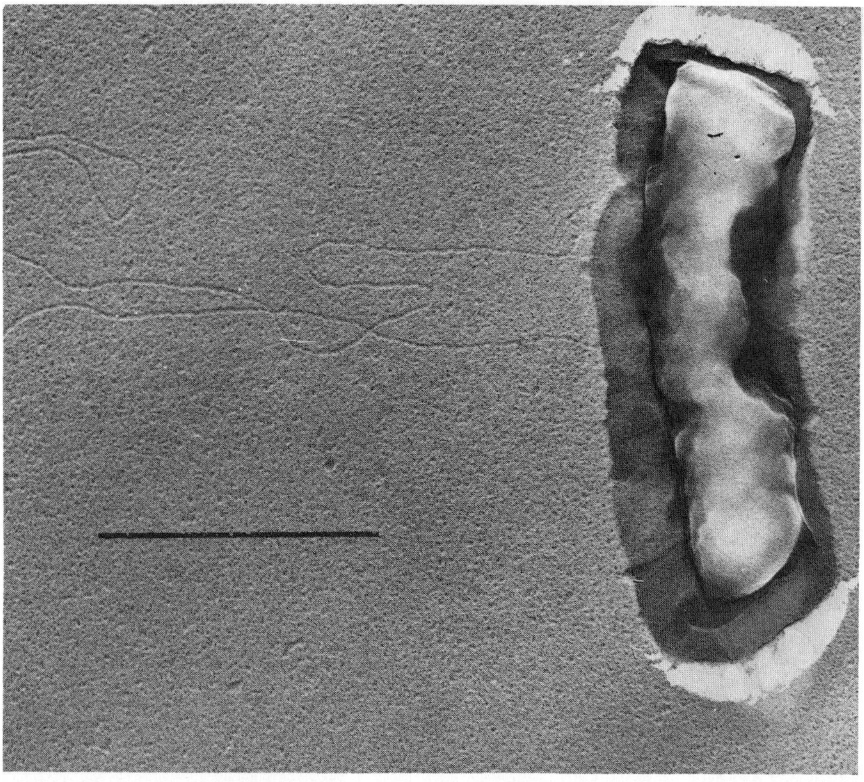

min (Levine and Strauss, 1965; Dubnau and Cirigliano, 1972c). These results appear to define a succession of surface-bound forms. Additional kinetic studies have established that the shear-sensitive form is a precursor of the shear-resistant form and that the latter is a precursor of the DNase-resistant material (Dubnau and Cirigliano, 1972c; Davidoff-Abelson and Dubnau, 1973b). It is reasonable to suppose that it is the double-strand cleavage event that renders the donor DNA relatively shear-resistant. It is possible that this is not the entire explanation for the transition to shear resistance. Possibly the cleavage fragments become less extended and hug the cell surface more closely. A similar fragmentation step occurs in *S. pneumoniae* (Morrison and Guild, 1973a,b). In this organism, single-strand nicks appear to precede the double-strand cleavage (Lacks and Greenberg, 1976; Lacks, 1979). It is not known whether this is true in *B. subtilis*. The observation that bound, intact donor DNA can be recovered immediately after binding shows that binding and double-strand cleavage are not concomitant events and implies that the binding receptor and the cleavage enzyme are distinct. This is consistent with the observation that the noncompetent cell surface can interact reversibly with DNA to yield double-strand cleavages without retaining these fragments (Haseltine and Fox, 1971).

The distribution of cleavage events has also been studied using T7 [^{3}H]DNA (MW ~25 × 106) and a preparation of *B. subtilis* chromosomal [^{3}H]DNA (MW ~ 125 × 10^6) (Dubnau, 1976). The T7 DNA recovered from the competent cell surface after 60 and 180 sec (at 37°C) had similar sedimentation distributions. The results showed that each molecule had suffered an average of one double-strand break, since the number average molecular weight was about 12 × 10^6 (weight average of 21 × 10^6). These breaks were distributed randomly throughout the length of the molecule, except that there appeared to be a bias against breakage within 4–5 × 10^6 from the molecular termini. Thus, the probability of formation of a cleavage product of given molecular weight is equal within the range 5–20 × 10^6. In the 180-sec sample no more than about 15–20% of the mass of bound DNA was in the form of intact T7-sized molecules. When the DNA of 125 × 10^6 molecular weight was tested, additional features were observed. Material with a weight average molecular weight of about 21 × 10^6 again accumulated with time, but the bulk of the bound DNA was distributed between this size and the starting size, with a progressive broadening of this distribution and a shift toward smaller sizes occurring between 60 and 180 sec. Thus, in contrast to the DNA of 25 × 10^6 molecular weight, which underwent (on the average) a single rapid cleavage event per molecule, this material suffered successive time-dependent cleavages but tended to reach a size limit similar to that reached by the T7 DNA.

Two plausible models can be invoked to explain these results and are consistent with the observations described above. In the model we favor, the donor DNA first binds to the cell surface without double-strand cleavage. This initial binding facilitates cleavage by interaction of the bound DNA with an endonuclease fixed elsewhere on the cell surface. This is consistent with the existence of distinct binding and cleavage sites as suggested above. The spacing between the sites of binding and of cleavage *on the DNA molecule* would be determined both by the stiffness of the DNA and by the spacing between the cell surface sites. The probability of multiple cleavages occurring on the molecule and the average spacing between these cleavages would be similarly determined. If binding and cleavage are both random with respect to sequence, then the probability of cleavage at a particular point will reflect the total likelihood of binding on either side of this point. However, DNA stiffness and the spacing of binding and cleavage sties will result in a diminished likelihood of cleavage of the DNA molecule near the site of binding. Therefore, the probability of cleavage near the molecular termini should be about half that observed for internal regions of the molecule, since binding can occur only on one side of the terminus. This is essentially what has been observed (Dubnau, 1976). Finally, the probability of multiple cleavages will increase with increasing molecular weight and should occur in a random, time-dependent fashion, as described above for the sample of 125×10^6 molecular weight. A second model can be envisioned in which the DNA binds at a given point and then moves along the cell surface (by diffusion?) past an endonuclease, with a probability of cleavage equal to about once per T7 length. In the first model physical parameters determine the distribution of cleavage events, and in the second these are determined by the frequency of endonuclease cutting per enzyme molecule.

It is not known what enzyme is responsible for cleavage. A surface-localized Mn^{2+}-dependent endonuclease is present in *B. subtilis,* but no mutant studies have been carried out (Scher and Dubnau, 1975). This endonuclease is present in both competent and noncompetent cells. This is not unreasonable. A similar enzyme activity has been detected in a donor DNA–protein complex recovered from transformed cells (Pieniazek *et al.,* 1977).

Kinetic studies, in which competent cells are exposed to DNA for 1 min and further binding is arrested by the addition of heterologous DNA, show that most of the mass of the cell surface-bound cleavage products is retained on the cell surface and is on the main uptake pathway. It is reasonable to suggest that at least one of the newly formed molecular termini resulting from cleavage at the cell surface remains attached and participates in the next step, which is uptake.

V. Uptake

Uptake is detected and therefore defined operationally, by measuring the increase in DNase-resistant radioactive DNA or transformants. This definition carries the nonrigorous but plausible inference that DNA becomes DNase-resistant because it has crossed a barrier to DNase permeation, presumably the cell membrane or cell wall. Piechowska and Fox (1971) first demonstrated high-molecular-weight DNA of donor origin in transformed cells. Dubnau and Cirigliano (1972b) showed independently that single-strand DNA was present in transformed cells and that the kinetics of the appearance of this material were consistent with its role as either a precursor of integrated material or a degradation intermediate. Their material, however, had a very low molecular weight ($\sim 10^5$). The isolation procedure of Piechowska and Fox employed high concentrations of lysozyme, a basic protein later shown to protect the high-molecular weight single-strand material from degradation during isolation (Davidoff-Abelson and Dubnau, 1973a). Evidence was then presented showing that fragmented DNA of donor origin, which was the only cell-associated product present at 30 sec, was converted to single-strand fragments after 1–2 min. The mass of single-strand material present corresponded to roughly one-half of the prior load of double-strand material (Davidoff-Abelson and Dubnau, 1973b). The rest was lost to the medium as acid-soluble material which consisted of 5′-mononucleotides, nucleosides, and free bases (Dubnau and Cirigliano, 1972b). The single-strand fragments were shown to be completely resistant to exogenous DNase and had thus been internalized. This single-strand material was nearly quantitatively (72%) converted to integrated material within a few minutes when the cells were incubated in the presence of DNase (to eliminate surface-bound double-strand DNA and to prevent further binding). Thus, it was established that donor DNA was bound to the cell surface in fragments (with retention of at least most of the fragment mass), that about one-half the mass of these retained fragments was internalized with concomitant conversion to single strands, and that these single-strand molecules were direct precursors of integrated material. The weight average molecular weight of the single-strand precursor was about 5×10^6 (Davidoff Abelson and Dubnau, 1973b).

The major features of the DNA uptake process in *B. subtilis* are very similar to those described in *S. pneumoniae*. In the latter organism also, double-strand DNA-fragments bound to the surface are internalized as single-strand material with the loss of about half of the mass as acid-soluble material (Lacks, 1962; Lacks *et al.*, 1967). In *S. pneumoniae* the latter material appears in the medium as oligonucleotides (Lacks and

Greenberg, 1973; Morrison and Guild, 1973a,b). Lacks and co-workers have shown that certain mutants (*noz*), which are deficient in an endonuclease, are also transformation-deficient and are deficient in uptake, although they can bind DNA (Lacks, 1974; Lacks *et al.,* 1974, 1975). Rosenthal and Lacks (1980) have shown that this nuclease exists as a "discrete structural complex in the membrane."

Lacks (1979) has shown that double-strand DNA bound to the *S. pneumoniae* cell surface is first nicked in one strand upon binding (see also Lacks and Greenberg, 1976). This event does not involve the membrane nuclease, since it proceeds in *noz* mutants. A break opposite this nick results in double-strand cleavage. Lacks postulates that this is followed by a processive action of the fixed membrane-bound nuclease on one strand. He believes that the nuclease also provides a hydrophillic "pore" for entry of the resulting (complementary) single strand. The principal features of this attractive model are consistent with the *B. subtilis* data, although no evidence for nicking at the binding site has been presented in the latter organism.

Several workers have reported, in *B. subtilis* and *S. pneumoniae,* that the kinetics of the appearance of DNase-resistant transformants for individual markers, compared to linked marker pairs, are consistent with this notion of linear transport across the membrane, perhaps initiated at a molecular terminus (Strauss, 1965, 1966; Gabor and Hotchkiss, 1966). The *Hemophilus* system provides a striking contrast to this picture. Not only are binding and uptake sequence-specific in this organism, but no single-strand donor material has been detected (Stuy, 1965; Notani and Goodgal, 1966; Deich and Smith, 1980). Instead, double-strand DNA, with single-strand regions probably located at the molecular termini, is observed in the cells following transformation (Sedgwick and Setlow, 1976).

VI. Integration and Subsequent Steps

A. Structure of Recombinant Molecules

Before briefly discussing the integration mechanism, it is appropriate to describe the structure of the end product, the recombinant molecule. In *B. subtilis* (as in *S. pneumoniae* and *Hemophilus*) the product of recombination is a heteroduplex in which a donor strand is paired to a complementary resident strand (Bodmer and Ganesan, 1964; Strauss, 1970; Dubnau and Davidoff-Abelson, 1971; Dubnau and Cirigliano, 1972a). Transformation, like all homology-dependent recombination processes, relies on recognition by base pairing and represents an extreme case in which one entire parental moiety is present in a heteroduplex. Other recombination processes involve heteroduplex overlap at the joint between duplex parental entities (Radding, 1974).

The average size of the integrated single-strand donor moiety has been measured by physical methods and found to have a molecular weight of 2.8×10^6 (weight average) or to be ~8.5 kb (Dubnau and Cirigliano, 1972a). This is consistent with the finding that the average single-strand precursor molecule has a molecular weight of $\sim 5 \times 10^6$. A similar average value for the integrated segment ($1.5–3 \times 10^6$) has been reported in *S. pneumoniae* (Gurney and Fox, 1968), and a somewhat higher value (6×10^6) in *Hemophilus* (Notani and Goodgal, 1966). The fragmentation phenomenon, the retention of fragments on the cell surface, and the high efficiency of integration, led to the inescapable conclusion that single-strand precursor fragments derived from one donor molecule must often insert near one another in *B. subtilis,* leading to a clustering of integration events (Dubnau and Cirigliano, 1972c). This prediction was confirmed by buoyant density studies using deuterium-labeled DNA (Dubnau and Cirigliano, 1972a) and later by electron microscopy (Fornili and Fox, 1977). The latter study also yielded an average size for the integrated segment of $2.9–3.9 \times 10^6$ molecular weight in reasonable agreement with the aforementioned estimate.

B. The State and Physical Location of Precursor Single Strands

Eisenstadt *et al.* (1975) have described a single-strand binding protein which appears to be competence-specific. Several workers have reported that intracellular donor DNA exists in a complex with protein (Piechowska and Fox, 1971; Soltyk *et al.,* 1975). Pieniazek *et al.* (1977) isolated such a complex from CsCl gradients after gentle lysis. The single-strand donor DNA was complexed with cell proteins. The complex exhibited cell-lytic activity and endonucleolytic activity toward double-strand linear DNA which was similar in its properties to the cell surface-localized Mn^{2+}-dependent endonuclease studied by Scher and Dubnau (1975). The donor DNA in this complex was partially protected from the action of DNase I. Similar observations have been reported in *S. pneumoniae* (Morrison, 1977).

C. Joint Molecules and Their Formation

A stable noncovalent association of donor and resident DNA has been described (Dubnau and Cirigliano, 1973a). When freed by treatment at pH 12, which separates strands of double-strand DNA, the donor moiety was found to have a weight average molecular weight of 2.5–3.0 $\times$ 10^6 in excellent agreement with the sizes reported for both the precursor and the integrated moieties. The structure of this joint molecule is not known, although both a branched triple-strand (redundant) structure in which only a portion of the donor strand is base-paired to the recipient, and a completely base-paired structure are plausible, and both structures may have been present in these preparations. Arwert and Venema (1973) have suggested that nicks (rather than gaps) separate the donor and resident moieties in these joint molecules, since DNA ligase increased recombinant-transforming activity of such a preparation as much as ligase plus DNA polymerase.

Strand assimilation (Radding, 1974, 1978) is certainly the most attractive proposal for integration. According to this concept, initial pairing occurs via invasion of a resident duplex by a donor single strand (Holloman *et al.,* 1975; Beattie *et al.,* 1977; Wiegand *et al.,* 1977) or, less attractively, by pairing at a preexisting gap. The paired length might then grow by exchanging with the resident redundant strand—a process known as branch migration (Lee *et al.,* 1970). Alternatively, the resident strand might be degraded, resulting in a facilitated, or driven, strand assimilation process (Cassuto and Radding, 1971).

Buitenwerf and Venema (1977, 1978) have reported that, at 17°C, an intermediate accumulates in which the donor moiety is partially single-strand (susceptible to S1 nuclease) and which has a higher buoyant density in netural CsCl than duplex DNA. However, this donor material was already covalently bound to resident DNA. It is plausible to postulate that the low temperature slowed down strand assimilation, although allowing ligation of a paired donor terminus to proceed. Popowski and Venema (1978) have also described a labile association of donor single strands with recipient DNA, which is stabilized by ultraviolet (UV) irradiation or by the chemical cross-linking agent 8-trimethylpsoralen. This labile complex presumably involves base pairing of the donor and recipient moieties.

Transformation is highly dependent on base sequence homology (Dubnau *et al.,* 1965; Harris-Warrick and Lederberg, 1978a). *Bacilius globigii* DNA, for example, transforms *B. subtilis* at 10^{-4} the frequency observed with homologous DNA. Melting temperature measurements for *B. globigii–B. subtilis* heteroduplex DNA suggest that there is about a 5% sequence mismatch between these DNA molecules (Chilton and McCarthy, 1969; Lomant and Fresco, 1975). A model of integration that involves merely annealing of a donor strand with an unpaired portion of the resident duplex would presumably not show this degree of discrimination. Strand assimilation, on the other hand, might be slowed down or arrested by mismatch. Interestingly, Cunningham *et al.* (1981) have reported that the *E. coli* recA protein can distinguish perfect from near-perfect homologous pairing. The additional contribution of mismatch repair to discrimination has not been properly assessed.

The intial pairing (synapsis) of donor and recipient DNA is extremely rapid. The lag between uptake and the formation of the first detectable stable complex of donor and recipient DNA is about 1.0 min at 37°C (Davidoff-Abelson and Dubnau, 1973b). It is not known whether initial pairing precedes or follows introduction of the first covalent discontinuity in the recipient DNA. In fact, virtually nothing is known about changes in the chromosome during transformation, although it has been reported that the chromosome of competent cells is highly "gapped" (Harris and Barr, 1971). This lack of knowledge, as noted above, probably derives from the large size of the chromosome. However, Collins and Guild (1972) and Shoemaker and Guild (1972) detected a step preceding integration in *S. pneumoniae* that they postulated, based on temperature dependence studies, to represent the initial formation of base pairs between donor and recipient DNA. The high temperature dependence of this process was interpreted as indicating that it is rate-limited by the unstacking of several recipient base pairs. Beattie *et al.* (1977) have pointed out that the temperature dependence of the rate-limiting step in their model studies on single-strand DNA uptake by superhelical DNA in solution is very similar to that observed by Guild and his co-workers. They interpret the rate-limiting step in their system as due to the unstacking of base pairs in superhelical DNA. This is consistent with the occurrence in transformation of pairing *before* covalent continuity is interrupted and with a role for superhelicity in synapsis.

To summarize, although several investigators have presented suggestive evidence that single-strand invasion and strand assimilation are the route followed by transforming DNA during integration, little direct evidence has been presented.

D. Resolution of the Heteroduplex

Heteroduplex recombinant molecules may be resolved either by semiconservative replication, to yield mixed clones, or by mismatch repair followed by replication, to yield pure clones. Transformation in *B. subtilis* is no exception, and the probability of mismatch repair (determined by clonal analysis) has been shown to depend on the nature of the mismatch (Bresler *et al.*, 1968; Spatz and Trautner, 1970). The probability of simultaneous correction of two mutations decreases with increasing distance between markers. An interesting approach to the problem of mismatch repair involves the study of heterologous transformation (Matsumoto *et al.*, 1978). DNA from *Bacillus amyloliquefaciens* was used to transform a *B. subtilis trp* mutant to prototrophy. DNA from this strain (termed an intergenote) (Wilson and Young, 1972) transformed *B. subtilis* to Trp^+ at a frequency 10- to 20-fold below that of homologous DNA. However, when strains carrying certain *uvr* mutations were used as recipients, the frequency of this transformation event rose strikingly. The authors suggested that this occurred because a mismatch correction mechanism, which could result in the loss of heteroduplex molecules, was inactivated by these mutations. Interestingly, when *B. amyloliquefaciens* rather than intergenotic DNA was used, no enhancement of heterologous transformation for several markers was observed in the same *uvr* strains. The authors point out that the frequency of transformation in this heterologous cross is normally depressed 10^2- to 10^4-fold, whereas the frequency using the heterogenote is depressed only 10-fold. Thus, base sequence discrimination at the integration step may predominate in the heterologous case, whereas the effect of mismatch correction may predominate in the more homologous intergenotic cross. The same group has recently isolated an S1-like endonuclease from *B. subtilis,* with a specificity for single-strand and superhelical DNA (Matsumoto *et al.*, 1979). They report that this enzyme appears to act preferentially on mismatched DNA.

In *S. pneumoniae,* heteroduplex mismatch repair does not occur for some markers, termed high efficiency (HE) (Guerrini and Fox, 1968a,b). Some mutations, termed low efficiency (LE), are transforned at a lower efficiency. A third class (very high efficiency, VHE) has also been described (Ephrussi-Taylor, 1966; Ephrussi-Taylor and Gray, 1966; Lacks, 1966; Tiraby and Sicard, 1973). These effects have long been attributed to mutation-specific mismatch repair causing a loss of LE markers. Certain mutants (*hex*) allow the transformation of LE markers at high efficiency and presumably lack this correction system (Lacks, 1970; Tiraby and Fox, 1973). Recently, Claverys *et al.* (1980) have shown, using heteroduplex transforming molecules prepared *in vitro,* that the *hex* system operates to eliminate LE genetic information carried on either strand. The VHE marker information is not removed from either strand by the *hex* system. In the case of each HE marker tested, a particular strand is preferentially eliminated. Whatever the precise molecular mechanism involved in this discrimination, the results argue strongly for the involvement of mismatch correction of a heteroduplex structure by the *hex* system of *S. pneumoniae*. No LE or HE markers have been reported in *B. subtilis*. Possibly a more exhaustive search would reveal such effects. It will be interesting to compare the properties of the *hex* system with the correction system now being investigated by the Japanese group (Matsumoto *et al.*, 1979).

VII. Linkage

Transformation has been widely used in genetic mapping of *B. subtilis*, and it is therefore worthwhile commenting on the factors that contribute to cotransformation frequencies in this system. No attempt will be made to develop a precise quantitative description. Since only 10–20% of the cells in a competent culture are able to take up DNA, and since transformation is efficient, the probability of cotransformation of two unlinked markers appears anomalously high. This phenomenon, termed congression, is useful, since it permits the ready construction of strains with new marker configurations. It is also a potential pitfall, and care must be taken to test suspected cotransformation events at low (subsaturating) DNA concentrations.

Dislinkage is obviously a consequence of two factors: separation of markers before binding (shear during isolation) and after binding. The probability of dislinkage during DNA isolation varies directly with the intermarker distance and therefore contributes strongly to apparent recombination frequencies. It has been shown that very high-molecular-weight DNA reveals previously unsuspected linkages and lower recombination frequencies between close markers (Kelly, 1967a). A contributory factor to this type of effect is probably the concentration of competent cells during incubation with transforming DNA, since a single DNA molecule can bind simultaneously to more than one cell (Kelly, 1967b).

Molecules bound to the competent cell surface undergo double-strand cleavage to produce fragments with a number average molecular weight of $10–20 \times 10^6$. These molecules enter the cell and are integrated independently with high efficiency, producing the clustering effect noted above. However, as a consequence of fragmentation, dislinkage may also be produced in several ways. First, the probability of integration is probably not unity for each fragment, although it is certainly at least 0.5 (Bodmer, 1966; Dubnau and Cirigliano, 1972c; Somma and Polsinelli, 1970). Second, integration (at least on the population level) is not strand-specific. Thus, linkage of markers carried on separate fragments is most likely reduced by a factor of 2 following replication and segregation of daughter chromosomes. Third, although it has been reported that at least the majority of competent cells are uninucleate (Singh and Pitale, 1968), Vermeulen and Venema (1974a) have found that about half of the competent cells have two nuclear bodies. Thus, integration of individual fragments into two separate chromosomes or into newly replicated arms may contribute to dislinkage. Thus, fragmentation per se may in several ways lead to a reduction in the cotransformation frequency of markers located on adjacent fragments. This type of dislinkage would be detectable by clonal analysis, since it would lead to mixed colonies with complementary allelic configurations.

Although transformational recombination has been formally analyzed in terms of a "switch frequency" (Guild *et al.*, 1968; Contente and Dubnau, 1979b), it is not certain how to interpret switching in molecular terms. It is possible that interruption of strand assimilation corresponds to one possible switching event. Very little detailed fine structure genetic analysis of multifactor transformation crosses has been carried out in *B. subtilis*. However, it appears that, in some genetic regions at least, few multiple crossovers occur; transformational integration appears to occur with infrequent loss of internal markers (Darlington and Bodmer, 1968). It is very likely that whatever multiple crossovers do occur are due to mismatch correction (gene conversion).

VIII. Recombination-Deficient Mutants

Many *rec* mutants of *B. subtilis* have been isolated, mostly as mitomycin C and as methylmethanesulfonate (MMS)-sensitive strains. At least 12 loci have been described that affect recombination. These have been reviewed recently (Mazza and Galizzi, 1978) and will not be described here in any detail. The nomenclature of these strains is somewhat confused, since various authors have used different designations for the same strains.

Several of the mutants allow uptake and processing of DNA by competent cells, but formation of stable donor-recipient DNA association is blocked (Dubnau *et al.*, 1973). In this category are *recA1*, *recE4* (renamed *recK* by Mazza and Galizzi), and *recD3* (renamed *recB* by Mazza and Galizzi). Another mutant (*recB2*) formed normal amounts of stable complex, but the DNA containing the donor moiety failed to replicate (Davidoff-Abelson and Dubnau, 1971). It is not known whether this was due to failure of the transformed chromosome to replicate or to cell death.

Mutations at two loci have been reported to depress ATP-dependent nuclease activity and transformation frequency. The *recE5* mutant isolated by Doly *et al.* (1974) is reportedly transformable at about 10% of the wild-type frequency and has no detectable ATP-dependent DNase activity. However, no data documenting the Rec$^-$ phenotype of this strain have been published. In our hands the *recE5* mutant was transformation-deficient, as reported, but was also poorly viable, possibly explaining the apparent transformation deficiency (J. Abramowitz and D. Dubnau, unpublished). The *recH* mutant (quoted in Mazza and Galizzi, 1978) was also reported to lack ATP-dependent DNase (Chestukhin *et al.*, 1972). We found this strain to be at least a double mutant (J. Abramowitz and D. Dubnau, unpublished). One mutation led to MMS sensitivity and was correlated with reduced transformability, although no studies on viability or DNA uptake were performed. Another seemed to result in resistance to lysis by lysozyme. When the *recH* strain was treated with lysozyme and then briefly sonicated, normal ATP-dependent nuclease levels were observed. The lysis-resistant and MMS-sensitive phenotypes were separable by transformation. It is very doubtful that ATP-dependent DNase is required for transformational recombination in *B. subtilis*. A similar conclusion has been reached in *Hemophilus,* in which the apparent deficiency in transformation of the mutant lacking ATP-dependent DNase activity was attributed to poor viability (Wilcox and Smith, 1975). In *S. pneumoniae* a mutant deficit in this enzyme appeared to be somewhat transformation-deficient (Vovis and Buttin, 1970). A later study showed that, in this mutant strain, the integration of donor DNA appeared to proceed normally (Vovis, 1973).

IX. Plasmid Transformation

A. Reasons for Studying Plasmid Transformation

Recent interest in the use of plasmids as vectors for molecular cloning has stimulated interest in plasmid transformation (Chapter 10). However, plasmid transformation is important in its own right. First, the use of defined CCC molecules can provide insight into the binding and uptake process. Second, the development of a plasmid model system promises to obviate the problems mentioned above (Section I,C) as inherent in the use of chromosomal DNA. We might imagine a plasmid, bearing a genetic marker, established in a competent strain and lacking homology with the chromosome of the host cell. DNA from the same plasmid, but with a different allelic configuration, can be used to transform this strain. The plasmid donor DNA may be linear, CCC, open circular, etc. If uptake and *rec*-dependent recombination can occur between the donor and resident plasmid, then the diffculties that derive from the heterogeneity of the donor DNA and the size of the recipient replicon will be eliminated.

The study of plasmid transformation in *B. subtilis* was made feasible by Ehrlich's (1977) discovery that several R plasmids from *Staphylococcus aureus* were capable of replication and expression of antibiotic resistance in *B. subtilis*.

B. Transformation of a Plasmid-Free Recipient by Plasmid DNA

With the use of native CCC plasmid DNA from several plasmids, it has been shown that the development of competence for this DNA closely parallels the development of competence for chromosomal DNA (Contente and Dubnau, 1979a). This, and the fact that plasmid CCC and chromosomal DNA compete for uptake (S. Contente and D. Dubnau, unpublished), demonstrate that at least part of the binding and uptake machinery is utilized in common during both types of transformation. In support of this conclusion are the observations of Lacks (1979) and of deVos *et al.* (1981) that most, if not all, CCC DNA is processed by *S. pneumoniae* and *B. subtilis* exactly like linear duplex transforming DNA.

The efficiency of plasmid transformation, with respect to DNA added or DNA taken up, is very low (Contente and Dubnau, 1979a). 10^3 to 10^4 plasmid molecules must be taken up per transformation event. However, the efficiency of plasmid transformation shows a marked dependence on molecular weight (Table I). This is not due entirely to a molecular weight dependence on uptake, since the efficiency per unit uptake of pSA2100 (MW 4.7×10^6, is about 10-fold greater than that of pUB110 (MW 3.0×10^6) Contente and Dubnau, 1979a). We will return to this observation below.

Over a considerable range, transformation using CCC DNA shows a first-order dependence on DNA concentration (Contente and Dubnau,

TABLE I Molecular Weight Dependence of Plasmid Transformation[a]

Plasmid	Molecular weight	Selected marker	Transformants/ml
pE5	1.750	Em^r	3.9×10^3
pC194	2.000	Cm^r	2.0×10^4
pUB110	3.000	Km^r	2.1×10^5
pSA2100	4.800	Cm^r	1.26×10^6
pBD64 (*spo0H*)[b]	6.200	Cm^r	2.5×10^6

[a] CCC plasmid DNA at a saturating concentration (5 μg/ml) was used to transform a BD170 (*trpC2 thr-5*) competent culture. Unpublished data of D. Dubnau and T. J. Gryczan.

[b] This plasmid was constructed by E. Dubnau, *et al.* (1981) by inserting a *B. licheniformis* 3.0-MDalton *Bgl*II fragment into the *Bgl*II site of pBD64. It complements *spo0H* mutants of *B. subtilis*.

1979a). This strongly indicates that the limiting step in plasmid transformation involves interaction of the cell with a single plasmid molecule. In this respect, plasmid transformation is similar to chromosomal transformation but different from transfection which shows a greater than first-order dependence on DNA concentration. In accordance with this, transfection requires Rec^+ cells. Presumably, recombinational repair takes place between damaged molecules, permitting formation of an intact phage molecule. Chromosomal transformation obviously also requires Rec^+ cells.

A most remarkable finding concerning plasmid transformation of competent cells is that of Canosi *et al.* (1978). These workers observed that plasmid pC194 CCC monomer DNA was inactive in transformation. The active form consisted of CCC multimeric molecules which are present in native plasmid preparations. This finding has been confirmed and extended to other plasmids (Mottes *et al.,* 1979; Gryczan *et al.,* 1980). Like monomers, dimeric molecules are virtually inactive, whereas trimers and higher forms are active in transformation. These observations hold in both Rec^+ and *recE4* competent cultures.

This requirement for multimers may be less than absolute. In *S. pneumoniae* monomer CCC DNA transforms with measurable frequency, although with higher than first-order concentration dependence (C. W. Saunders and W. R. Guild, personal communication). Multimeric forms transform, as in *B. subtilis,* with first-order kinetics. Undoubtedly related to this is the observation that linear pAM77 plasmid DNA transforms *S. pneumoniae,* although at a reduced frequency (Barany and Tomasz, 1980). In view of the overall similarity between *S. pneumoniae* and *B. subtilis* transformation, it will be interesting to reexamine the *B. subtilis* system for highly concentration-dependent transformation by monomer DNA.

The apparent requirement for multimeric DNA forces a reevaluation of some earlier data on the effect of plasmid DNA conformation on transformation (Contente and Dubnau, 1979a). The observation that linear DNA has no transforming activity may be a consequence of the monomerization that necessarily accompanies cleavage by single-site restriction endonuclease. This question remains unanswered, although it is susceptible to experimental resolution. The inactivation kinetics by DNase I were interpreted as indicating a requirement for CCC DNA. In fact, the inactivation of transforming activity proceeded about three times *more rapidly* than the conversion of CCC monomer to the open circular form. This is consistent with the notion that the active molecular form is larger in size than monomer. Therefore it is not known whether a single nick inactivates the active form, since the inactivation rate of transforming activity is consistent with single-nick inactivation of trimer or of more than one nick per molecule inactivating a mixture of trimer and higher forms. This question can be answered by studying the inactivation kinetics of single oligomeric species isolated from electrophoretic gels. Although it is therefore not known whether absolute continuity is required for the transformation of competent cells, superhelicity is clearly not required. Relaxed CCC DNA transforms as well as the native (superhelical) form (Contente and Dubnau, 1979a).

Whatever the mode of processing of multimeric plasmic DNA, these molecules are converted to monomer by a *rec*-independent process (S. Contente and D. Dubnau, unpublished). Trimeric CCC DNA of pUB110, a 3.0-MDalton kanomycin resistance plasmid, was purified from gels and used to transform Rec^+ and *recE4* competent cells to Km^r. Single colonies were picked and grown overnight in 1-ml cultures; the crude lysates were then prepared and analyzed electrophoretically. The ethidium bromide fluorescence patterns of the lysates from *recE4* and Rec^+ cells were indistinguishable from one another and from the pattern observed with native plasmid DNA. The predominant forms in each case were superhelical CCC monomer, then open circular monomer, then CCC superhelical dimer, etc. Thus, whatever the mechanism of conversion of multimer to monomer, it is *rec*-independent, unlike the situation in *E. coli* transformation (Potter and Dressler, 1977).

The observations concerning the transformation of plasmid-free competent cultures by plasmid DNA are compatible with recently published models (Dubnau *et al.*, 1980; deVos *et al.*, 1981). This model is illustrated in Fig. 2 in slightly altered form and is based on the supposition that plasmid transformation proceeds by the same uptake and processing pathway as linear duplex DNA. This assumption not only possesses the virtue of simplicity but is supported by the available evidence; competence for chromosomal and plasmid DNA develops with similar (if not identical) kinetics, and chromosomal and plasmid DNA compete with one another. In *S. pneumoniae* it has been shown that PM2 DNA is processed to form intracellular single strands (Lacks, 1979) and that *noz* mutants are not transformed by plasmid DNA (Barany and Tomasz, 1980; Saunders and Guild, 1980). In *B. subtilis* as well, intracellular single strands are derived from plasmid DNA (deVos *et al.*, 1981). The model therefore attempts to explain the properties of plasmid transformation on the basis of the usual mechanism of DNA uptake by competent cells and from the requirements of plasmid replication, notably the need to establish CCC molecules. In further support of this line of reasoning is the observation that protoplasts of *B. subtilis* can be transformed by CCC monomer and open circular monomer with the same frequency as

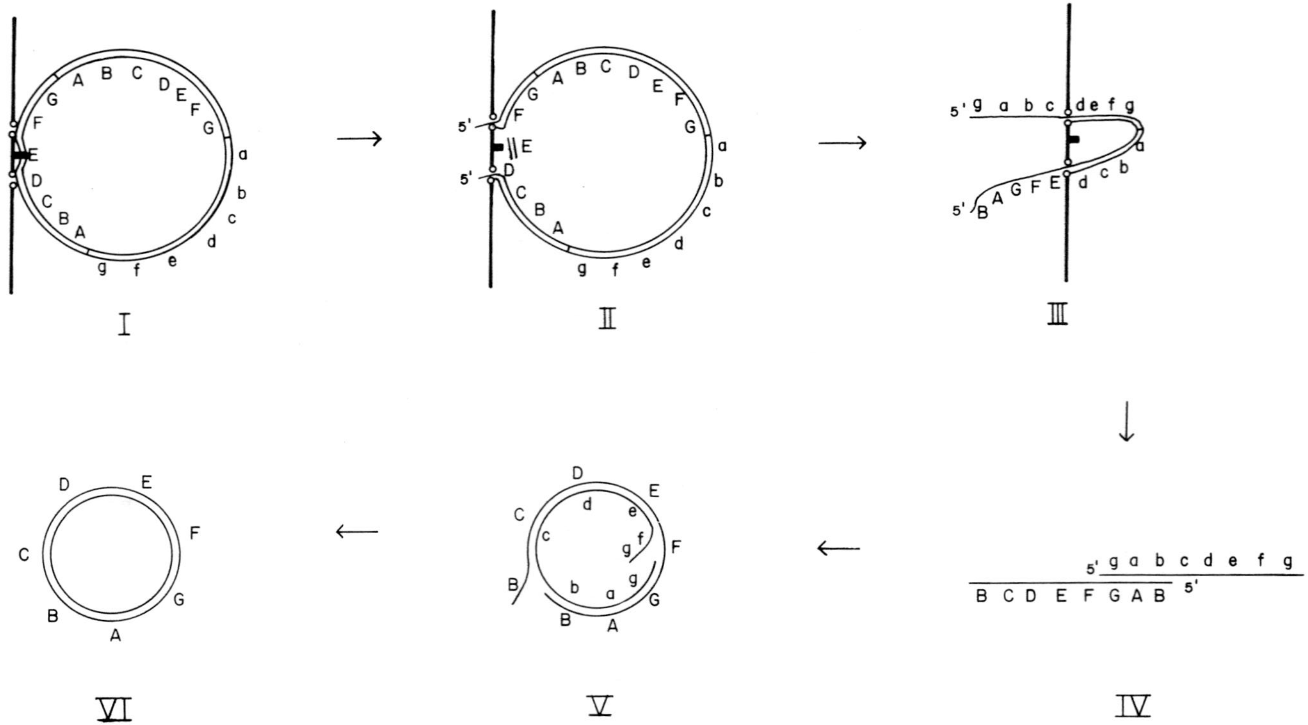

Fig. 2. Model of plasmid transformation. (I) A trimeric CCC molecule of plasmid DNA is bound to the competent cell surface at a receptor site (solid rectangle). (II) Double-strand cleavage occurs at two neighboring cleavage sites (open circles). Attachment of one terminus resulting from each cleavage occurs, with the loss of an intervening segment. (Alternatively, a single cleavage may occur, with binding of both termini.) (III) Independent entry of the complementary single strands take place with the same polarity. This is shown here as a $5 \rightarrow 3$ entry for illustrative purposes. Degradation of the nonentering strands accompanies this process, and some material is lost from one or both ends of each strand taken up. (IV) Complementary strands anneal. (V) Circularization occurs as a result of the presence of redundant single-strand termini. (VI) Repair to yield CCC monomer takes place. Upper- and lowercase letters indicate complementarity.

by multimers (Gryczan *et al.*, 1980). In fact, linear DNA also transforms protoplasts, but at a reduced frequency (Chang and Cohen, 1979). Thus, the requirement for multimericity appears to be a consequence of the competence mechanism.

The model supposes that a single trimeric (or higher multimer) molecule, in order to yield a transformant, must be bound at a receptor site and then cleaved at two neighboring cleavage sites on the competent cell surface. Following double-strand cleavage, "opposite" single strands enter with the same (unknown) chemical polarity. Eventually, sufficiently long complementary sequences may be internalized to permit annealing and circularization. Thus, monomer molecules are regenerated by a *rec*-independent process, as observed. The model predicts first-order kinetics and is consistent with the observed low efficiency of plasmid transformation (see below). It is also consistent with the observed sensitivity to nicking, since covalent discontinuity will decrease the probability of achieving sufficient intracellular complementarity if a nick impedes further uptake. The model is also consistent with the inactivity of dimers, since it is difficult, if not impossible, to generate sufficient complementarity from dimer molecules.

Finally, the model can be used, together with a concept introduced earlier (Section IV), to explain the striking dependence of plasmid transformation on the molecular weight of the plasmid used if we assume that two cleavages must occur to ensure the uptake of two complementary strands (Fig. 2). In our discussion of the fragmentation kinetics of linear duplex DNA on the competent cell surface, we suggested that the length distribution of cleavage depended on the average spacing of cleavage and binding sites on the cell surface and on the rigidity of the DNA, so that the probability of a cleavage within a certain distance from a binding site is low. If the uptake of complementary strands from a transforming plasmid molecule is subject to the same constraints, it is easy to see that lowering the plasmid molecular weight would decrease the probability of achieving intracellular complementarity, since the two postulated cleavages might then occur within different members of a multimer. Higher-molecular-weight plasmids would have an advantage, since they would possess an enhanced probability of suffering the two required cleavages within one member of a multimer. It is interesting that the data obtained with T7 DNA suggest that cleavage is infrequent within about 4–5 $\times$ 10^6 of a binding site. The data in Table I show that the transforming frequency rises steeply above a plasmid molecular weight of 2–3 $\times$ 10^6 and is still increasing at 6 $\times$ 10^6. These observations are consistent with this model.

At this point, we will briefly discuss transfection, which will not otherwise be dealt with in this chapter. "Transfection" is the term used for transformation with bacteriophage DNA. As a result of this process, infectious centers are formed, indicating that phage DNA has been taken up and processed and that a phage growth cycle has been completed with the liberation of infectious particles. Transfection possesses several features that can be instructively compared with plasmid transformation. Transfection with mature DNA from most bacteriophages shows a higher than first-order concentration dependence in *B. subtilis* and *S. pneumoniae* (Trautner and Spatz, 1973; Porter and Guild, 1978). When replicating forms, shown to be concatamers, or multimers prepared *in vitro* using T4 DNA ligase are used, the concentration dependence is seen to be first order (Flock, 1978; Porter and Guild, 1978). This latter case is directly reminiscent of plasmid transformation by multimers. The higher order concentration-dependent transfection observed with mature (monomer) phage DNA is *rec*-dependent (Rutberg and Rutberg, 1971; Spatz and Trautner, 1971). When DNA preparations from two genetically marked ϕ105 strains were used in mixed transfection experiments, the frequency of wild-type recombinants was not dependent on DNA concentration (Flock and Rutberg, 1974). This was interpreted as indicating that recombination is required for infectivity (see also Spatz and Trautner, 1971).

It is reasonable to assume that the transfecting particle is processed by the usual competence pathway, for easons exactly analogous to those presented above for plasmid transformation. The observed high-order kinetics and *rec* dependence have been explained as follows (Porter and Guild, 1978). Fragmentation on the cell surface and single-strand uptake yield independently transported intracellular single-strand fragments derived from several phage DNA molecules and from several regions of the same molecule. These anneal in overlapping fashion to produce gapped duplex structures of greater than unit length. Repair replication and recombination combine to produce a duplex bacteriophage genome with complete genetic information. This is a plausible model and is consistent with the data described above. It is also consistent with the observed low efficiency of transfection and with the finding that single-strand bacteriophage DNA can yield infectious centers (Dishman, 1972).

Some interesting reports have appeared on the susceptibility of transformation and transfection by unmodified DNA to restriction by r^+ cells. Chromosomal transformation is not affected by restriction, whereas transfection is affected Trautner *et al.*, 1974; Bron *et al.*, 1980a,b). This was interpreted as indicating that unmodified single strands of chromosomal DNA are integrated to yield heteroduplexes (m^+/m^-) which are resistant to the restriction endonuclease, whereas single strands of transfecting DNA anneal with one another to yield restriction-sensitive homoduplexes. Tanaka (1979) has reported that transformation of a plasmid marker is insensitive to restriction when the plasmid can integrate at a region of homology on the chromosome but is restricted when the plasmid must enter and replicate autonomously (in a *rec* recipient). Prozorov *et al.* (1980) have likewise reported that pUB110 transformation is restriction sensitive. We have also observed that, like transfection, plasmid transformation by unmodified plasmid DNA is restriction-sensitive, with a sensitivity that depends on the number of cleavage sites on the plasmid (S. Contente and D. Dubnau, unpublished). However, these data may also be interpreted as suggesting that the entry of duplex (m^-/m^-) DNA is required for both transfection and plasmid transformation. Since these processes are inherently inefficient, such "escape" entry of duplex DNA is difficult to rule out, however, unattractive as a hypothesis.

In conclusion, similar models of entry and primary processing are currently in vogue to explain transfection and transformation with chromosomal or plasmid DNA. The differences in kinetics, *rec* dependence, and efficiency between these three systems are most likely not due to fundamental differences in the way the molecules are initially processed but rather to differences in the intracellular interactions required to generate the final products. In the case of chromosomal transformation, this is an interaction between donor single strands and the resident chromosome. In the case of plasmid transformation we believe that the interaction is between complementary single strands derived from separate regions of a multimer, and in the case of transfection it is between gapped duplex molecules. One observation that remains unexplained is the apparent failure of plasmid monomers to transform *B. subtilis* competent cells by a transfection-like, second-order process such as has been found for plasmid transformation in *S. pneumoniae* (Saunders and Guild, 1981). This may be a quantitative rather than a qualitative failure; perhaps the experiment has not been performed at high enough DNA concentration.

C. Transformation of a Recipient Carrying a Homologous Plasmid

From the foregoing discussion it seems obvious that inclusion of a homologous plasmid in a competent culture might allow the occurrence of interactions analogous with chromosomal transformation. This prediction has been tested using chimeric plasmids constructed, for the most part, *in vitro* (Contente and Dubnau, 1979b). In these experiments a cointegrate plasmid of the form A · B (where A and B are DNA segments conferring resistance to antibiotics a and b) has been used to transform a competent culture carrying only plasmid A or plasmid B. Thus, the resident plasmid is completely homologous with a portion of the donor. In this situation, transformation to a^r or to b^r is augmented 10- to 20-fold by the presence of B or A in the recipient. An unrelated plasmid (C) has no augmenting effect. When linear A · B is used to transform a culture carrying A, transformation to b^r is observed at high frequency when the linearizing cleavage is within the A moiety of the donor. (It should be recalled that transformation is not detectable using linear plasmid DNA in a plasmid-free recipient.) When the B moiety is cleaved, no transformation is seen unless B is present in the recipient and selection is for a^r. This "marker rescue" phenomenon is *rec*-dependent and behaves like classic transformation into a deletion; the selected marker must be flanked by segments that are homologous with the resident plasmid (compare Harris-Warrick and Lederberg, 1978b). The quantitative effect of varying the position of the single cleavage site within the portion of the donor chimera that is homologous with the recipient has also been examined. Based on these observations evidence was obtained for a single-strand "excluded length" of 0.15 MDalton occurring at the end(s) of the transforming DNA. This length can be characterized as DNA that has been removed or is otherwise unavailable for recombination. An excluded length of similar size had earlier been inferred for *S. pneumoniae* and *B. subtilis,* based on very different experiments (Guild *et al.*, 1968). The plasmid data also yielded an estimate for the "exchange frequency" during recombination of 0.2–0.5 MDalton^{-1}. Thus, the probability of more than two exchanges per integrated segment of about 3×10^6 may be low, in accordance with published measurements of multiple crossover frequency (Darlington and Bodmer, 1968). Canosi *et al.* (1981) have performed analogous experiments using a hybrid plasmid carrying a segment of *B. subtilis* DNA. Transformation occurs using the monomeric form of this plasmid, but in a *rec*-dependent fashion. It is likely that these results are interpretable in the same way as those of Contente and Dubnau (1979b). Various properties of this marker rescue system have been explored and are beyond the scope of this chapter. Two consequences have resulted. First, the marker rescue effect has been exploited to generate a highly efficient molecular cloning system for heterologous DNA in *B. subtilis* (Gryczan *et al.*, 1980). Second, and more relevant to this chapter, the transformation of plasmid-bearing strains by linear homologous plasmid DNA appears to be similar in every respect to classic chromosomal transformation, except that internalized products of a unique duplex molecule, pair and recombine with a small recipient replicon. Thus, this system holds great promise for increasing our understanding of the molecular basis of transformation.

Acknowledgments

The author thanks C. Cirigliano, R. Davidoff-Abelson, B. Scher, S. Contente, and T. J. Gryczan for their contributions to the study of transformation in his laboratory over the years. He is indebted to I. Smith, E. Dubnau, A. G. Shivakumar, J. Hahn, Y. Kozlov, and G. Grandi for discussions, to F. Barany, A. Tomasz, C. W. Saunders, W. R. Guild, J. P. Claverys, M. Roger, A. M. Sicard, S. Lacks, and A. L. Rosenthal for providing unpublished information, and to A. Howard for expert secretarial assistance.

During the preparation of this review the author's laboratory was supported by grants from the National Institutes of Health, AI-10311, and from the American Cancer Society, VC300.

References

Anagnostopoulos, C., and Spizizen, J. (1961). *J. Bacteriol.* **81,** 741–746.
Arwert, F., and Venema, G. (1973). *In* "Bacterial Transformation" (L. J. Archer, ed.), pp. 203–214. Academic Press, New York.
Barany, F., and Tomasz, A. (1980). *J. Bacteriol.* **144,** 698–709.
Beattie, K. L., Wiegand, R. C., and Radding, C. M. (1977). *J. Mol. Biol.* **116,** 783–803.
Bodmer, W. F. (1966). *J. Gen. Physiol.* **49,**(6), Part 2, 233–258.
Bodmer, W. F., and Ganesan, A. T. (1964). *Genetics* **50,** 717–738.
Bresler, S. E., Kreneva, R. A., and Kushev, V. V. (1968). *Mol. Gen. Genet.* **102,** 257–268.
Bron, S., Luxen, E., and Trautner, T. A. (1980a). *Mol. Gen. Genet.* **179,** 111–117.
Bron, S., Luxen, E., Venema, G., and Trautner, T. A. (1980b). *Mol. Gen. Genet.* **179,** 103–110.
Buitenwerf, J., and Venema, G. (1977). *Mol. Gen. Genet.* **156,** 145–155.
Buitenwerf, J., and Venema, G. (1978). *Mol. Gen. Genet.* **160,** 67–75.
Cahn, F. H., and Fox, M. S. (1968). *J. Bacteriol.* **95,** 867–875.
Canosi, U., Morelli, G., and Trautner, T. A. (1978). *Mol. Gen. Genet.* **166,** 259–267.
Canosi, U., Iglesias, I., and Trautner, T. A. (1981). *Mol. Gen. Genet.* **181,** 434–440.
Cassuto, E., and Radding, C. M. (1971). *Nature (London) New Biol.* **229,** 13–16.
Chang, S., and Cohen, S. N. (1979). *Mol. Gen. Genet.* **168,** 111–115.
Chestukhin, A. V., Shemyakin, M. F., Kalinina, N. A., and Prozorov, A. A. (1972). *FEBS Lett.* **24,** 121–125.
Chilton, M.-D., and McCarthy, B. J. (1969). *Genetics* **62,** 697–710.
Ciferri, O., Barlati, S., and Lederberg, J. (1970). *J. Bacteriol.* **104,** 684–688.
Claverys, J. P., Roger, M., and Sicard, A. M. (1980). *Mol. Gen. Genet.* **178,** 191–202.
Collins, C. J., and Guild, W. R. (1972). *J. Bacteriol.* **109,** 266–275.
Contente, S., and Dubnau, D. (1979a). *Mol. Gen. Genet.* **167,** 251–258.
Contente, S., and Dubnau, D. (1979b). *Plasmid* **2,** 555–571.
Cunningham, R. P., Wu, A. M. Shibata, T., Dasgupta, C., and Radding, C. M. (1981). *Cell* **24,** 213–223.

Darlington, A. J., and Bodmer, W. F. (1968). *Genetics* **60,** 681–684.
Davidoff-Abelson, R., and Dubnau, D. (1971). *Proc. Nat. Acad. Sci. USA* **68,** 1070–1074.
Davidoff-Abelson, R., and Dubnau, D. (1973a). *J. Bacteriol.* **116,** 146–153.
Davidoff-Abelson, R., and Dubnau, D. (1973b). *J. Bacteriol.* **116,** 154–162.
Deich, R. A., and Smith, H. O. (1980). *Mol. Gen. Genet.* **177,** 369–374.
deVos, W., Venema, G., Canosi, U., and Trautner, T. A. (1981). *Mol. Gen. Genet.* **181,** 424–433.
Dishman, B. (1972). Ph.D. Thesis, Univ. of Texas.
Doly, J., Sasarman, E., and Anagnostopoulos, C. (1974). *Mutat. Res.* **22,** 15–23.
Dooley, D. C., Hadden, C. T., and Nester, E. W. (1971). *J. Bacteriol.* **108,** 668–679.
Dubnau, D. (1976). *In* "Microbiology, 1976" (D. Schlessinger, ed.), pp. 14–27. American Society for Microbiology, Washington, D.C.
Dubnau, D., and Cirigliano, C. (1972a). *J. Bacteriol.* **111,** 488–494.
Dubnau, D., and Cirigliano, C. (1972b). *J. Mol. Biol.* **64,** 9–29.
Dubnau, D., and Cirigliano, C. (1972c). *J. Mol. Biol.* **64,** 31–46.
Dubnau, D., and Cirigliano, C. (1973a). *Mol. Gen. Genet.* **120,** 101–106.
Dubnau, D., and Cirigliano, C. (1973b). *J. Bacteriol.* **113,** 1512–1514.
Dubnau, D., and Cirigliano, C. (1974). *In* "Mechanisms in Recombination" (R. F. Grell, ed.), pp. 167–178. Plenum Press, New York.
Dubnau, D., and Davidoff-Abelson, R. (1971). *J. Mol. Biol.* **56,** 209–221.
Dubnau, D., Smith, I., Morell, P., and Marmur, J. (1965). *Proc. Nat. Acad. Sci. USA* **54,** 491–498.
Dubnau, D., Davidoff-Abelson, R., Scher, B., and Cirigliano, C. (1973). *J. Bacteriol.* **114,** 273–286.
Dubnau, D., Contente, S., and Gryczan, T. J. (1980). *In* "DNA-Recombination Interactions and Repair" (S. Zadrazil and J. Sponar, eds.), pp. 365–386. Pergamon, Oxford.
Dubnau, E., Ramakrishna, N., Cabane, K., and Smith, I. (1981). *J. Bacteriol.* **147,** 622–632.
Ehrlich, S. D. (1977). *Proc. Nat. Acad. Sci. USA* **74,** 1680–1682.
Eisenstadt, E., Lange, R., and Willecke, K. (1975). *Proc. Nat. Acad. Sci. USA* **72,** 323–327.
Ephrussi-Taylor, H. (1966). *Genetics* **54,** 211–222.
Ephrussi-Taylor, H., and Gray, T. C. (1966). *J. Gen. Physiol.* **49**(6), Part 2, 211–231.
Flock, J.-I. (1978). *Mol. Gen. Genet.* **163,** 7–15.
Flock, J.-I., and Rutberg, L. (1974). *Mol. Gen. Genet.* **131,** 301–311.
Fornili, S. L., and Fox, M. S. (1977). *J. Mol. Biol.* **113,** 181–191.
Fox, M. S. (1978). *Ann. Rev. Genet.* **12,** 249–287.
Fox, M. S., and Hotchkiss, R. D. (1957). *Nature (London)* **179,** 1322–1324.
Gabor, M., and Hotchkiss, R. D. (1966). *Proc. Nat. Acad. Sci. USA* **56,** 1441–1448.
Garcia, E., Lopez, P., Ureña, M. T. P., and Espinosa, M. (1978). *J. Bacteriol.* **135,** 731–740.
Gryczan, T., Contente, S., and Dubnau, D. (1980). *Mol. Gen. Genet.* **177,** 459–467.
Guerrini, F., and Fox, M. S. (1968a). *Proc. Nat. Acad. Sci. USA* **59,** 429–436.
Guerrini, F., and Fox, M. S. (1968b). *Proc. Nat. Acad. Sci. USA* **59,** 1116–1123.
Guild, W. R., Cato, A., Jr., and Lacks, S. (1968). *Cold Spring Harbor Symp. Quant. Biol.* **33,** 643–645.
Gurney, T., Jr., and Fox, M. S. (1968). *J. Mol. Biol.* **32,** 83–100.
Hadden, C., and Nester, E. W. (1968). *J. Bacteriol.* **95,** 876–885.
Harris, W. J., and Barr, G. C. (1971). *Mol. Gen. Genet.* **113,** 316–330.
Harris-Warrick, R. M., and Lederberg, J. (1978a). *J. Bacteriol.* **133,** 1237–1245.
Harris-Warrick, R. M., and Lederberg, J. (1978b). *J. Bacteriol.* **133,** 1246–1253.

Haseltine, F. P., and Fox, M. S. (1971). *J. Bacteriol.* **107,** 889–899.
Holloman, W. K., Wiegand, R., Hoessli, C., and Radding, C. M. (1975). *Proc. Nat. Acad. Sci. USA* **72,** 2394–2398.
Joenje, H., and Venema, G. (1975). *J. Bacteriol.* **122,** 25–33.
Joenje, H., Gruber, M., and Venema, G. (1972). *Biochim. Biophys. Acta* **262,** 189–199.
Kelly, M. S. (1967a). *Mol. Gen. Genet.* **99,** 333–349.
Kelly, M. S. (1967b). *Mol. Gen. Genet.* **99,** 350–361.
Kleinschmidt, A. K. (1968). *Methods Enzymol.* **12,** Part B, 361–377.
Lacks, S. (1962). *J. Mol. Biol.* **5,** 119–131.
Lacks, S. (1966). *Genetics* **53,** 207–235.
Lacks, S. (1970). *J. Bacteriol.* **101,** 373–383.
Lacks, S. (1974). *In* "Mechanisms in Recombination" (R. F. Grell, ed.), pp. 217–223. Plenum Press, New York.
Lacks, S. (1977). In "Microbial Interactions, Receptors and Recognition." (J. L. Reissig, ed.), pp. 179–232. Chapman and Hall, London.
Lacks, S. (1979). *J. Bacteriol.* **138,** 404–409.
Lacks, S., and Greenberg, B. (1973). *J. Bacteriol.* **114,** 152–163.
Lacks, S., and Greenberg, B. (1976). *J. Mol. Biol.* **101,** 255–275.
Lacks, S., Greenberg, B., and Carlson, K. (1967). *J. Mol. Biol.* **29,** 327–347.
Lacks, S., Greenberg, B., and Neuberger, M. (1974). *Proc. Nat. Acad. Sci. USA* **71,** 2305–2309.
Lacks, S., Greenberg, B., and Neuberger, M. (1975). *J. Bacteriol.* **123,** 222–232.
Lee, C. S., Davis, R. W., and Davidson, N. (1970). *J. Mol. Biol.* **48,** 1–22.
Levin, B. C., and Landman, O. E. (1973). *In* "Bacterial Transformation" (L. J. Archer, ed.), pp. 217–240. Academic Press, New York.
Levine, J. S., and Strauss, N. (1965). *J. Bacteriol.* **89,** 281–287.
Lomant, A. J., and Fresco, J. R. (1975). *Prog. Nucleic Acid Res. Mol. Biol.* **15,** 185–218.
Loveday, K. S. (1978). *J. Bacteriol.* **135,** 1158–1161.
Low, K. B., and Porter, D. D. (1978). *Ann. Rev. Genet.* **12,** 47–68.
Matsumoto, K., Takahashi, H., Saito, H., and Ikeda, Y. (1978). *Mol. Gen. Genet.* **162,** 229–235.
Matsumoto, K., Ando, T., Saito, H., and Ikeda, Y. (1979). *J. Biochem.* **86,** 627–637.
Mazza, G., and Galizzi, A. (1978). *Microbiologica* **1,** 111–135.
McCarthy, C., and Nester, E. W. (1967). *J. Bacteriol.* **94,** 131–140.
Morrison, D. A. (1971). *J. Bacteriol.* **108,** 38–44.
Morrison, D. A. (1977). *J. Bacteriol.* **132,** 576–583.
Morrison, D. A., and Guild, W. R. (1973a). *J. Bacteriol.* **115,** 1055–1062.
Morrison, D. A., and Guild, W. R. (1973b). *Biochim. Biophys. Acta* **299,** 545–556.
Mottes, M., Grandi, G., Sgaramella, V., Canosi, U., Morelli, G., and Trautner, T. A. (1979). *Mol. Gen. Genet.* **174,** 281–286.
Nester, E. W., and Stocker, B. A. D. (1963). *J. Bacteriol.* **86,** 785–796.
Notani, N., and Goodgal, S. H. (1966). *J. Gen. Physiol.* **49**(6), Part 2, 197–209.
Notani, N. K., and Setlow, J. K. (1974). *Prog. Nucl. Acid Res. Mol. Biol.* **14,** 39–100.
Piechowska, M., and Fox, M. S. (1971). *J. Bacteriol.* **108,** 680–689.
Piechowska, M., Soltyk, A., and Shugar, D. (1975). *J. Bacteriol.* **122,** 610–622.
Pieniazek, D., Piechowska, M., and Venema, G. (1977). *Mol. Gen. Genet.* **156,** 251–261.
Popowski, J., and Venema, G. (1978). *Mol. Gen. Genet.* **166,** 119–126.
Porter, R. D., and Guild, W. R. (1978). *J. Virol.* **25,** 60–72.
Potter, H., and Dressler, D. (1977). *Proc. Nat. Acad. Sci. USA* **74,** 4168–4172.

Prozorov, A. A. (1965). *Dokl. Akad. Nauk SSR* **160,** 472–474.
Prozorov, A. A. (1980). "Genetic Transformation and Transfection." Academy of Science of the U.S.S.R., Moscow (in Russian).
Prozorov, A. A., Belova, T. S., and Surikov, N. N. (1980). *Mol. Gen. Genet.* **180,** 135–138.
Radding, C. M. (1974). *Ann. Rev. Genet.* **7,** 87–111.
Radding, C. M. (1978). *Ann. Rev. Biochem.* **47,** 847–880.
Rosenthal, A. L., and Lacks, S. A. (1980). *J. Mol. Biol.* (in press).
Rutberg, B., and Rutberg, L. (1971). *J. Virol.* **8,** 919–921.
Scher, B., and Dubnau, D. (1975). *J. Bacteriol.* **126,** 429–438.
Saunders, C. W., and Guild, W. (1980). *Mol. Gen. Genet.* **180,** 573–578.
Saunders, C. W., and Guild, W. (1981). *J. Bacteriol.* **146,** 517–526.
Sedgwick, B., and Setlow, J. K. (1976). *J. Bacteriol.* **125,** 588–594.
Shoemaker, N. B., and Guild, W. R. (1972). *Proc. Nat. Acad. Sci. USA* **69,** 3331–3335.
Singh, R. N. (1972). *J. Bacteriol.* **110,** 266–272.
Singh, R. N., and Pitale, M. P. (1967). *Nature* (*London*) **213,** 1262–1263.
Singh, R. N., and Pitale, M. P. (1968). *J. Bacteriol.* **95,** 864–866.
Sisco, K. L., and Smith, H. O. (1979). *Proc. Nat. Acad. Sci. USA* **76,** 972–976.
Smith, H. O., Danner, D. B., and Deich, R. A. (1981). *Ann. Rev. Biochem.* **50,** 41–68.
Soltyk, A., Shugar, D., and Piechowska, M. (1975). *J. Bacteriol.* **124,** 1429–1438.
Soltyk, A., Piechowska, M., Shugar, D., and Doskočil, J. (1976). *Mol. Gen. Genet.* **148,** 307–313.
Somma, S., and Polsinelli, M. (1970). *J. Bacteriol.* **101,** 851–855.
Spatz, H. Ch., and Trautner, T. A. (1970). *Molec. Gen. Genet.* **109,** 84–106.
Spatz, H. Ch., and Trautner, T. A. (1971). *Molec. Gen. Genet.* **113,** 174–190.
Spizizen, J. (1958). *Proc. Nat. Acad. Sci. USA* **44,** 1072–1078.
Strauss, N. (1965). *J. Bacteriol.* **89,** 288–293.
Strauss, N. (1966). *J. Bacteriol.* **91,** 702–708.
Strauss, N. (1970). *Genetics* **66,** 583–593.
Stuy, J. H. (1965). *J. Mol. Biol.* **13,** 554–570.
Stuy, J. H., and Stern, D. (1964). *J. Gen. Microbiol.* **35,** 391–400.
Tanaka, T. (1979). *Mol. Gen. Genet.* **175,** 235–237.
Tichy, P., and Landman, O. E. (1969). *J. Bacteriol.* **97,** 42–51.
Tiraby, J.-G., and Fox, M. S. (1973). *Proc. Nat. Acad. Sci. USA* **70,** 3541–3545.
Tiraby, J.-G., and Sicard, M. A. (1973). *J. Bacteriol.* **116,** 1130–1135.
Trautner, T. A., and Spatz, H. Ch. (1973). *Current Topics Microbiol. Immunol.* **62,** 61–88.
Trautner, T. A., Pawlek, B., Bron, S., and Anagnostopoulos, C. (1974). *Mol. Gen. Genet.* **131,** 181–191.
Venema, G. (1979). *Adv. Microbiol. Physiol.* **19,** 245–331.
Vermeulen, C. A., and Venema, G. (1974a). *J. Bacteriol.* **118,** 334–341.
Vermeulen, C. A., and Venema, G. (1974b). *J. Bacteriol.* **118,** 342–350.
Vovis, G. F. (1973). *J. Bacteriol.* **113,** 718–723.
Vovis, G. F., and Buttin, G. (1970). *Biochim. Biophys. Acta* **224,** 42–54.
Wiegand, R. C., Beattie, K. L., Holloman, W. K., and Radding, C. M. (1977). *J. Mol. Biol.* **116,** 805–824.
Wilcox, K. W., and Smith, H. O. (1975). *J. Bacteriol.* **122,** 443–453.
Wilson, G. A., and Bott, K. F. (1968). *J. Bacteriol.* **95,** 1439–1449.
Wilson, G. A., and Bott, K. F. (1970). *Biochim. Biophys. Acta* **199,** 464–475.
Wilson, G. A., and Young, F. E. (1972). *J. Bacteriol.* **111,** 705–716.

6
Sporulation Genes and Their Regulation

RICHARD LOSICK

I. Introduction

An intriguing biological problem posed by the bacilli is their ability to undergo cellular differentiation. When deprived of nutrients, vegetative

The Molecular Biology
of the Bacilli

ISBN 0-12-222701-8

cells of spore-forming bacilli cease normal cell division and form instead polar septa that partition the sporulating cells into sporangia and forespore compartments (Fitz-James and Young, 1969). After engulfment by the sporangium, the forespore matures into a dormant endospore which is ultimately liberated by lysis of the mother cell. Although among the most primitive examples of cellular differentiation, sporulation is nevertheless a complex developmental process. The formation of endospores is dependent upon the products of at least 30 genes and involves extensive alterations in the biochemistry, physiology, and morphology of the developing cell (Hoch, 1976; Piggot and Coote, 1976).

Bacterial sporulation is an attractive model system for studying cellular differentiation because of the facility with which bacilli and in particular *Bacillus subtilis* can be manipulated genetically and biochemically. Sporulation genes (*spo* genes) have been the subject of intensive investigation in classic transformation genetics for over two decades (Schaeffer and Ionesco, 1960). The recent introduction from *Staphylococcus* and *Streptococcus* of cloning vectors (Chapter 10, this volume) and transposons (P. Youngman, personal communication) into *B. subtilis* has provided powerful new tools for the study of sporulation genes. Moreover, as bacilli sporulate efficiently and synchronously in liquid culture, large quantities of sporulating cells can be obtained for biochemical analysis. Thus, endospore formation is one of the most tractable biological systems for the study of developmental genes and their regulation during cellular differentiation.

II. Sporulation Genes

A. The *spo* Loci

At least 30 genetic loci are known whose products are apparently required for sporulation (Fig. 1). These *spo* loci are defined by mutations that do not affect vegetative growth but impair the sporulation process (reviewed by Piggot and Coote, 1976). The map of *spo* genes on the *B. subtilis* chromosome is not saturated; based on the frequency with which randomly isolated mutations fall within known *spo* loci, Hranueli *et al.* (1974) estimate that the total number of sporulation loci is 40–50.

The *spo* mutations at particular loci block sporulation at characteristic stages of development (Hoch, 1976; Piggot and Coote, 1976). Thus, mutations in the *spo0A* locus are arrested at the earliest stage of endospore formation (stage 0), whereas mutations in the *spoIIE* locus form

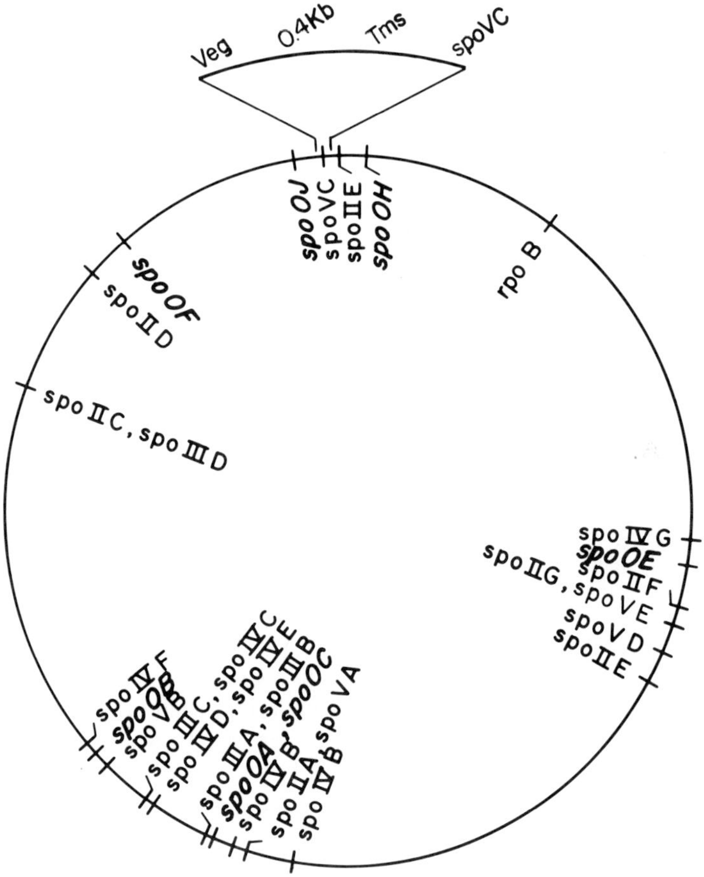

Fig. 1. Location of *spo* genes on the *B. subtilis* chromosome. The seven *spo0* genes are highlighted in boldface. The *rpoB* gene encodes the β subunit of RNA polymerase (Halling *et al.,* 1977). The location and orientation of the *veg–0.4 kb–tms–spoVC* region is shown in the expansion at the top of the map. The origin of DNA replication is at the top of the map, just to the right of *spo0J* (Trowsdale *et al.,* 1979).

a septum (stage II) but are unable to proceed with the formation of a detached forespore in the mother cell (stage III).

The developmental stage at which a *spo* mutant is blocked is not, however, necessarily coincident with the time of expression of the *spo* gene; a *spo* gene could be turned on well before the time that its product functions in development. For example, in the one case in which the transcript of a *spo* locus has been identified (*spoVC,* see below) transcription is activated several hours before the developmental stage (stage IV–V) at which a SpoVC mutant arrests.

The *spo0* genes are of special interest, as their products apparently initiate the sporulation process (Hoch *et al.*, 1978). Mutations at the *spo0* loci block development before the earliest morphological events of endospore formation. For this reason one or more of the *spo0* gene products are thought to be regulatory proteins that turn on *spo* genes early in development. At least seven *spo0* loci are known. These are located at widely scattered sites on the *B. subtilis* genetic map (highlighted in boldface in Fig. 1). The products of at least some (if not all) of these genes are presumably vegetative proteins, as mutations in certain *spo* loci detectably alter the properties of growing cells. For example, mutations in the *spo0A* and *spo0B* genes are known to affect the sensitivity of vegetative cells to certain phages and surface-active antibiotics (Hoch, 1976; Hoch *et al.*, 1978). Moreover, mutations in the *spo0A, spo0B, spo0E, spo0F,* and *spo0H* genes alter the pattern of protein synthesis in exponential-phase cells (Brehm *et al.*, 1975; Linn and Losick, 1976). A model of how the *spo0* gene products might be involved in the initiation of development will be considered at the end of this chapter.

Recently, it has become possible to isolate by recombinant DNA techniques the *B. subtilis spo0A, spo0B, spo0C, spo0F* and *spo0H* genes (Ferrari *et al.*, 1981; Kawamura *et al.*, 1981; Dubnau *et al.*, 1981). The *spo0B* gene has been correlated with a 39,000 dalton polypeptide (Kobayashi *et al.*) whereas *spo0H* appears to encode a 27,000 dalton product (Dubnau *et al.*). As expected from genetic considerations discussed above, both the *spo0B* and *spo0H* genes are expressed in vegetative as well as in early sporulating cells.

Another important class of *spo* mutants is altered in the *B. subtilis* transcriptional and translational machinery (reviewed by Sonenshein and Campbell, 1978; Piggot and Coote, 1976). These were discovered as a frequent class among mutants resistant to specific RNA polymerase and ribosomal inhibitors. For example, certain RNA polymerase mutations [mapping in the *rpoB* gene (Fig. 1), which encodes the β subunit of polymerase (Halling *et al.*, 1977)] that confer resistance to rifampin impair the sporulation process without affecting vegetative growth (e.g., Sonenshein *et al.*, 1974; Leighton, 1973; Sumida-Yasumoto and Doi, 1977; Linn *et al.*, 1975). In at least some cases, these drug-resistant mutants are blocked at an early stage in development, fail to undergo sporulation-associated changes in RNA polymerase (Sonenshein *et al.*, 1974), and fail to activate a gene whose transcription is known to be under *spo0* control (Segall and Losick, 1977). Similarly, ribosomal mutants that are resistant to a variety of antibiotics including erythromycin (Leighton, 1974; Tipper *et al.*, 1977), spectinomycin (Cannon and Bott, 1979), and streptomycin (Campbell *et al.*, 1978) are often defective in the sporulation process (Chapter 4, this volume). These findings suggest that RNA polymerase and ribosomes play a special, perhaps regulatory, role in the sporulation process, a possibility that will be considered below.

B. Phage Conversion of *spo* Mutants

An intriguing phenomenon that may bear on the special role of RNA polymerase in spore formation is the conversion of certain Spo mutants to a Spo^+ phenotype by pseudolysogenic phage PMB12. Bramucci, Keggins and Lovett (1979) and Kinney and Bramucci (1981) have discovered that phage PMB12 restores the ability of both Spo^- Rfm^r and Spo^- Str^r mutants to sporulate. This conversion is dependent upon three phage genes known as *scn*. Apparently, the *scn* gene products circumvent an early sporulation pathway that somehow interacts with the transcription and translation machinery in the cell. This pathway is at least in part distinct from the *spo0* gene action because PMB12 does not enable most *spo0* mutants to form spores.

Like its close relatives PBS1 (Rima and Takahashi, 1973) and PBS2 (Price and Frabotta, 1972), phage PMB12 carries out its entire developmental cycle in the presence of rifampin (Bramucci *et al.*, 1977). This suggests that RNA synthesis by this class of phages is either controlled by phage-specific RNA polymerase or a novel form of the host cell polymerase that is insensitive to rifampin. In the best studied case of phage PBS2, Clark (1978) has distinguished two temporal classes of phage-directed RNA synthesis. The "late" class of transcription is directed by a phage-induced multisubunit RNA polymerase entirely distinct from the *B. subtilis* transcriptase (Clark *et al.*, 1974). Although the transcribing enzyme for early genes has not been identified, genetic experiments by Osburne and Sonenshein (1980) suggest that RNA synthesis early after PBS2 infection may be directed by a novel phage-modified form of the bacterial enzyme that has been converted to rifampin resistance but which retains sensitivity to another RNA polymerase inhibitor, lipiarmycin.

It is unlikely, however, that either a PMB12 induced modification or replacement of RNA polymerase could explain the capacity of this phage to convert Spo mutants. First, PBS2 is not a converting phage and PMB12 converts both RNA polymerase and ribosome mutants. Second, *scn* mutants prevent conversion but are not otherwise defective in the PMB12 life cycle. So the mechanism of conversion remains an intriguing but unexplained mystery.

C. Compartmentalization of *spo* Gene Expression

Septation at stage II of development partitions the sporulating cell into two compartments: the mother cell or sporangium and the smaller forespore compartment which is destined to become the dormant endospore (Fitz-James and Young, 1969). Thus, two cell types are involved in endospore formation. Each cell type contains a copy of the chromosome, and both chromosomes are transcriptionally active during development (Ryter and Whitehouse, 1978).

Does each cell type exhibit a distinct program of *spo* gene expression? To investigate this question Lencastre and Piggot (1979) devised an indirect but clever genetic procedure for determining whether a *spo* locus is expressed in the mother cell or the forespore. Asporogenous mutants were rendered capable of forming spores by transformation with wild-type DNA at the start of sporulation. Lencastre and Piggot (1979) then examined the genotype of cells arising from the heat-resistant spores. For some mutants, such as *spoVB*, a significant proportion of the heat-resistant spores were genetically asporogenous. This indicated that only the mother cell genome need have been transformed to Spo^+ in order for endospore formation to proceed normally. Thus, the *spoVB* locus must be expressed in the mother cell compartment. Heat-resistant spores arising from transformation of certain other mutants such as *spoVA* were almost always genetically Spo^+. The *spoVA* locus must therefore be expressed in the forespore compartment.

Lencastre and Piggot's experiments strongly suggest that the products of various *spo* loci function in different compartments. This, in turn, could mean that the expression of certain *spo* genes is compartmentalized. This need not be the case, however, for it could be imagined that a *spo* locus might be expressed in both cell types but that its product is *required* only in one or the other compartment. A direct demonstration of compartmentalized gene expression thus requires purification of the *spoVA* and *spoVB* genes in order to probe for their transcription in the mother cell and in the forespore.

D. Temporal Programming of Gene Expression

Endospores are formed over the course of several hours in an orderly progression of morphological and biochemical events. At least five morphological stages can be distinguished. These morphological stages are associated with characteristic biochemical events such as antibiotic and protease secretion early in development and dipicolinic acid synthesis at an intermediate stage of spore formation.

Does this developmental progression reflect a temporally ordered program of gene activation? It is tempting to believe that sets of sporulation genes are turned on at different stages of development, but little experimental evidence is as yet available to support this. Although sporulating bacteria are known to synthesize RNA sequences that are absent in vegetative cells (Sumida-Yasumoto and Doi, 1974; DiCioccio and Strauss, 1973; Pero *et al.*, 1975), hybridization–competition studies on global transcription have provided little information about the timing of gene activation. High-resolution one-dimensional and two-dimensional gel electrophoresis of pulse-labeled proteins, on the other hand, has revealed detailed changes in the program of protein synthesis during the course of spore formation (Linn and Losick, 1976; Campbell *et al.*, 1978; Arnaud *et al.*, 1980). At least five classes of polypeptides could be distinguished on the basis of their time of appearance during sporulation (Linn and Losick, 1976). Unfortunately, however, changes in the pattern of pulse-labeled proteins need not reflect changes in gene expression. At least one spore coat polypeptide, for example, appears during the fourth hour of sporulation, but Cheng and Aronson (1977) and Munoz *et al.* (1978) believe that coat protein is derived by proteolytic cleavage from a larger precursor polypeptide synthesized early in development. A direct demonstration of programmed gene expression requires the isolation of cloned probes for individual sporulation loci.

III. Cloned Genes That Are Actively Transcribed during Sporulation

A. A Cloned Gene Cluster

To study sporulation gene expression, this laboratory has employed radioactively labeled sporulation RNA as a hybridization probe for identifying and cloning segments of *B. subtilis* DNA that are actively transcribed during sporulation. With this approach we have identified within an 8-kb segment of the *B. subtilis* chromsome four genes that are among the most actively transcribed sequences in sporulating cells (Segall and Losick, 1977; Ollington *et al.*, 1981). These three genes are known as *veg, 0.4 kb, ctc,* and *0.3 kb*. The location and direction of transcription of three of these genes are indicated on the endonuclease restriction map in Fig. 2. Although clustered within a small segment of DNA, these genes exhibit different patterns of transcription.

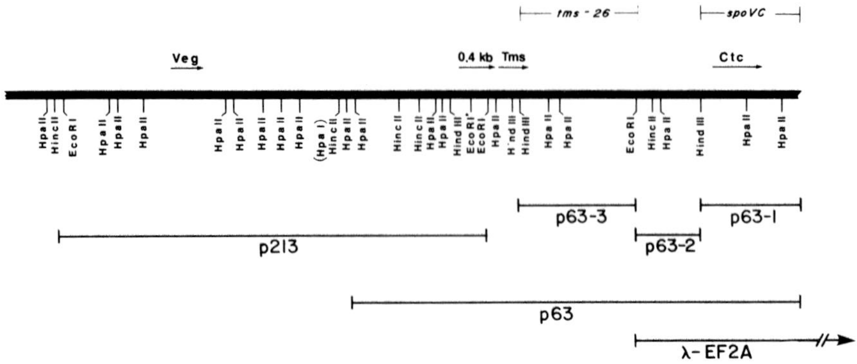

Fig. 2. Physical map of the *veg–0.4 kb–tms–spoVC* gene region. The arrows above the endonuclease restriction map show the location and direction of transcription of the *veg, 0.4 kb, tms,* and *ctc* genes; the right-hand termini of the *tms* and *ctc* transcription units are uncertain. The boundaries within which the mutation *tms-26* lies are shown at the top of the figure (Haldenwang *et al.*, 1980). The boundaries within which the *spoVC* mutation lies were determined by Moran *et al.* (1980) and C. Moran and M. Stephens (Table II). Cloned DNA segments are indicated below the endonuclease restriction map and correspond to the DNAs employed in the transformation experiment in Table II. p213, p63, p63-3, p63-2, and p63-1 are described in Segall and Losick (1977) and Ollington *et al.* (1981). The phage λ hybrid EF2A was a gift from E. Ferrari and J. Hoch.

1. The *veg* Gene

This gene encodes an RNA of about 350 bases and is actively transcribed both in growing cells and in cells undergoing sporulation (Ollington *et al.*, 1981; Ollington and Losick, 1981).

2. The *0.4 kb* Gene

This gene (as its name implies) encodes an RNA of about 400 bases. Segall and Losick, 1977; Ollington *et al.*, 1981). The *0.4 kb* gene is transcribed at only a low rate in vegetative cells but is actively transcribed at both early and late stages of sporulation. As summarized in Table I, several lines of evidence show that the turn-on of the *0.4 kb* gene is specifically under the control of regulatory events at the onset of sporulation. First, transcription of the *0.4 kb* gene is dependent upon the products of five of the seven *spo0* loci; mutations at *spo0A, spo0B, spo0E, spo0F,* and *spo0H* markedly restrict *0.4 kb* RNA synthesis. (Mutations in the genes *spo0C, spo0J,* and *spoIIA,* in contrast, do not impair *0.4 kb* transcription.) Second, treatment of wild-type vegetative cells with 0.7 *M* ethanol, a procedure that mimics the effect of a *spo0* mutation but does not impair vegetative growth (Bohin *et al.*, 1976),

TABLE I Modification of RNA Polymerase and Activation of the *0.4 kb* Gene

	RNA polymerase modification	Ref.	*0.4 kb* Transcription	Ref.[a]
Growth phase				
Vegetative	–	*a–e*	–	*h,i*
Sporulation	+	*a–e*	+	*h,i*
Mutants				
spo0A	–	*a,d*	–	*h,i*
spo0B	–	*a,b d,e*	–	*h,i*
spo0C	–	*a*	+	*i*
spo0E	n.d.		–	*i*
spo0F	n.d.		–	*i*
spo0H	–	*a*	–	*i*
spo0J	n.d.		+	*i*
Rfm[r] Spo[–]	–	*c,e*	–	*h*
spoIIA	+	*a,d,e*	+	*h*
spoIIIC	+	*a*	n.d.	
spoIVE	+	*a*	n.d.	
Cell treatment				
0.7 *M* ethanol	–	*f*	–	*h*
Chloramphenicol	–	*g*	–	*h*

[a] Cells in which RNA polymerase was found to be modified as measured by inhibition of σ factor activity or in which the *0.4 kb* gene was turned on are indicated by a "+", and cells in which polymerase was unmodified or the *0.4 kb* gene was off are indicated by a "–"; "n.d." indicates not determined. Treatment of cells with 0.7 *M* ethanol produces a *spo0* phenocopy without interfering with vegetative growth (see the text). Chloramphenicol treatment restores σ factor activity to sporulating cells (see the text). The key to the reference numbers is: [a]Brevet and Sonenshein (1972); [b]Nishimoto and Takahashi (1974); [c]Sonenshein *et al.* (1974); [d]Tjian and Losick (1974); [e]Tjian and Losick (1975); [f]Bohin *et al.* (1976); [g]Segall *et al.* (1974); [h]Segall and Losick (1977); [i]Ollington *et al.* (1981).

blocks the turn-on of the *0.4 kb* gene. Third, *0.4 kb* RNA synthesis is prevented in certain Rfm[r] Spo[–] mutants that are blocked early in development. Fourth, treatment of sporulating cells with chloramphenicol, a procedure known to reverse a sporulation-induced modification of RNA polymerase, strongly inhibits *0.4 kb* transcription (Segall *et al.*, 1974).

3. The *ctc* Gene

This gene encodes an RNA of about 600 bases (Ollington *et al.*, 1981; N. Lang, unpublished results). Like the *0.4 kb* gene, the *ctc* gene is activated by nutritional deprivation in the sporulation medium. Unlike the *0.4 kb* transcription, however, *ctc* RNA synthesis is not dependent upon the *spo0* gene products. Hence the *ctc* gene is under catabolite control but not specifically under sporulation control. Nevertheless, as we shall see, the *ctc* gene apparently plays a specific role in spore development.

4. The *0.3 kb* Gene

The gene, which is located immediately to the right of *veg* (Fig. 1), encodes a transcript of 300 bases (Ollington and Losick, 1981). Although its transcription is under *spo0* control, the *0.3 kb* gene differs from both the *0.4 kb* and *ctc* sequences in that its expression is not activated until an intermediate stage of development (stage III to IV). Thus, *0.3 kb* is the first example of a gene whose transcription is under temporal control during the course of sporogenesis.

Although new cloning methods promise to provide a variety of *spo* genes, the *veg, 0.4 kb, ctc,* and *0.3 kb* genes are as yet the only probes for studying RNA synthesis in sporulating cells. I shall therefore focus on these genes in the remainder of this chapter as a model system for sporulation gene expression. It is by no means certain, however, that these genes will prove to be representative of *spo* loci in general.

B. Mapping the Gene Cluster

The *veg, 0.4 kb, ctc,* and *0.3 kb* genes were identified as transcriptional units, hence their location on the *B. subtilis* genetic map, and their possible functions in sporulation were not inmediately apparent. To determine the map position of these genes, Haldenwang *et al.* (1980) developed a method for locating the chromosomal site homologous to any particular cloned *B. subtilis* DNA. These workers constructed a mapping vector known as p1949 that carries a chloramphenicol resistance determinant and is incapable of autonomous replication in *B. subtilis*. When a segment of *B. subtilis* DNA is inserted into p1949, the resulting hybrid plasmid can integrate into the *B. subtilis* chromosome by recombination at a site of homology with cloned DNA. Thus, the genetic location of a cloned *B. subtilis* DNA can be determined simply by mapping the integrated drug resistance marker.

To map the cloned gene cluster, Haldenwang *et al.* (1980) inserted into p1949 a DNA fragment from the promoter proximal side of the *0.4 kb* gene. This recombinant plasmid was then introduced into *B. subtilis* by transformation, and drug-resistant transformants were then selected. The integrated *cam* determinant in these transformants mapped in the *purA–cysA* region of the chromosome, a region that contains several sporulation genes and the origin of DNA replication (Trowsdale *et al.*,

1979). Integration of the *cam* determinant at a second site in the *purA–cysA* region (by insertion into p1949 of a DNA fragment from the promoter *distal* side of the *0.4 kb* gene) established the physical orientation of the cloned gene cluster on the *B. subtilis* genetic map as shown in Fig. 1.

p1949 should provide a general method for mapping and orienting other cloned *B. subtilis* DNAs for which precise genetic loci are not known. Several examples are rRNA and tRNA genes, the origin of DNA replication and cloned genes that specify important sporulation proteins such as spore coat protein (Aronson and Fitz-James, 1976), and the germination polypeptides (Setlow, 1975).

C. Identification of the *tms-26* Locus in Cloned DNA

The identification of a genetic locus from the *purA–cysA* region in cloned DNA provided a precise map position for the *veg, 0.4 kb,* and *ctc* genes. *tms-26* is a temperature-sensitive mutation that apparently affects the synthesis or integrity of the bacterial cell wall (Copeland and Marmur, 1968). An interesting feature of this locus is its striking genetic conservation as demonstrated by Copeland and Marmur (1968) in interspecies transformation experiments. Haldenwang *et al.* (1980) showed that a strain harboring the *tms-26* mutation could be converted to temperature resistance by transformation with cloned DNA.

Transformation with DNA restriction fragments positioned the *tms-26* locus between the *0.4 kb* gene and the *ctc* gene, within the 1.3-kb interval identified in Fig. 2. This poses a dilemma, as the *tms-26* locus must lie outside both the *0.4 kb* and *ctc* transcribed sequences and within a region in which transcription *in vivo* has not been detected. Presumably the *tms* gene is transcribed at too low a level to be detected by the Southern hybridization techniques employed by Ollington *et al.* (1981). Just to the right of the *0.4 kb* gene, however, is a promoter that is recognized *in vitro* by RNA polymerase holoenzyme (Haldenwang and Losick, unpublished results). Polymerase initiating at this site transcribes into the region containing the *tms-26* locus. Thus, this region of *in vitro* RNA synthesis we refer to as the *tms* gene could correspond to the *tms-26* locus (see Fig. 2).

D. Proximity of the *spoVC* Locus to the *ctc* Gene

The *spoVC* locus is defined by a temperature-sensitive mutation (*spoVC285*) in the *purA–cysA* region of the chromosome that causes a blockage at a late stage (IV or V) of sporulation (Young, 1976; Piggot and Coote, 1976). Two- and three-factor crosses have placed the *spoVC* locus just to the right of the *tms-26* mutation (Moran, Losick, and Sonenshein, 1980), a finding that suggested that *spoVC* could lie within the *ctc* gene (Fig. 2). Table II summarizes the evidence for this. C. Moran and M. Stephens (unpublished) tested a series of cloned DNAs from the *ctc* region of the chromosome for the ability to convert competent *spoVC* mutant cells to a Spo^+ phenotype. The physical locations of these cloned DNAs are shown in Fig. 2. Only DNAs that contained the *ctc* sequence were able to convert the *spoVC* mutant. These DNAs defined a 1-kb interval containing both the *ctc* sequences and the *spoVC* locus. Thus, the *spoVC* locus must lie within or very near the *ctc* gene.

This is an unexpected finding, as the *ctc* gene is an example of a gene that is turned on during sporulation but whose transcription is not dependent upon the *spo0* gene products. If *ctc* is in fact the *spoVC* gene, then this could mean that the *spo0* gene products are required only for turning on a subset of sporulation genes.

The discovery that the *ctc* gene could correspond to the *spoVC* locus also raises an apparent contradiction between the time of *ctc* RNA synthesis and the phenotype of the *spoVC* mutant. The *ctc* sequence is actively transcribed both at early and late times in sporulation (Ollington *et al.*, 1981). Yet, the *spoVC* mutation arrests development at the fourth to fifth stage of spore formation. "Shift-up" and "shift-down" experiments with the *spoVC* mutant have shown, however, that the heat-sensitive period in this Spo^{ts} mutant precedes by several hours the stage at

TABLE II Transformation of *spoVC* by Cloned DNA[a]

DNA	Met^+ transformants	Met^+ Spo^+ transformants
p213	800	0
p63	3500	202 (6%)
p63-3	3500	0
p63-2	1500	0
p63-1	3500	48 (1.4%)
λ-EF2A	500	39 (7.8%)

[a] Competent cells of a Met^- $SpoVC^-$ mutant were transformed with a mixture of *met*$^+$ *spoVC*$^-$ chromosomal DNA and the indicated cloned DNA as described by Moran *et al.* (1980). Since in a population of *B. subtilis* cells only a small proportion are capable of undergoing transformation, competent cells, which are capable of undergoing multiple transformation events, were selected by growth on medium lacking methionine. Spo^+ transformants were then identified among these Met^+ transformants by incubation on Difco sporulation medium. The cloned DNAs are described in Fig. 2. This experiment was performed by C. Moran and M. Stephens.

which development is blocked (Young, 1976). One interpretation of this is that the *spoVC* gene product (the *ctc* gene product) must accumulate early in sporulation in order for development to proceed past stage V. If this interpretation is correct, then the stage of blockage of a *spo* mutation may not necessarily correlate with the time of expression of the *spo* gene.

[Another interpretation is that the *spoVC* gene product does not play an essential role in the sporulation process but that the *spoVC285* allele produces an altered product that *interferes* with a late developmental stage. The isolation of other alleles of the *spoVC* locus (or the development of a system for creating partial diploids at the *spoVC* locus) should provide a test of this.]

Despite these complications, I will assume here that *ctc* is in fact the *spoVC* gene, although this remains to be proven by nucleotide sequence analysis. The *spoVC* gene is, then, the first example of a genetically defined locus whose product is apparently involved in sporulation and whose physical structure, time of expression, and regulation can now be determined directly.

E. Construction of a Deletion Mutation in the *0.4 kb* Gene

Although a previously known *spo* locus has not been mapped to the *0.4 kb* gene, Rosenbluh *et al.* (1981) recently removed by recombinant DNA techniques a small DNA segment from within the *0.4 kb* gene. This "deletion" mutation was then inserted into the *B. subtilis* chromosome by transformation of competent cells. This resulted in transformants that were defective in spore formation at a late developmental stage, a finding that indicates that the *0.4 kb* gene, like the *ctc* gene, plays a functional role in the sporulation process and identifies the *0.4 kb* gene as a new sporulation locus, which we designate *spoVG*.

IV. Transcription by Modified RNA Polymerases

A. RNA Polymerase Modification during Sporulation

Bacillus subtilis DNA-dependent RNA polymerase undergoes changes in transcription specificity and subunit composition during the course of endospore formation (reviewed by Losick and Pero, 1976a,b). RNA polymerase holoenzyme from vegetative cells consists of a core enzyme containing the subunits $\beta'\beta\alpha_2$ that catalyzes the polymerization of RNA chains and a σ subunit of 55,000 daltons (herein referred to as σ^{55}) that governs the recognition of promoters (see Chapter 3, this volume). The onset of sporulation is associated with a marked decrease in the activity of subunit σ^{55} (Losick and Sonenshein, 1969). Evidence for a sporulation-specific inhibitor of σ^{55} has come from an examination of RNA polymerase in crude extracts of sporulating cell (Brevet and Sonenshein, 1972; Linn *et al.*, 1973; Tjian and Losick, 1974, 1975; Tjian *et al.*, 1975). Although sporulating bacteria contain as much σ^{55} as vegetative cells, RNA polymerase in crude extracts of sporulating *B. subtilis* exhibits little σ^{55} activity, as measured both by transcription of σ-dependent templates and by sensitivity to anti-σ antibodies. The σ^{55} in extracts of sporulating cells appears to be only weakly bound to RNA polymerase, as indicated by its failure to copurify efficiently with the core enzyme during conventional purification procedures or during precipitation with antibodies to core enzyme. The observation that inhibition of protein synthesis restores σ^{55} activity to sporulating cells suggests that the inhibitor of σ^{55} could be unstable *in vivo* (Segall *et al.*, 1974). This inhibitor has not yet been purified (Tjian *et al.*, 1975), hence the evidence for its existence remains indirect.

Nevertheless, as summarized in Table I, an examination of σ^{55} activity in a variety of mutants by several investigators strongly suggests that this RNA polymerase modification is a sporulation-specific event. In most *spo0* mutants tested, as well as in certain Rfm^r Spo$^-$ mutants, RNA polymerase fails to undergo the loss of σ^{55} activity characteristic of the enzyme in wild-type cells. The loss of σ^{55} activity is also blocked in wild-type cells that have been converted to a *spo0* phenocopy by treatment with 0.7 *M* ethanol (Bohin *et al.*, 1976). Since RNA polymerase modification in *spoII, spoIII,* and *spoIV* mutants occurs normally, the loss of σ^{55} activity is apparently under the control of genetic events occurring at the onset of sporulation. As discussed below, Table I also reveals a striking correlation between the loss of σ^{55} activity and the turn-on of the *0.4 kb* gene.

Another modification of RNA polymerase that occurs during sporulation is the acquisition of new subunits. Several modified forms of RNA polymerase have been isolated that lack σ^{55} and that contain one or more sporulation-induced subunits of about 85,000, 36,000, 29,000, or 20,000 daltons (Linn *et al.*, 1975; Nakayama *et al.*, 1978; Nishimoto and Takahashi, 1974; Haldenwang and Losick, 1979; Chapter 3, this volume). A characteristic feature of these modified polymerases is that they bind more tightly to DNA–cellulose than unmodified RNA polymerase holoenzyme. Doi and co-workers (Nakayama *et al.*, 1978) have termed the sporulation-induced components of polymerase δ (differentiation) factors; however, to avoid confusion with a vegetative subunit of polymerase of the same name [a 21,000-dalton polypeptide that enhances specific transcription by phage-modified RNA polymerase (Pero *et al.*, 1975; Tjian *et al.*, 1977)] and to avoid prejudging their function, I prefer the designations P^{85}, P^{36}, P^{29}, and P^{20}.

What is the significance of these sporulation-induced modifications? Losick and Sonenshein (1969) have proposed that the transcription of sporulation genes could be controlled, at least in part, by replacing σ^{55} on polymerase with a sporulation-induced subunit(s). Indeed, *B. subtilis* phage SP01 (and its close relative SP82) is known to code for σ-like regulatory proteins that replace σ^{55} on the *B. subtilis* RNA polymerase and direct transcription of phage middle and late genes (Fox *et al.*, 1976; Duffy and Geiduschek, 1977; Talkington and Pero, 1978; Whiteley *et al.*, 1976; Chapter 7, this volume). The hypothesis that sporulation gene expression is controlled in a like manner poses two experimental questions: Are sporulation genes, in fact, transcribed by modified forms of RNA polymerase and, if so, do these modified polymerases contain new σ-like factors that direct the recognition of sporulation promoters? The isolation by recombinant DNA techniques of two sporulation-related genes (*0.4 kb* (*spoVG*) and *spoVC*) has provided the first opportunity to investigate these questions directly.

B. A Novel RNA Polymerase σ Factor

Using cloned DNAs as templates for *in vitro* RNA synthesis, Haldenwang and Losick (1979) showed that unmodified RNA polymerase holoenzyme failed to transcribe either the *0.4 kb* gene or the *spoVC* (*ctc*) gene *in vitro*, although it actively copied RNA from the *veg* gene. These workers have, however, isolated a modified form of *B. subtilis* RNA polymerase that transcribes both developmentally controlled genes in cloned DNA. The modified RNA polymerase copied RNA selectively and asymmetrically, initiating at sites close to or identical with the apparent starting points for transcription of the *0.4 kb* and *spoVC* genes *in vivo* (Moran *et al.*, 1981).

The modified RNA polymerase lacks σ^{55} but contains a newly identified subunit of 37,000 daltons termed P^{37} (Haldenwang and Losick, 1979). In a reconstitution experiment P^{37} polypeptide purified from urea-denatured enzyme conferred on core RNA polymerase the ability to transcribe the *0.4 kb* and *ctc* genes (Haldenwang and Losick, 1980). Thus, P^{37} is itself responsible for the novel transcriptional specificity of modified RNA polymerase. P^{37} apparently acts at the level of promoter recognition, since in a reconstitution experiment purified P^{37} polypeptide enabled core polymerase to form stable binary and ternary (initiation) complexes with endonuclease restriction fragments containing the *0.4 kb* and *spoVC* genes. Thus, P^{37} is itself a σ-like protein, and hereafter we will refer to this transcriptional determinant as σ^{37}. This is the first example within a single bacterium of distinct σ factors (not associated with bacteriophage) that differ in their promoter recognition specificities.

Even though σ^{37} directs the transcription of two genes whose expression *in vivo* appears to be associated with sporulation, this novel transcriptional determinant can be detected in vegetative cells as well as in sporulating bacteria (Haldenwang *et al.*, 1980). Preliminary experiments indicate, however, that the synthesis of σ^{37} or its association with polymerase is nutritionally controlled; little σ^{37} enzyme can be detected in glucose-grown cells, although modified polymerase is relatively abundant in cells growing on such complex carbon sources as are found in Difco sporulation medium (Haldenwang, unpublished results). The presence of σ^{37} in sporulating cells also appears to be regulated. The σ^{37} enzyme is abundant at an early stage of sporulation, but little of the modified enzyme can be detected after the second hour of spore formation. The loss of σ^{37} at intermediate stages of sporulation appears to be a sporulation-specific event, as stationary-phase cells of a *spo0A* mutant actually accumulate σ^{37} polymerase. In this regard, the behavior of σ^{37} parallels that of σ^{55} whose association with polymerase has also been shown to be under *spo0* gene control (as discussed above). In summary, σ^{37} is associated with the early stages of sporulation and under certain circumstances can actually be present in growing cells. With these observations in mind I will consider below a specific model for the involvement of σ^{37} in the regulation of *0.4 kb* and *spoVC* transcription *in vivo*.

C. A Sporulation-Induced σ Factor

As noted above, several different forms of RNA polymerase from sporulating cells have been described. One especially abundant form contains, in addition to the subunits of core RNA polymerase, a subunit of about 27,000–29,000 daltons, herein referred to as P^{29} (Linn *et al.*, 1975; Nakayama *et al.*, 1978). Unlike σ^{37}, P^{29} appears to be a sporulation-specific component of RNA polymerase, as it appears during the second hour of sporulation in wild-type cells and fails to appear in stationary-phase cells of a *spo0A* mutant. Haldenwang *et al.* (1980) have recently found that P^{29}-containing RNA polymerase, like σ^{37} enzyme, is capable of transcribing both the *0.4 kb* and *spoVC* genes in cloned DNA. Indeed, in an RNA "runoff" experiment with a truncated DNA template, P^{29}-containing RNA polymerase initiated transcription of these genes at sites close to or identical with the initiation points recognized by σ^{37}-containing

RNA polymerase. It seems likely, therefore, that P^{29}, like σ^{37}, is a σ-like transcriptional determinant, and I shall refer to this factor henceforth as σ^{29}. The presence of σ^{29} at intermediate and late stages of sporulation could account for the continued transcription of *0.4 kb* and *spoVC* at times when σ^{37}-containing RNA polymerase can no longer be detected.

D. Nucleotide Sequence of Sporulation Promoters

The recognition of promoters by bacterial RNA polymerases is largely determined by hexanucleotide sequences centered approximately 10 and (the −10 or Pribnow box sequence) and 35 (the −35 or recognition sequence) base pairs upstream from the startpoint of transcription. In the best studied case of *E. coli* promoters, the preferred −10 and −35 hexanucleotides are TTGACA and TATAAT, respectively (Rosenberg and Court, 1979). Although only a few examples have been documented, *B. subtilis* σ^{55}-controlled promoters, including the *veg* and *tms* promoters (C. Moran and N. Lang, unpublished results), are highly homologous to the consensus sequences established for *E. coli* RNA polymerase. The *0.4 kb* (Moran *et al.*, 1981a) and *spoVC* promoters (Moran *et al.*, 1981b), on the other hand, do not conform to this pattern, as their nucleotide sequences differ significantly from previously studied bacterial promoters. Thus, the novel promoter recognition specificity of σ^{37} and σ^{29} may reflect corresponding novel nucleotide sequences in the sporulation promoters that these σ factors regulate.

V. A Positive Control Model for Regulation by the *spo0* Gene Products

DNA–RNA hybridization experiments have distinguished two categories of *spo0* genes. Mutations in genes *spo0A, spo0B, spo0E, spo0F,* and *spo0H* severely restricted transcription of the *0.4 kb* gene, whereas mutations in genes *spo0C* and *spo0J* had no measurable effect on the pattern of gene transcription. This implies that the requirements for *0.4 kb* gene transcription are complex and involve the products of at least five genetic loci at scattered sites on the *B. subtilis* chromosome. Yet, *in vitro,* the *0.4 kb* gene was transcribed with apparent fidelity by polymerase that contained simply σ^{37} in addition to the subunits of core RNA polymerase. How, then, do the *spo0A, spo0B, spo0E, spo0F,* and *spo0H* gene products influence expression of the *0.4 kb* gene?

Two general models are plausible. In a negative control model, transcription of the *0.4 kb* gene by modified RNA polymerase is blocked by a repressor. In response to nutrient deprivation, the *spo0* gene products

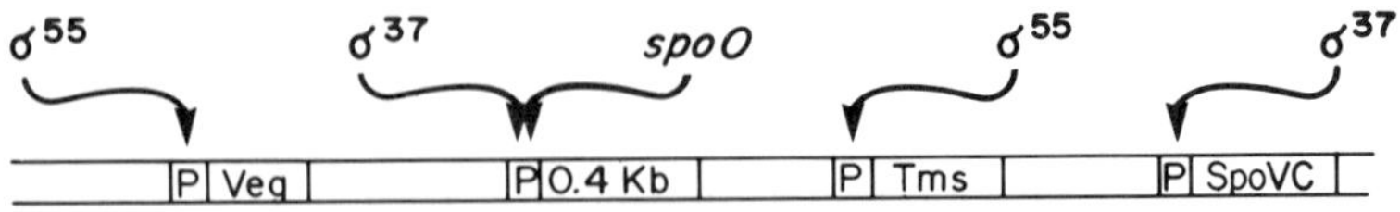

Fig. 3. A positive control model for regulation by the *spo0* gene products. The position of the genes and their promoters (P) is indicated schematically. In this model (see the text), a *spo0* gene product(s) activates the *0.4 kb* gene either by interaction at its promoter and/or by interaction with σ^{37}-containing RNA polymerase.

turn on *0.4 kb* RNA synthesis by inactivating this repressor. In a positive control model (Fig. 3), one or more of the *spo0* gene products could act as a positive regulator in activating *0.4 kb* RNA synthesis. In this case, I would suppose that the *0.4 kb* gene starting site is a weak promoter and that active *in vivo* transcription of the *0.4 kb* gene requires a *spo0* gene product(s) (or a product under *spo0* control) in addition to σ^{37}. Consistent with this model, a promoter competition experiment has shown that the *0.4 kb* gene initiation site is a much weaker promoter for σ^{37}-containing RNA polymerase than the promoter for the *spoVC* gene, a gene whose transcription *in vivo* was not dependent upon the *spo0* gene products (Ollington *et al.*, 1981). Additional evidence in favor of a positive control model is the finding (Banner, 1980) that amplification of the *0.4 kb* gene promoter on a plasmid that replicates with a high copy number in *B. subtilis* strongly inhibits sporulation at an early stage, an observation consistent with the idea that multiple copies of the *0.4 kb* gene control region titrate a regulatory protein(s) (a *spo0* gene product?) required at an early developmental stage.

I favor, therefore, a positive control model and propose that the five *spo0* gene products are components of a pathway (or pathways) that senses nutrient deprivation in *B. subtilis* and translates this environmental signal into transcriptional activation of the *0.4 kb* gene and certain other early sporulation genes. In this view, one or more of the *spo0* gene products must be present before the onset of sporulation, hence I believe that they are "vegetative" proteins. Indeed, as discussed above, mutations in the *spo0A*, *spo0B*, *spo0E*, *spo0F*, and *spo0H* genes are known to have pleiotropic effects on the properties of growing cells.

It is pertinent to note here a striking correlation between the loss of RNA polymerase σ factor activity and the turn-on of the *0.4 kb* gene as determined under a variety of growth conditions and in various mutants. As summarized in Table I, the *0.4 kb* gene was turned on in cells in which RNA polymerase had undergone modification (as measured by inhibition of σ factor activity). Conversely, with only one exception (*spo0C*), the *0.4 kb* gene was off in cells in which RNA polymerase had not undergone modification. Thus, the products of the *spo0* genes appear to be intimately involved in both the inhibition of σ factor activity and the turn-on of the *0.4 kb* gene by σ^{37}-containing RNA polymerase. It is tempting to suppose that both events are controlled by common underlying mechanisms and that one or more of the *spo0* gene products interact directly with RNA polymerase.

The next question that must be addressed is how the *spoVC* gene is turned on, as *spoVC* is an example of a gene whose transcription is not dependent upon the *spo0* gene products. Although at least some σ^{37} can be detected in vegetative cells, preliminary experiments indicate that the amount of σ^{37}-containing RNA polymerase is strongly influenced by the growth medium. Conceivably, *spoVC* transcription is regulated at the level of the synthesis of σ^{37} or its association with RNA polymerase. Alternatively, σ^{37} may account only for a low level of *spoVC* RNA synthesis in growing cells, and its replacement by other regulatory proteins such as σ^{29} could be responsible for the strong stimulation of *spoVC* transcription during sporulation.

Finally, I consider the question of how the process of sporulation is triggered. Freese and co-workers (Mitani *et al.*, 1978; Freese *et al.*, 1978) have shown that sporulation can be induced in the presence of excess glucose, ammonia, and phosphate (conditions under which spore formation is normally blocked) by depriving leaky purine-requiring mutants of purine or by the addition of inhibitors of AMP or GMP synthesis. Limitation of purine synthesis does not, however, overcome the inability of *spo0* mutants to sporulate. Freese thus suggests that the triggering signal for sporulation nay be a reduction in the concentration of purine nucleotides or some closely related event that is dependent upon purine synthesis. In my model for activation of the *0.4 kb* and *spoVC* genes, I presume that one or more of the *spo0* gene products responds directly to a change in the intracellular concentration of purine nucleotide or some closely related molecule whose synthesis is dependent upon purine nucleotides. Alternatively (or additionally), low purine nucleotides could be a signal for controlling the synthesis of σ^{37} or its association with RNA polymerase.

VI. Control of Sporulation Genes: Unanswered Questions

Although the *veg–0.3 kb–0.4 kb–spoVC* gene cluster appears to offer a promising model system, many of the most basic questions about the control of sporulation genes remain unanswered. Perhaps the three most important questions are: What is the nature of the *spo0* gene products and what role do they play in the initiation of sporulation? How is the program of sporulation gene expression controlled and which *spo* loci regulate different temporal classes of sporulation genes? What are the distinguishing features of sporulation genes and their promoters and by what components of the *B. subtilis* transcriptional machinery are they recognized?

The *spo0* gene products have not all been identified, but genetic and physiological experiments suggest that the *spo0A, spo0B, spo0E, spo0F,* and *spo0H* genes are expressed in vegetative cells and presumably encode vegetative proteins. As a working hypothesis, I have proposed a positive control model for how the *spo0* gene products could control *0.4 kb* gene transcription. Little information is available as yet, however, to test this model rigorously or to examine its validity for other *spo0*-controlled genes.

Spore formation proceeds according to a well-defined temporal program. Presumably, the developmental program is controlled, *in part,* by the activation of sporulation genes or groups of sporulation genes in a temporal sequence, but cloned DNA probes are not yet available to test this directly. Mutations in *spo* genes are pleiotropic and characteristically block development at particular stages. Presumably, the products of certain *spo* genes are regulatory proteins that turn on, at particular times in development, subsets of *spo* genes, whereas the products of other *spo* genes are structural proteins or enzymes directly involved in spore morphogenesis. Cloned probes for genes activated at different stages of sporulation should make it possible to distinguish regulatory genes from morphogenetic genes. It should be noted, however, that the stage at which a *spo* gene is expressed need not correlate with the stage at which its product functions; the *spoVC* and *0.4 kb* (*spoVG*) genes are transcribed well before the time at which their products act in spore morphogenesis.

Ultimately, an understanding of sporulation gene expression requires a determination of the regulatory signals for *spo* genes and identification of the regulatory proteins with which they interact. I assume here that *spo* genes are largely regulated at the level of transcription, as there are few well-documented examples of posttranscriptional control mechanisms in prokaryotes and little evidence for posttranscriptional control during sporulation. Thus far, the only examples of sporulation-controlled genes we have to draw upon are the cloned probes described herein which, as we have seen, require novel forms of RNA polymerase for their transcription. As cloned probes for a variety of *spo* loci become available, an examination of their promoters and the forms of RNA polymerase that control them will be of special interest.

Acknowledgments

I thank members of the Losick–Pero group for helpful discussion and critical readings of the manuscript. I also thank C. Moran and M. Stephens for permission to report the experiment of Table II. This work was supported by a grant from the NIH.

References

Arnaud, M., Mahler, I., Halvorson, H. O., Boschwitz, H., and Keynan, A. (1980). *J. Bacteriol.* **142,** 1045–1048.
Aronson, A., and Fitz-James, P. (1976). *Bacteriol. Rev.* **40,** 360–402.
Banner, C. D. B. (1980). *Abstr. Ann. Meeting Am. Soc. Microbiol.* p. 100.
Bohin, J. P., Rigomier, P., and Schaeffer, P. (1976). *J. Bacteriol.* **127,** 934–940.
Bramucci, M. G., Keggins, K. K., and Lovett, P. S. (1977). *J. Virol.* **22,** 194–202.
Brehm, S., Le Hegarat, F., and Hoch, J. (1975). *J. Bacteriol.* **124,** 977–984.
Brevet, J., and Sonenshein, A. L. (1972). *J. Bacteriol.* **112,** 1270–1274.
Campbell, K., Leventhal, J., and Chambliss, G. H. (1978). *In* "Spores VII" (G. Chambliss and J. C. Vary, eds.), pp. 250–253. American Society for Microbiology, Washington, D.C.
Cannon, J., and Bott, K. (1979). *Mol. Gen. Genet.* **174,** 149–162.
Cheng, Y., and Aronson, A. (1977). *Proc. Nat. Acad. Sci. USA* **74,** 1254–1258.
Clark, S. (1978). *J. Virol.* **25,** 224–237.
Clark, S., Losick, R., and Pero, J. (1974). *Nature (London)* **252,** 21–24.
Copeland, J., and Marmur, J. (1968). *Bacteriol. Rev.* **32,** 302–312.
DiCioccio, R., and Strauss, N. (1973). *J. Mol. Biol.* **77,** 325–336.
Dubnau, E., Ramakrishna, N., Cabane, K., and Smith, I. (1981). *J. Bacteriol.* **147,** 622–632.
Duffy, J., and Geiduschek, E. P. (1977). *Nature (London)* **270,** 28–32.
Ferrari, E., Henner, D., and Hoch, J. (1981). *J. Bacteriol.* **146,** 430–432.
Fitz-James, P., and Young, I. (1969). *In* "The Bacterial Spore" (G. Gould and A. Hurst, eds.), pp. 39–72. Academic Press, New York.
Fox, T., Losick, R., and Pero, J. (1976). *J. Mol. Biol.* **101,** 427–433.
Freese, E., Heinze, J., Mitani, T., and Freese, E. B. (1978). *In* "Spores VII" (G. Chambliss and J. C. Vary, eds.), pp. 277–285. American Society for Microbiology, Washington, D.C.
Haldenwang, W. G., Banner, C. D. B., Ollington, J. F., Losick, R., Hoch, J. A., O'Connor, M. B., and Sonenshein, A. L. (1980). *J. Bacteriol.* **142,** 90–98.
Haldenwang, W. G., Lang, N., and Losick, R. (1981). *Cell* **23,** 615–624.
Haldenwang, W. G., and Losick, R. (1979). *Nature (London)* **282,** 256–260.
Haldenwang, W. G., and Losick, R. (1980). *Proc. Nat. Acad. Sci. USA* **77,** 7000–7004.
Haldenwang, W. G., Ollington, J. F., and Losick, R. (1980). *In* "RNA Polymerase, tRNA, and Ribosomes: Their Genetics and Evolution" (S. Osawa, H. Ozaki, H. Uchida, and T. Yura, eds.), pp. 127–134. Univ. of Tokyo Press, Tokyo.
Halling, S., Burtis, K., and Doi, R. (1977). *J. Biol. Chem.* **252,** 9024–9031.
Hoch, J. (1976). *Adv. Genet.* **18,** 69–98.
Hoch, J. A., Shiflett, M. A., Trowsdale, J., and Chen, S. M. H. (1978). *In* "Spores VII" (G. Chambliss and J. C. Vary, eds.), pp. 127–130. American Society for Microbiology, Washington, D.C.
Hranueli, D., Piggot, P., and Mandelstam, J. (1974). *J. Bacteriol.* **119,** 684–690.

Kawamura, F., Shimotsu, H., Saito, H., Hirochika, H., and Kobayashi, Y. (1981). *In* "Sporulation and Germination (Proceedings of the Eighth International Spore Conference)" (H. S. Levinson, A. L. Sonenshein and D. J. Tipper, eds.), pp. 109–113. American Society for Microbiology, Washington, D.C.

Kinney, D. and Bramucci, M. (1981). *J. Bacteriol.* **145,** 1281–1285.

Kobayashi, Y., Hirochika, H., Kawamura, F., and Saito, H. (1981). *In* "Sporulation and Germination (Proceedings of the Eighth International Spore Conference)" (H. S. Levinson, A. L. Sonenshein and D. J. Tipper, eds.), pp. 114–118). American Society for Microbiology, Washington, D.C.

Leighton, T. (1973). *Proc. Nat. Acad. Sci. USA* **70,** 1179–1183.

Leighton, T. (1974). *J. Mol. Biol.* **86,** 858–863.

Lencastre, H., and Piggot, P. (1979). *J. Gen. Microbiol.* **114,** 377–389.

Linn, T., and Losick, R. (1976). *Cell* **8,** 103–114.

Linn, T., Greenleaf, A., Shorenstein, R., and Losick, R. (1973). *Proc. Nat. Acad. Sci. USA* **70,** 1865–1869.

Linn, T., Greenleaf, A., and Losick, R. (1975). *J. Biol. Chem.* **250,** 9256–9261.

Linn, T., Losick, R., and Sonenshein, A. L. (1975). *J. Bacteriol.* **122,** 1387–1390.

Losick, R., and Pero, J. (1976a). *Adv. Enzymol.* **44,** 165–185.

Losick, R., and Pero, J. (1976b). *In* "RNA Polymerase" (R. Losick and M. Chamberlin, eds.), pp. 227–246. Cold Spring Harbor Laboratory, Cold Spring Harbor, New York.

Losick, R., and Sonenshein, A. L. (1969). *Nature* (*London*) **224,** 35–37.

Mitani, T., Heinze, J., and Freese, E. (1978). *Bíochem. Biophys. Res. Commun.* **77,** 1118–1125.

Moran, C. P., Jr., Lang, N., Banner, C. D. B., Haldenwang, W. G., and Losick, R. (1981). *Cell* **25,** 783–791.

Moran, C. P., Jr., Lang, N., and Losick, R. (1981). *Nucleic Acids Research* **10**, in press.

Moran, C. P., Jr., Losick, R., and Sonenshein, A. L. (1980). *J. Bacteriol.* **142,** 331–334.

Munoz, L. E., Nakayama, T., and Doi, R. H. (1978). *In* "Spores VII" (G. Chambliss and J. C. Vary, eds.), pp. 213–219. American Society for Microbiology, Washington, D.C.

Nakayama, T., Williamson, V., Burtis, K., and Doi, R. (1978). *Eur. J. Biochem.* **88,** 155–164.

Nishimoto, H., and Takahashi, I. (1974). *Can. J. Biochem.* **52,** 966–973.

Ollington, J. F., Haldenwang, W. G., Huynh, T. V., and Losick, R. (1981). *J. Bacteriol.* **147,** 432–442.

Ollington, J. F., and Losick, R. (1981). *J. Bacteriol.* **147,** 443–451.

Osburne, M., and Sonenshein, A. L. (1980). *J. Virol.* **33,** 945–953.

Pero, J., Nelson, J., and Fox, T. (1975a). *Proc. Nat. Acad. Sci. USA* **72,** 1589–1593.

Pero, J., Nelson, J., and Losick, R. (1975b). *In* "Spores VI" (P. Gerhardt, R. N. Costilow, and H. L. Sadoff, eds.), pp. 202–212. American Society for Microbiology, Washington, D.C.

Piggot, P., and Coote, J. (1976). *Bacteriol. Rev.* **40,** 908–962.

Price, A., and Frabotta, M. (1972). *Biochem. Biophys. Res. Commun.* **48,** 1578–1585.

Rima, B., and Takahashi, I. (1973). *Can. J. Biochem.* **51,** 1219–1224.

Rosenberg, M., and Court, D. (1979). *Ann. Rev. Genet.* **13,** 319–353.

Rosenbluh, A., Banner, C. D. B., Losick, R., and Fitz-James, P. (1981). *J. Bacteriol.* **148,** 341–351.

Ryter, A., and Whitehouse, R. (1978). *Arch. Microbiol.* **118,** 27–37.

Schaeffer, P., and Ionesco, H. (1960). *C. R. Acad. Sci. Paris* **251,** 3125–3127.

Segall, J., and Losick, R. (1977). *Cell* **11,** 751–761.

Segall, J., Tjian, E., Pero, J., and Losick, R. (1974). *Proc. Nat. Acad. Sci. USA* **71,** 4860–4863.

Setlow, P. (1975). *J. Biol. Chem.* **250,** 631–637.

Sonenshein, A. L., and Campbell, K. M. (1978). *In* "Spores VII" (G. Chambliss and J. C. Vary, eds.), pp. 179–192. American Society for Microbiology, Washington, D.C.

Sonenshein, A. L., Cami, B., Brevet, J., and Cote, R. (1974). *J. Bacteriol.* **120,** 253–265.

Sumida-Yasumoto, C., and Doi, R. (1974). *J. Bacteriol.* **117,** 775–782.

Sumida-Yasumoto, C., and Doi, R. (1977). *J. Bacteriol.* **129,** 433–444.

Talkington, C., and Pero, J. (1978). *Proc. Nat. Acad. Sci. USA* **75,** 1185–1189.

Tipper, D., Johnson, W. C., Ginther, C., Leighton, T., and Wittmann, G. (1977). *Mol. Gen. Genet.* **150,** 147–159.

Tjian, R., and Losick, R. (1974). *Proc. Nat. Acad. Sci. USA* **71,** 2872–2876.

Tjian, R., and Losick, R. (1975). *In* "Organization and Expression of the Eukaryotic Genome" (*Proc. Meeting F.E.B.S., 10th, Paris*) (G. Bernardi and F. Gros, eds.), pp. 277–287. Elsevier/North-Holland Publ., Amsterdam.

Tjian, R., Stinchcomb, D., and Losick, R. (1975). *J. Biol. Chem.* **250,** 8824–8828.

Tjian, R., Losick, R., Pero, J., and Hinnebush, A. (1977). *Eur. J. Biochem.* **74,** 149–154.

Trowsdale, J., Chen, S. M. H., and Hoch, J. A. (1979). *Mol. Gen. Genet.* **173,** 61–70.

Whiteley, H. R., Spiegelman, G. B., Lawrie, J. M., and Hiatt, W. R. (1976). *In* "RNA Polymerase" (R. Losick and M. Chamberlin, eds.), pp. 587–600. Cold Spring Harbor Laboratory, Cold Spring Harbor, New York.

Young, M. (1976). *J. Bacteriol.* **126,** 928–936.

7

Regulatory Mechanisms in the Development of Lytic Bacteriophages in *Bacillus subtilis*

E. PETER GEIDUSCHEK AND JUNETSU ITO

I. Introduction

In this chapter, we analyze the development of two classes of lytic DNA viruses in *Bacillus subtilis*: the hydroxymethyluracil(hmU)-containing phages, of which SPO1 and SP82 are the most studied examples, and the small, linear DNA phages with covalently bound terminal protein,

The Molecular Biology
of the Bacilli

ISBN 0-12-222701-8

of which φ29 is the most studied. Our plan is to present the available information on the genomes, describe sequences of gene activity during viral development and then focus on the regulation of this gene action. Of course, our understanding of this regulation is quite incomplete: Although much is already known and certain generalizations do emerge, it is our hope that the reader will gather some impression not only of the current status but also of unanswered questions.

The two classes of viruses we discuss here have entirely different genomes and rather different features of viral development. Moreover, what is known about the two classes of viruses is, to some degree, complementary. With its small genome, φ29 has been genetically much more thoroughly analyzed than the larger phage SPO1. The replication of φ29 DNA involves mechanisms that compel attention: the covalently bound terminal protein of the linear φ29 genome is essential for DNA replication. The involvement of this terminal protein, which recalls the adenoviruses and parvoviruses (Winnacker, 1978; Revie *et al.*, 1979), is currently under intensive study in several laboratories. Despite its small genome, the φ29 virion is structurally a relatively complex object. Accordingly, φ29 has been an attractive subject of studies on viral morphogenesis (reviewed by Anderson and Reilly, 1976; Wood and King, 1979). In contrast, SPO1 DNA replication has been relatively little studied, and the first really detailed analyses of the virion and its morphogenesis have only recently been completed (Parker, 1979; Eiserling, 1979).

Neither φ29 nor SPO1 completely shuts off gene expression of the infected bacterium. In the case of φ29, infection has very little effect on host cell functions (Schachtele *et al.*, 1972), and viral protein synthesis constitutes a relatively small fraction of the total throughout the infectious cycle. The case of the hmU-containing large phages is less extreme, but it takes several minutes for the synthesis of virus-coded protein to dominate the total activity of the cell. In neither case is the bacterial host's chromosome degraded.

The expression of φ29 and SPO1 viral genes in infected cells is regulated in orderly sequences. The larger phage, with its larger collection of genes, appears to elicit the more complex sequence. It is these sequences and the transcriptional mechanisms by which they are generated that constitute the core material of this chapter. Here also, there is a certain degree of complementarity in our presentation. The analysis of transcription of the larger SPO1 and SP82 genomes presents a more complex task and is less complete than for the φ29 genome. On the other hand, the availability of *in vitro* systems for positively regulated transcription of hmU DNA genomes has advanced this aspect of hmU phage regulation relative to that of φ29.

In the following discussion we first present the smaller and then the larger of these phages, following much the same sequence in each section: The virion and the genome are described first. Then, after brief comments on the relatedness of the phages of each group, the general features of viral development are outlined and the discussion focuses on the analysis of regulated transcription during the viral lytic cycle.

II. Phage φ29

A. The Virion

Phage φ29 is the prototype of a group of small but morphologically very complex viruses. The φ29 virion consists of a prolate head, a neck, and a short, noncontractile tail (Anderson *et al.*, 1966). Figures 1 and 2 show, respectively, an electron micrograph of a negatively stained preparation of φ29 and a schematic diagram of φ29 (Anderson *et al.*, 1966). Somewhat different representations of φ29 have also been constructed (Viñuela *et al.*, 1976; Eiserling, 1979).

The head is about 40 nm long and 30 nm wide, containing numerous protein fibers which are characteristic of this phage. The neck assembly consists of two collars, an upper one and a lower one. The upper collar is attached to, and is a part of, the head. Twelve spindle-shaped appendages, which are attached to the lower collar, appear to be the organelles of adsorption (Tosi and Anderson, 1973). The noncontractile tail is about 30-nm long and is enlarged at the distal end.

The molecular mass of a φ29 particle is 18×10^6 daltons (Rubio *et al.*, 1974). Seven structural proteins have been resolved by sodium dodecyl sulfate (SDS)–polyacrylamide gel electrophoresis (Mendez *et al.*, 1971; Pène *et al.*, 1973), but two of these may be the products of a single gene: The minor head protein may be generated from the major head protein by nonspecific cleavage (Anderson and Reilly, 1976), no separate structural gene for the minor head protein having so far been found.

B. Genome Structure

The φ29 genome is a linear, double-stranded DNA molecule of about 12×10^6 daltons (Anderson and Mosharrafa, 1968). The unique features of the φ29 genome are that its DNA contains a short, inverted 6-bp-long

Fig. 1. An electron micrograph of φ29. (From Anderson *et al.*, 1966.)

terminal repetition (Yoshikawa *et al.*, 1981; Escarmis and Salas, 1981) and that a protein is linked covalently to the 5′-ends of each DNA strand (Ortin *et al.*, 1970; Salas *et al.*, 1978; Yehle, 1978; Ito, 1978; Hermoso and Salas, 1980). This DNA terminal protein is essential for φ29 DNA replication (Yanofsky *et al.*, 1976; Hagen *et al.*, 1976; Carrascosa *et al.*, 1976; Mellado *et al.*, 1980). The DNA of φ29 is blunt-ended (Yoshikawa *et al.*, 1981; Escarmis and Salas, 1981), so that the general structure is

Protein 5′-AAAGTA........................TACTTT-3′

3′-TTTCAT........................ATGAAA-5′ Protein

The φ29 DNA–protein linkage appears to be a phosphodiester bond

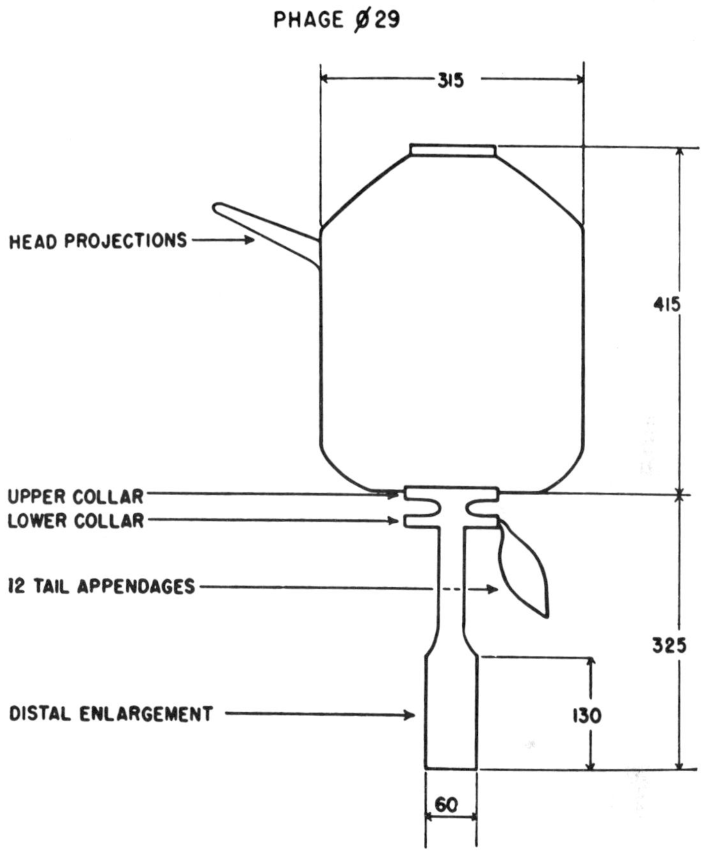

Fig. 2. Schematic drawing of ϕ29. Distances are marked in angstrom units. (From Anderson *et al.*, 1966.)

between the β-OH of a serine, and the 5′-P of the terminal deoxyadenylate (Hermoso and Salas, 1980).

ϕ29 DNA is resistant to cleavage by many restriction endonucleases such as *Hae*III, *Bam*HI, *Pst*I, *Sal*I, *Sma*I, *Bgl*II, and *Bal*I. This lack of susceptibility is due to the absence of the nucleotide sequences in ϕ29 DNA recognized by these enzymes, rather than the presence of modified nucleotides (Ito and Roberts, 1979). The G + C content of ϕ29 DNA is 36%, lower than that of host DNA (42.5%; Rubio *et al.*, 1974). The two complementary strands of ϕ29 DNA can be separated by CsCl density gradient centrifugation in the presence of poly (UG), which bind preferentially to the H strand of ϕ29 DNA (Mosharrafa *et al.*, 1970).

Since the genome of ϕ29 contains approximately 18,000 bp, this phage can code for about 660,000 daltons of protein, assuming that the average amino acid has a molecular weight of 110 and all potential codons are used only once. Eighteen genes have been identified and arranged on a linear genetic map (Mellado *et al.*, 1976; Reilly *et al.*, 1977; Fig. 3). A list of these genes and of the functions that are known, is presented in Table I.

Proteins specified by genes *1–3*, *5*, and *6* are required for ϕ29 DNA replication (Talavera *et al.*, 1972; McGuire *et al.*, 1974, 1977; Carrascosa *et al.*, 1976; Harding and Ito, 1976; Hagen *et al.*, 1976). The product of gene *4* appears to be a positive control element, since very little late protein is synthesized in cells infected with gene-*4* mutant phage (Anderson and Reilly, 1976; Carrascosa *et al.*, 1976). As described below, mutants in gene *4* are defective in synthesizing late ϕ29 mRNA.

The identification of ϕ29 nonstructural proteins has been carried out by infecting ultraviolet (UV)-irradiated *B. subtilis* with wild-type and mutant phage in the presence of radioactive amino acids and subseguently separating the isolated proteins on SDS–polyacrylamide gels (Carrascosa *et al.*, 1973; Hawley *et al.*, 1973; Pène *et al.*, 1973). However, to date little is known about the molecular weights and functions of ϕ29 early

TABLE I *ϕ29 Genes, Proteins, and Functions*

Gene	Time of expression	Protein or phenotype
1	Early	DNA synthesis (?)[a]
2	Early	DNA synthesis
3	Early	Terminal protein
4	Early	Control of late transcription
5	Early	DNA synthesis
6	Early	DNA synthesis
7	Late	Head morphogenesis
8	Late	Major head protein
8.5	Late	Head fiber protein
9	Late	Tail protein
10	Late	Upper collar protein
11	Late	Lower collar protein
12	Late	Precursor of the neck appendage protein
13	?	Morphogenesis
14	?	Lysis
15	Late	Morphogenesis
16	Late	Morphogenesis
17	Early	?[a]

[a] Some *sus* mutants in genes *1* and *17* synthesize ϕ29 DNA, and others do not (Hagen *et al.*, 1976; Harding and Ito, 1976; Pène, personal communication).

gene products. Perhaps the most interesting of these is the gene-*3* protein, which is the DNA terminal protein. It has been reported that the protein specified by gene *3* has a molecular weight of 27,000 (McGuire *et al.*, 1974 Carrascosa *et al.*, 1976). On the other hand, Anderson and Reilly (1976) observed that a nonsense mutant in φ29 gene *3* failed to synthesize two proteins with molecular weights of about 28,000 and 29,000, called AF and BF. It is possible that the terminal protein is processed during φ29 development.

Sogo *et al.* (1976) have shown that very little late RNA is synthesized in φ29 *sus* gene *4*-infected cells, suggesting that the product of gene *4* is required for late transcription. When UV-irradiated cells were infected either with *sus* or with *ts* gene-*4* mutants under nonpermissive conditions and RNA was analyzed by polyacrylamide gel electrophoresis, essentially only early φ29 RNA species could be detected (Kawamura and Ito, unpublished results).

The products of the late φ29 genes, *7–12, 15,* and *16* have been identified, as shown in Table I (Anderson and Reilly, 1974; Camacho *et al.*, 1974; McGuire *et al.*, 1974; Hagen *et al.*, 1976; Carrascosa *et al.*, 1976). The product of gene *12* is a precursor protein which is cleaved to generate the neck appendage (Anderson and Reilly, 1974; Carrascosa *et al.*, 1974; Tosi *et al.*, 1975). Genes *13* and *14* also appear to be late genes. There is evidence suggesting that the product of gene *13* is required for phage morphogenesis (Hagen *et al.*, 1976). It is also known that *sus* gene-*14* mutants produce normal phage after artificial lysis (Carrascosa *et al.*, 1976).

C. Small Bacillus Phages Related to φ29

Several other small phages that are morphologically similar to φ29 have been described (Hemphill and Whiteley, 1975; Reilly, 1976). These include φ15, φ21, Nf, M2Y, SF5, and GA-1 (Reilly, 1965; Ikeda *et al.*, 1965; Bradley, 1965; Shimizu *et al.*, 1970; Rima *et al.*, 1971; Kawamura and Ito, 1977c). All these phages contain linear, duplex DNA with molecular weights of 11–13 $\times$ 10^6, and the infectivity of this DNA is sensitive to proteolytic enzymes (Hirokawa, 1972; Ito *et al.*, 1973; Arwert and Venema, 1974; Kawamura and Ito, 1977c; Takagi and Hirokawa, 1979). Some properties of these phages are presented in Table II. The phages fall into three serological groups: Phages φ29 and φ15 form one group, phages M2 and Nf are similarly related to each other, and GA-1 is not related to any of the others. Furthermore, the genome of GA-1 is larger than those of other phages. On the other hand, phage M2Y

TABLE II Comparison of the Small Bacillus Phages

Group	Phage	Molecular weight of genome	Terminal inverted repeat	Molecular weight of terminal protein	Remarks
I	φ29	12×10^6	5′-AAAGTA-	29,000	—
	φ15	12×10^6	5′-AAAGTA-	29,000	—
II	NF	11.7×10^6	5′-AAAGTAAG-	30,000	—
	M2Y	11.4×10^6	5′-AAAGTAAG-	30,000	No head fibers
III	GA-1	13.5×10^6	5′-AAATAGA-	28,000	No head fibers

is the smallest known bacteriophage genome composed of linear duplex DNA (Yoshikawa and Ito, 1981a).

These phages are similar in that their DNA contains short, inverted terminal repeats and terminal proteins at the 5′-ends (Yoshikawa and Ito, 1981a). The related phages, φ29 and φ15, have identical sequences at the left end for the first 29 bases and at the right end for the first 26 bases. On the other hand, phages M2Y and Nf have identical sequences at both ends for at least the first 200 bases. Phages φ29 and M2Y exhibit a high degree of homology for the first 50 bases at the termini, although they are not related serologically. Even GA-1, which is serologically as well as chemically distinct from φ29, appears to have some homology with φ29 in the terminal DNA sequences, suggesting an evolutionary relationship.

D. General Features of Viral Development

φ29 can infect some strains of *Bacillus subtilis* and varieties of *B. licheniformis, B. pumilus,* and *B. amyloliquefaciens* (Reilly, 1976). Young (1967) has shown that glucosylated teichoic acid is essential for φ29 adsorption to *B. subtilis*. However, φ29 DNA can transfect competent bacterial cells unable to adsorb this phage (Reilly and Spizizen, 1965). Recently, it has been shown that *B. subtilis* protoplasts are susceptible to φ29 DNA transfection in the presence of poly(ethylene glycol) (Stahly and Ito, 1981). Transfection of φ29 DNA either by competent bacterial cells or protoplasts is sensitive to proteolytic enzymes (Hirokawa, 1972; Stahly and Ito, 1981).

The latent period of φ29 in *B. subtilis* at 37°C is 55–60 min, and the burst size, which can reach 200 (Reilly, 1965), varies greatly under different physiological conditions (Kawamura and Ito, 1975).

φ29 DNA synthesis at 37°C starts 8–10 min after phage infection, increases until about 35–40 min, and then levels off until lysis (Schachtele *et al.*, 1970; Talavera *et al.*, 1972; Ivarie and Pene, 1973; Harding and Ito, 1976).

Biochemical and electron microscopic experiments indicate that φ29 DNA replication proceeds via a strand displacement mechanism: Replication starts at either end of the φ29 linear double-stranded DNA and proceeds by displacing one of the parental strands. Synthesis of the complementary strand is then initiated in the opposite direction (Ito *et al.*, 1979; Inciarte *et al.*, 1980; Harding and Ito, 1980). A similar mechanism characterizes the replication of adenovirus DNA, which also contains terminal protein (Winnacker, 1978). However, little is known about the mechanism by which the ends of linear, nonredundant DNA molecules replicate. All known DNA polymerases require a 3′-OH group as a primer in addition to a template and elongate the product DNA chain in the 5′-to-3′ direction (Kornberg, 1974). Since φ29 DNA contains short, inverted terminal repeats, the two termini of the genome can form neither circular intermediates in infected cells as λ phage do (Young and Sinsheimer, 1964), nor concatemeric intermediates as T7 phage do (Kelly and Thomas, 1969). A model has been proposed in which the DNA terminal protein acts as a primer for the initiation of DNA replication (Rekosh *et al.*, 1977; Inciarte *et al.*, 1980; Harding and Ito, 1980).

Infection of φ29 has very little effect on host cell function (Schachtele *et al.*, 1972). Thus, the rate of macromolecular synthesis after φ29 infection remains constant until very late in phage development. It has also been demonstrated that, when φ29 infects sporulating cells, phage multiplication is inhibited at a very early stage of cell development and φ29 DNA is incorporated into spores in a heat-stable form (Ito *et al.*, 1973; Kawamura and Ito, 1974). Expression of the incorporated phage genome resumes only after germination and outgrowth of the spore.

E. Transcription of the φ29 Genome

Mosharrafa *et al.* (1970) first observed that there are two temporally separable classes of φ29 mRNAs. RNA–DNA hybridization–competition experiments indicate that φ29 early RNA is transcribed exclusively from the L strand of φ29 DNA and that it is transcribed continuously throughout phage development (Loskutoff *et al.*, 1973; Schachtele *et al.*, 1973). This class of RNA is synthesized when cells are infected in the presence of chloramphenicol. In contrast, late mRNA, which is exclusively transcribed from the H strand of φ29 DNA, requires the action of the gene-*4* product. While its first appearance coincides with the onset of viral DNA replication, late transcription is independent of φ29 DNA replication (Schachtele *et al.*, 1973; Loskutoff *et al.*, 1973). Studies with rifamycin-sensitive and rifamycin-resistant host strains have shown that complete transcription of ϕ29 DNA requires the components of the host RNA polymerase, at least of the β subunit, that are sensitive to this antibiotic (Loskutoff *et al.*, 1973; Schachtele *et al.*, 1973). Evidently, late ϕ29 mRNA is transcribed by a system involving host RNA polymerase, rather than by a totally phage-specified RNA polymerase. No *in vitro* system for late ϕ29 transcription has yet been established. This contrasts with the situation of the hmU-containing phages and suggests the possibility that the ϕ29 gene-*4* product may be something other than a simple analog of the SPO1 gene-*33* and -*34* products or of *B. subtilis* σ. Holland and Whiteley (1973) found that a portion of RNA polymerase was associated with a different subunit in ϕ29-infected *B. amyloliquefaciens*. However, the function of this enzyme fraction in selective ϕ29 late gene transcription is not known.

Sogo *et al.* (1976) have determined the extent of *in vivo* ϕ29 transcription by using separated strands of DNA restriction fragments and have shown the existence of symmetric transcription within most of the *Eco*RI-B fragment and in about one-third of the *Eco*RI-D fragment. However, since no early genes are known in these regions of the ϕ29 genome, the biological significance of such symmetric transcription is not understood at present.

The average chemical half-life of ϕ29 mRNA is about 10 min, whereas the average half-life of host mRNA is about 2 min, indicating that ϕ29 mRNA is four to five times more stable than host bacterial mRNA (Loskutoff *et al.*, 1973). On the other hand, the chemical stabilities of individual species of ϕ29 early mRNA, as examined by polyacrylamide gel electrophoresis, differ greatly (Kawamura and Ito, 1977c). Thus, larger RNA species are unstable, whereas smaller RNA species are considerably more stable. It is known that the functional half-lives of phage T7 and ϕX174 mRNAs are not much different from the half-life of host cell mRNA, although the former are chemically more stable (Yamada *et al.*, 1974; Lica and Ray, 1976).

1. *Mapping of ϕ29 Transcripts*

Loskutoff *et al.* (1973; Loskutoff and Péne, 1973) have analyzed early and late ϕ29 mRNA by sucrose gradient centrifugation and polyacrylamide gel electrophoresis. Their results showed that early mRNA consisted of six components ranging in molecular weight from 0.04×10^6 to 0.75×10^6, and that late mRNA consisted of three components (MW 1.75×10^6, 0.93×10^6, and 0.07×10^6). However, when specific DNA fragments, generated by restriction endonucleases, were used to detect ϕ29 mRNA, it was found that even the sharp single band of RNA in a polyacrylamide gel may contain more than one transcript (Kawamura and Ito, 1977a). With *Eco*RI-cleaved ϕ29 DNA fragments, as many as 13 early RNA bands could be resolved by polyacrylamide gel electrophoresis. It is not known however, which, if any, of these RNA species are primary transcripts and which are products of processing or of degradation.

Transcription of ϕ29 DNA by purified *B. subtilis* RNA polymerase holoenzyme yields early mRNA which hybridizes with the *Eco*RI-A and -C fragments (Fig. 3; Inciarte *et al.,* 1976a; Kawamura and Ito, 1977a; Sogo *et al.,* 1979; Davison *et al.,* 1979, 1980). Since the latter are derived from the ends of ϕ29 DNA (Ito *et al.,* 1976; Inciarte *et al.,* 1976b) and since there is no evidence indicating that ϕ29 DNA becomes covalently closed and circular in infected cells, these results suggest that there are at least two efficient initiation sites for early transcription of the ϕ29 genome. The work of Inciarte *et al.,* (1976a) on *in vitro* transcription of ϕ29 DNA by purified *B. subtilis* RNA polymerase showed that the transcription of *Eco*RI-A and -C fragments was initiated preferentially with GTP and to a lesser extent with ATP. Kawamura and Ito (1977a) have used gel electrophoresis to analyze the size distribution of RNA molecules synthesized *in vitro.* Using [γ-^{32}P]GTP or [γ-^{32}P]ATP they showed that three major RNA products with molecular weights of 1.6×10^6, 0.8×10^6, and 0.5×10^6 were initiated with GTP and that one small RNA product with a molecular weight of 001×10^6 was initiated with ATP. It was further shown that the largest RNA initiated with GTP hybridized with the *Eco*RI-A fragment, whereas the other larger RNAs hybridized with both *Eco*RI-A and -C fragments and a small RNA initiated with ATP hybridized with the *Eco*RI-A fragment exclusively (Kawamura and Ito, 1977a). Similar results were obtained by Davison *et al.* (1979, 1980) who referred to these transcripts as G1, G2, G3b, and A1a, respectively. These investigators have observed that, when transcription is performed at ionic strengths higher than 50 m*M,* an additional small RNA (MW 4.3×10^4), referred to as A1b, is synthesized with ATP. (The letters A and G represent the 5′-terminal nucleotides of ϕ29 mRNA.)

Davison *et al.* (1980) have found that, under the same reaction conditions, *Escherichia coli* RNA polymerase holoenzyme can generate four additional transcripts (designated G3a, G4a, G4b, and G5). Using specific restriction endonuclease fragments and exonuclease III-truncated DNA, Davison *et al.* (1980) constructed an *in vitro* transcription map of the ϕ29 genome (Fig. 4). Their results indicate that ϕ29 DNA contains three termination sites for early transcription and that *E. coli* RNA polymerase can initiate transcription at three sites (G4a/b, G1a/G3a, and G5) not efficiently utilized by *B. subtilis* RNA polymerase. They have also demonstrated that transcription of ϕ29 DNA and of T7 DNA by *E. coli* core polymerase either with *B. subtilis* or with *E. coli* σ factors results in the

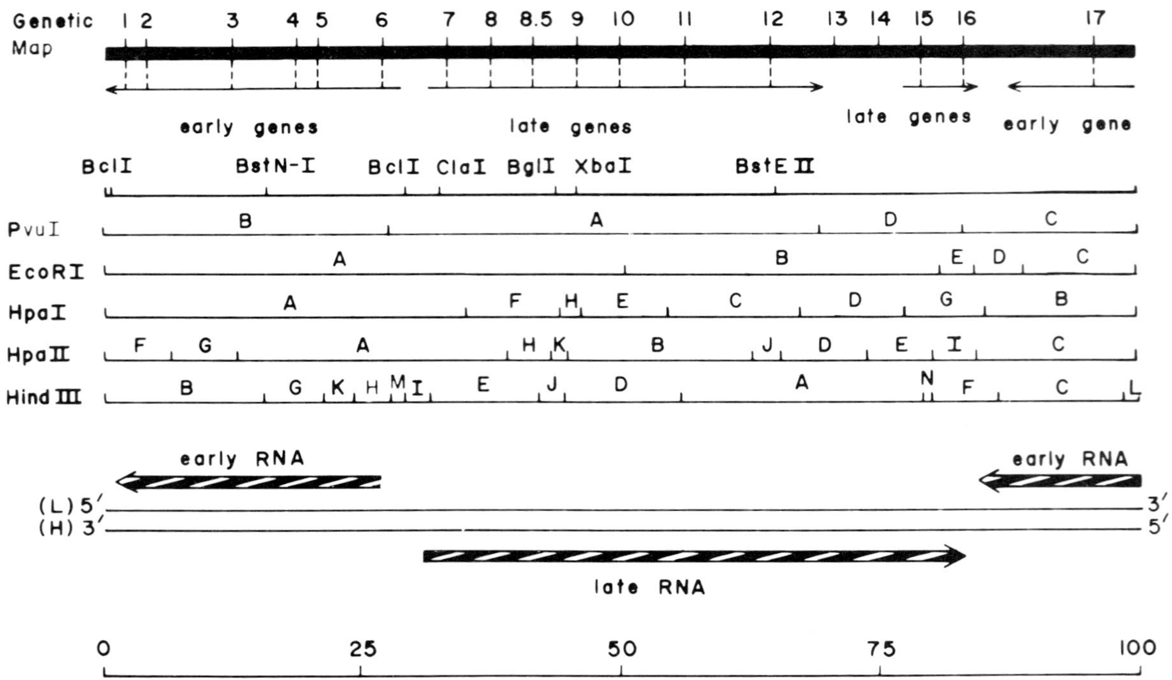

Fig. 3. Genetic, physical, and transcription (*in vivo*) maps of bacteriophage φ29. (Data compiled from Mellado *et al.* (1976); Reilly *et al.* (1977); Ito and Kawamura (1976); Ito *et al.* (1976); Inciarte *et al.* (1976); Kawamura *et al.* (unpublished); Kawamura and Ito, (1977a); Ito and Roberts (1979); Sogo *et al.* (1979); Davison *et al.* (1980); Yoshikawa and Ito (1981a).

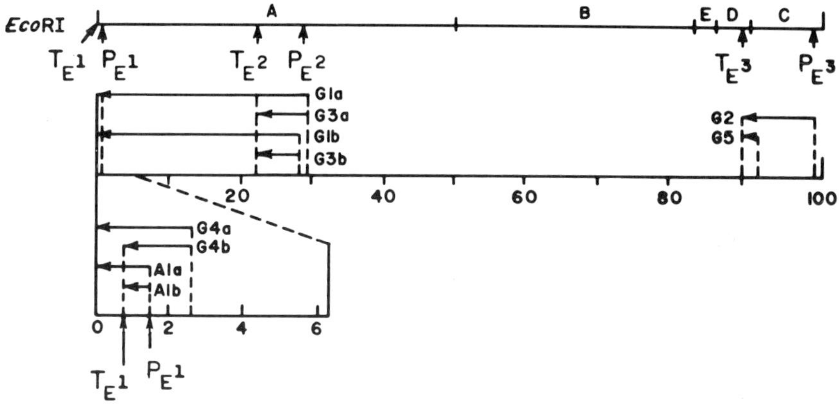

Fig. 4. *In vitro* transcription map of φ29. Data are from Davison *et al.* (1980). Designations of early promoters (P_E) and terminators (T_E) are tentative. P_E1 (A1a, A1b), P_E2 (G1b, G3b), and P_E3 (G2) are three major early φ29 promoters recognized by *B. subtilis* RNA polymerase.

synthesis of RNAs identical to those transcribed by *E. coli* holoenzyme. These results indicate an important role of polymerase core in promoter recognition and strength determination. In support of this notion they have shown that the efficiency of rifampicin-resistant complex formation on φ29 and T7 promoters is also dependent upon the source of RNA polymerase core (Davison *et al.*, 1980).

2. *RNA Polymerase-Binding Sites*

Sogo *et al.* (1979) have determined the specific interactions between φ29 DNA and *B. subtilis* RNA polymerase by electron microscopy. Seven specific binding sites have been visualized at positions 1.7 (±0.4), 25.5 (±0.5), 26.7 (±0.4), 59.4 (±1.2), 79.3 (±0.9), 91.3 (±0.6), and 99.3 (±0.4) (measured as percentage of total genome length from the left end). These sites have been designated A1, A2, A3, B1, B2, C1, and C2. Three sites (A1, A2, and A3) are located in fragment *Eco*RI-A, two (B1 and B2) in fragment *Eco*RI-B, and two (C1 and C2) in fragment *Eco*RI-C. Five of the locations of the initiation sites mapped by Davison *et al.* (1980) for φ29 early transcription correspond to these RNA polymerase-binding sites: G1a (A2), G2 (C2), G3b (A2), A1a (A1), and A1b (A1). Sogo *et al.* (personal communication) have also found that, under the same reaction conditions, *E. coli* RNA polymerase can bind at 13 specific sites on φ29 DNA. Six of them are located at the same positions as those found with *B. subtilis* RNA polymerase (only B2 is missing), and the remaining 7 appear as separate binding sites, 6 in the *Eco*RI-A fragment at positions 2.6 (A1I), 5.6 (A1II), 10.4 (A1III), 13.9 (A1IV), 25.8 (A2I), and 33.7 (A3I) and 1 in *Eco*RI-B at position 69.4 (B1I) (Sogo *et al.*, personal communication).

Dickel *et al.* (1980) determined which *Hin*dIII fragments of ϕ29 DNA were capable of binding *B. subtilis* RNA polymerase by making use of the face that ternary initiation and elongation complexes formed by RNA polymerase, a DNA fragment, and polynucleotide products are trapped on nitrocellulose filters (Hinkle and Chamberlin, 1972). Their results show that *B. subtilis* RNA polymerase forms initiation complexes with fragments *Hin*dIII-B, -C, -F, -H, and -L, in general agreement with the *in vitro* transcription map of Davison *et al.* (1980).

3. *Nucleotide Sequences of the ϕ29 Major Promoters*

As described above, the ϕ29 genome contains three strong promoters that are efficiently utilized by *B. subtilis* RNA polymerase. These are designated tentatively as P_E1, P_E2, and P_E3, respectively (Fig. 4). DNA fragments containing these regions have been isolated, and their nucleotide sequence have been determined (Yoshikawa and Ito, 1981b). As shown in Fig. 5, these sequences conform to the prototype sequence

Fig. 5A. ϕ29 major early promoter sequences and the *E. coli* promoter consensus sequence. All sequences are of the DNA strand parallel to mRNA (the coding strand). Data for the *E. coli* promoter consensus sequence are from Rosenberg and Court (1979), Siebenlist (1979), and Siebenlist *et al.*, (1980) and show the Pribnow box and the −35 region. The number of base pairs between the 3′-T of the Pribnow box and the initiating nucleotide varies from 4 to 8 (i.e., the Pribnow box occupies positions ranging from −10 to −5 at one extreme and −14 to −9 at the other). The number of base pairs between the 3′-A of the −35-region consensus sequence and the 5′-T of the Pribnow box varies from 16 to 18.

Fig. 5B. SPO1 early and middle promoter sequences. All sequences are of the DNA strand parallel to mRNA (the coding strand). Data are from Talkington and Pero (1979) and Lee *et al.* (1980b). Line 1: The consensus sequence of *E. coli* promoters repeated. Lines 2 and 3: An early promoter (P_E5) and a middle promoter (P_M) located in the 29–33 kbp region (fragment *Eco*RI-12). The latter sequence is written with U replacing T to denote the fact that this promoter functions *in vitro* only in hmU-containing DNA (Lee *et al.*, 1980a). Line 4: The consensus sequence of five middle SPO1 promoters (Lee *et al.*, 1980b). The number of base pairs between the 3′-end of the Pribnow-like box and the initiating nucleotide ranges from 4 to 6 (for four of the middle promoters). There are 15 or 16 base pairs between the 5′-U of the Pribnow-like box and the 3′-A of the upstream consensus sequence. Capital letters are used in cases of perfect concordance among the five promoters, and lowercase letters when the same nucleotide appears in four out of the five promoters; x denotes no concordance (or inadequate to satisfy the above criteria). Line 5: The 3′-end of *B. subtilis* 16 S RNA is complementary to the upstream consensus sequence of the SPO1 middle promoters.

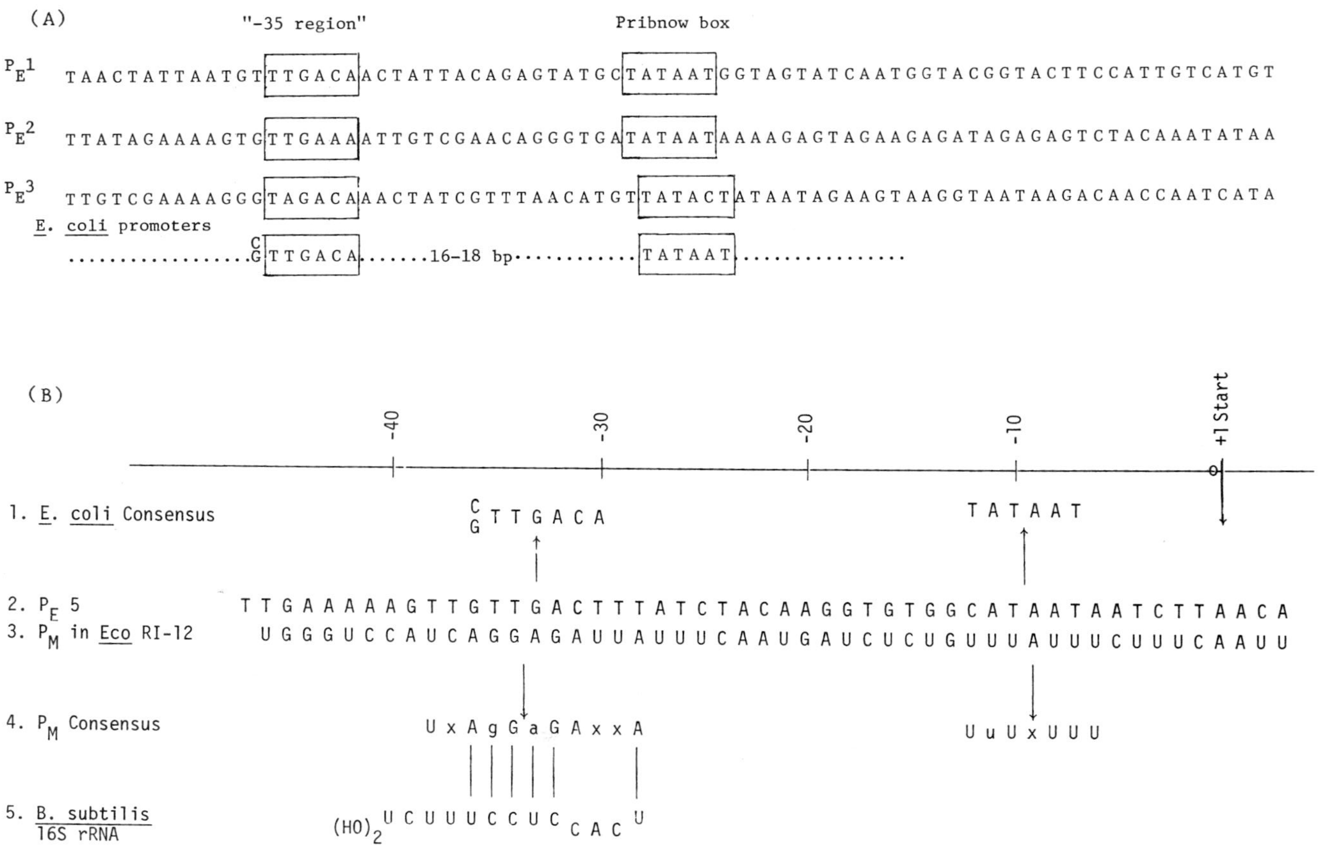

(A)
"-35 region"
Pribnow box
PE1 TAACTATTAATGTTTGACAACTATTACAGAGTATGCTATAATGGTAGTATCAATGGTACGGTACTTCCATTGTCATGT
PE2 TTATAGAAAAGTGTTGAAAATTGTCGAACAGGGTGATATAATAAAAGAGTAGAAGAGATAGAGAGTCTACAAATATAA
PE3 TTGTCGAAAAGGGTAGACAAACTATCGTTTAACATGTTATACTATAATAGAAGTAAGGTAATAAGACAACCAATCATA
E. coli promoters
.................C/G TTGACA.......16–18 bp............TATAAT................
(B)
-40
-30
-20
-10
+1 Start
1. E. coli Consensus
C/G T T G A C A
T A T A A T
2. PE 5
T T G A A A A A G T T G T T G A C T T T A T C T A C A A G G T G T G G C A T A A T A A T C T T A A C A
3. PM in Eco RI-12
U G G G U C C A U C A G G A G A U U A U U U C A A U G A U C U C U G U U U A U U U C U U U C A A U U
4. PM Consensus
U x A g G a G A x x A
U u U x U U U
5. B. subtilis 16S rRNA
(HO)2 U C U U U C C U C C A C U

TATAAT in the region frequently referred to as the "Pribnow box" and to TTGACA in the "−35 region" of the *E. coli* promoter (Rosenberg and Court, 1979; Siebenlist *et al.,* 1980). However, since the precise initiation sites of mRNA synthesis have not yet been determined, these sequences must be considered putative promoters of φ29. As described later, the nucleotide sequence of a phage SPO1 early promoter is also closely homologous to that of *E. coli* promoters (Talkington and Pero, 1979; Lee *et al.,* 1980b).

III. The Hydroxymethyluracil-Containing Phages

A. The Virion

The virions of the hmU phages have icosahedral heads and tails with elaborate baseplates (Eiserling and Romig, 1962; Truffaut *et al.,* 1970). The head of each SPO1 virion contains approximately 145 kbp of double-stranded DNA, which includes a terminal redundancy of 12.6 kbp, so that the complexity of the SPO1 genome is approximately 132 kbp. The DNA of the related SP82 and φe phages is of comparable size; SP82 has a terminally redundant segment of almost identical length (Lawrie, *et al.,* 1978b; Panganiban and Whiteley, 1981).

The recent work of Parker (1979) and Eiserling (1979) on SPO1 provides detailed information on the structure of the virion. Figure 6 shows an electron micrograph and an interpretive drawing of a SPO1 phage particle. The arrangement of morphological units on the surface of the 87-nm-diameter head suggests a $T = 16$ icosahedral surface lattice built of hexagonal morphological units. The primary component of this surface lattice is a 45.7-kdalton protein which is proteolytically cleaved from a slightly larger (47-kdalton) precursor. Proteolysis probably accompanies head maturation, as it does in phage T4 head assembly (Wood and King, 1979; Eiserling, 1979). The phage head and tail are connected by means of a neck, a connector structure composed of two washerlike disks, and a collar. Nineteen different proteins make up the head and neck.

The tail, with its central tube and contractile sheath, is 140 nm long; the sheath is composed of between 33 and 36 disks of protein so stacked on top of each other as to generate a sixfold helical array of sheath protein subunits. Six different proteins have thus far been identified in the tail and its sheath.

As is also the case with T-even phages, the SPO1 tailplate is by far the most complex subassembly of the virion. Composed of at least 28 proteins, it is seen in intact phage particles as a relatively ill-defined

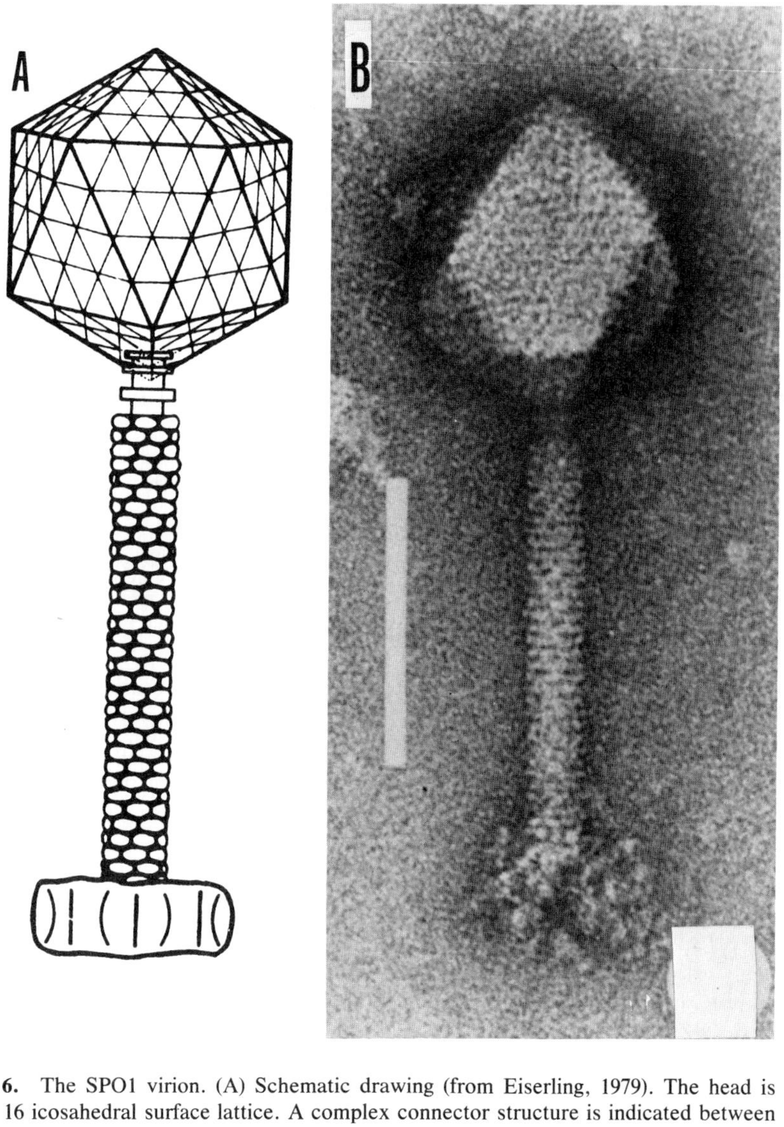

Fig. 6. The SPO1 virion. (A) Schematic drawing (from Eiserling, 1979). The head is a $T = 16$ icosahedral surface lattice. A complex connector structure is indicated between the head and tail. The tail subunit arrangement is derived fron analysis of electron micrographs made by Parker (1979). The baseplate structure is difficult to make out on extended tails but clearly has sixfold symmetry when converted to the expanded state after sheath contraction. (B) An electron micrograph (from Parker and Eiserling, 1981). The bar represents 100 nm.

structure approximately 25 nm thick and 60 nm wide. The tailplate, which is attached to the tailsheath, undergoes a major structural transition as the tail sheath contracts and is then seen as a complex, hexagonal double-plate structure approximately 33 nm thick and 90 nm wide (Eiserling, personal communication).

Parker has counted 53 different polypeptides in all the subassemblies of the SPO1 virion. The sum of the molecular weights of these proteins is approximately 2.2×10^6, requiring a corresponding coding complexity of 60 kbp of DNA. If there are no overlapping genes, and also making no allowance for control sequences, this corresponds to 46% of the complexity of the SPO1 genome (taken to be 132 kbp). Compared with φ29, it is clear that a substantial fraction of the eightfold greater complexity of the SPO1 genome is dedicated to the construction of a much more complex virion.

B. Genome Organization

Genetic maps have been constructed for phages SP82 and SPO1 (Green and Laman, 1972; Okubo *et al.*, 1972). Neither map is detailed. A partial physical genetic map of SPO1 having also been constructed, we select the latter for presentation and comment.

Okubo and collaborators located 34 SPO1 genes on a linear genetic map, primarily by two-factor crosses but with some three-factor crosses. They were unable to map mutations in two genes because these were unlinked from all others. Okubo and co-workers acknowledged the provisional nature of their map and cautioned that their assignments of gene order were tentative for several genes. They pointed to several sources of error including (1) dependence of recombination frequencies on total multiplicity of infection at a constant phage input ratio, (2) poor additivity of recombination frequencies in neighboring intervals, and (3) high negative interference. Based on argument by analogy with what was then known about phage T5, it was later suggested that the unlinked SPO1 genes might be located in a terminally redundant region (Rabussay and Geiduschek, 1977); such a region has, indeed, been shown to exist (Cregg and Stewart, 1978b). More recently, a restriction map has been constructed (Pero *et al.*, 1979), and mutations in several genes have been mapped to restriction fragments (Cregg and Stewart, 1978a; Stewart, personal communication). This genetic information is collected in Figs. 7 and 8. Figure 7 presents a restriction map of the SPO1 genome. The terminal redundancy of the linear DNA is accommodated in a circular map in which one end of the genome is placed at 12 o'clock and the other end of the genome between 1 and 2 o'clock. A less complete

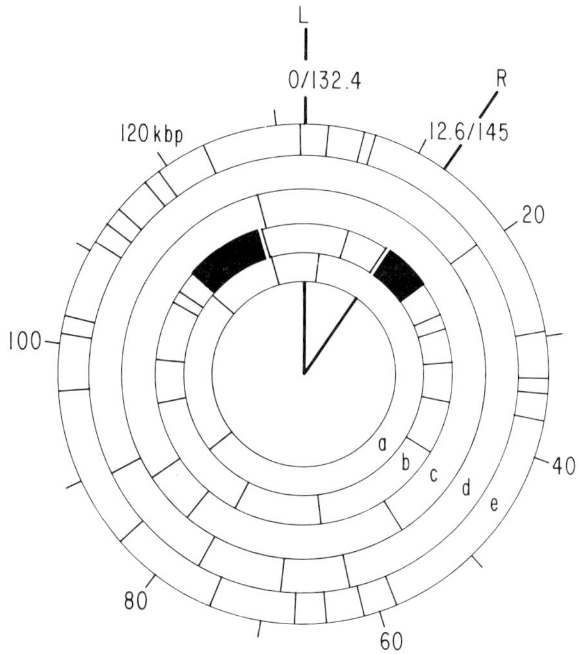

Fig. 7. Restriction map of the SPO1 genome. The terminally redundant, *linear* 145-kbp genome is represented as 1.095 clockwise turns around a cricular map starting at L (0/ 132.4 kbp) and finishing at R (12.6/145 kbp). Data for *Hae*III (ring a), *Sal*I (ring c), *Bgl*II (ring d), and *Eco*RI (ring e) are taken from Pero *et al.* (1979) with a redetermination of the sizes of certain large fragments based on further mapping (B. Chelm, personal communication). Data for *Kpn*I (ring b) are from Romeo *et al.* (1981), Chelm *et al.* (1981), and Chelm (personal communication). Additional *Kpn*I sites in the region of 12.3–24.3 and 111.2–126.4 kbp (shaded in the figure) have not yet been mapped.

restriction map for SP82 DNA has also been published (Lawrie and Whiteley, 1977). The genetic map is presented in two versions in Fig. 8: a conservative version containing only the primary physical genetic mapping data (ring I), and a version incorporating the recombination map of Okubo *et al.* into the physical map to provide estimates of the placements of other genes (ring II). Mutants in several complementation groups (*2, 7, 21, 23,* and *30*) can be subdivided on the basis of weak complementation (*2a–e, 7a* and *b, 21a* and *b, 23a* and b, *30*a–c; Okubo *et al.,* 1972). There is evidence that complementation groups *30a* and *c* code for different polypeptides (Reeve *et al.,* 1978); however, we guess that one is dealing here at least partly with intragenic complementation.

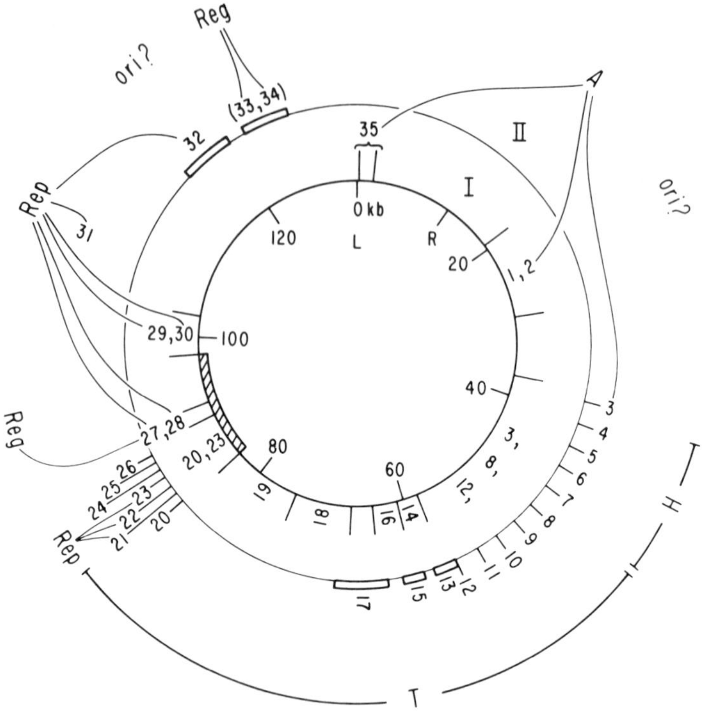

Fig. 8. A genetic map of phage SPO1. Ring I: Physical map of phage SPO1 genes primarily based on marker rescue–transfection experiments with *Eco*RI restriction fragments (Cregg and Stewart, 1978a). Gene *28* has been mapped by *in vitro* protein synthesis with restriction fragments to the 89–91 kbp region (Chelm *et al.*, 1981). Genes *20, 23,* and *27* have been mapped by marker rescue into the *Eco*RI fragment at 84–99 kbp (indicated by the shaded bar). The position of gene *14,* relative to genes *12, 16,* and *18* is the most probable location on the basis of recent, unpublished data of Stewart (personal communication) and differs from the original assignment. Ring II: Genetic map of phage SPO1 based on the recombination analysis of Okubo *et al.* (1972) and auxiliary information. In the region of 37.0–58.5 kbp, which holds genes *3, 8,* and *12* (ring I), genes *3–12* have been arbitrarily positioned at regular intervals. Genes *13, 15,* and *17* are located by interspersal between the closest physically mapped genes. The gene order *13, 14, 15, 16, 17, 18* is the most readily compatible with the three-factor crosses of Okubo *et al.* (1972). In the region of genes *20–28,* the order of genes *20, 23, 27,* and *28* is not fixed by the physical mapping shown in ring I. The placement of genes *20–28* is predicated on the gene order specified by Okubo *et al.* (1972) who regarded this assignment as tentative. Since gene *28* is located at 89–91 kbp, this gene order positions genes *20–27* in the 84–91 kbp segment. If the genes are correctly placed, then there must be considerable map expansion (i.e., higher recombination frequency) in the gene *20–28* region (84–91 kbp) relative to the gene *3–12* region (37–59 kbp). Genes *32, 33,* and *34* are placed on the map on the basis of the replication gradient analysis of Glassberg *et al.* (1977a) and of certain assumptions. The order of genes *33* and *34* is predicated on the data of Okubo *et al.* (1972). Genes *2, 7, 21, 23,* and *30* can each be subdivided into two or more complementation groups (see text). As noted in the

The defective morphogenesis of SPO1 mutants has been examined by Fujita (1971), Parker (1979), and Eiserling and Beard (unpublished observations) as a way of assigning normal morphogenetic functions to various phage genes. As is the case for so many other phage genomes, including φ29, related functions are clustered on the SPO1 genetic map (Fig. 8). Mutants affected in genes *4–20* are defective in head, tail, or baseplate morphogenesis. Mutants affected in genes *1–3* and *35* make all these subassemblies but fail to join heads to tails, fill heads, or otherwise completely traverse the morphogenetic pathways. Several of the originally isolated mutants in genes *2* and *3* are abnormal in DNA synthesis (Fujita, 1971; Okubo *et al.*, 1972), but one should be careful to note that these observations were made on original mutant isolates which might contain secondary mutations, rather than on extensively backcrossed stocks. Genes *3–20* are located on restriction fragments that cover approximately 45 kbp of SPO1 DNA, a large fraction of the 61 kbp that we have already estimated as required to code the hitherto identified proteins of the virion. The task of assigning SPO1 virion proteins to genes has been somewhat neglected up to this time.

The morphogenetic defects of phage SP82G *ts* mutants (Kahan, 1966, 1971) have been analyzed by Green and Laman (1972). Genes responsible for head formation (H), tail formation (T) and DNA replication (D) are linked in the order H-T-D (Hemphill and Whiteley, 1975), as they are in SPO1.

Phage SPO1 mutants in two gene clusters (genes *21–23* and *27–32*) are absolutely defective in synthesizing viral DNA. Mutants in two of these genes, genes *27* and *28*, are defective in the program of viral gene expression, as are mutants in genes *33* and *34* (Section III,E,2).

C. Genetic Relatedness of the Hydroxymethyluracil-Containing Phages

The genomes of the SP82, φe, and SPO1 phages are approximately the same size (Lawrie *et al.*, 1978). SP82 and SPO1 have terminally

---

text, while the circular representation is convenient for the terminal redundancy, the linkage map of genes *1–34* is linear.

Perimeter: Gene functions. The symbol "H" indicates genes whose products are involved in phage head morphogenesis. Similarly, "T" indicates tail and baseplate morphogenesis, "A" indicates unspecified steps in phage assembly (Fujita, 1971; Eiserling and Beard, 1972; Parker, 1979; Parker and Eiserling, 1981), "Rep" indicates DNA replication (Okubo *et al.*, 1972), and "Reg" indicates regulation of gene expression (Fujita *et al.*, 1971; Greene *et al.*, 1982; Heintz and Shub, 1981). Putative origins of replication (indicated as "ori?") are also shown (Glassberg *et al.*, 1977a).

redundant regions of almost identical length (Panganiban and Whiteley, 1981). It has just been mentioned that the H, T and D genes of SP82 and SPO1 are linked in the same order. Recent experiments further show that five individually corresponding D genes of SP82 and SPO1 are in the same order (Stewart and Franck, 1981). Thus, it appears likely that these two phages share a common plan of genome organization. A hybridization analysis of the hmU phages, performed several years ago (Truffaut *et al.,* 1970), showed SP8, SP82, ϕe, 2C, and SPO1 to be closely homologous. More extensive comparison of SPO1, ϕe, and SP82 shows appreciable nucleotide sequence divergence between SPO1 and SP82 and more between ϕe and both of the others. A comparative restriction enzyme analysis of these genomes, although incomplete, clearly shows that, of the 65–75 restriction sites for the *Hpa*I and *Hpa*II endonucleases, few, at best, are conserved in *all three* of these genomes (Lawrie *et al.,* 1978).

In mixed infections by phages SPO1 and SP82, the latter is predominant in the progeny. The region of the chromosome responsible for this property is located near gene *32,* where one of the origins of DNA replication is also located. Similar dominance relationships exist between T-even phages (Stewart and Franck, 1981).

D. General Features of Viral Development

The time scale of the developmental cycle of these phages, like that of the other bacteriophages, is in minutes: eclipse periods of about 25–35 min and latent periods of about 40–50 min at 37°C (Okubo *et al.,* 1964), but with significant variations possible if, for example, cell densities are greatly changed (Parker, 1979).

Host DNA replication is shut down soon after infection, but the bacterial DNA is not degraded (Lavi and Marcus, 1972; Wilson and Gage, 1971). Host protein synthesis is greatly decreased soon after infection, further decreasing progressively thereafter, but some host proteins continue to be made throughout infection (Shub, 1966; Lawrie *et al.,* 1978; Parker, 1979). Host rRNA synthesis continues for several minutes after infection but declines strongly at about the onset of viral DNA replication (Shub, 1966; Cocito, 1974).

Phage DNA replication starts about 10 min after infection (at 37°C). SPO1 DNA contains the abnormal pyrmidine, hmU, in place of thymine, and phage-coded enzymes of nucleotide biosynthesis are made after infection, as is a new DNA polymerase (Roscoe and Tucker, 1964, 1966; Nishihara *et al.,* 1969; Alegria *et al.,* 1968; Aposhian and Tremblay, 1966; Roscoe, 1969; Price *et al.,* 1972; Yehle and Ganesan, 1972; reviewed by Hemphill and Whiteley, 1975). Changes in the resistance to a DNA gyrase-inhibiting antibiotic (Sugino *et al.*, 1977) after infection (Gage and Fujita, 1969; Sarachu *et al.*, 1980) suggest that a virus-coded topoisomerase or virus-specific components of a topoisomerase may also be made after infection. SPO1 DNA is replicated in such a way as to generate linear concatmers from which mature viral DNA is packaged into virions (Levner and Cozzarelli, 1972). Gene *35*, which is located in the terminal redundancy of mature virion DNA, appears not to be redundant in concatemeric DNA. This suggests that *all* SPO1 genes might be unit copy in concatemeric DNA. In this respect, SPO1 would resemble T7 phage, for example (Cregg and Stewart, 1978b). What distinguishes the two phages is the size of the terminally redundant stretch of DNA. Short terminal redundancies like those in the T7 genome are thought to have a role in the replication of linear DNA via concatemeric intermediates (Watson, 1972). However, it would be astonishing if the hmU phages' enormous terminal redundancy of over 12 kbp fulfilled only this role.

In order to locate the sites on the viral genome at which DNA replication is initiated, Glassberg *et al.*, (1977a) performed density shift experiments during the first minutes of SPO1 DNA replication and combined these with marker rescue analysis of density-labeled DNA. They concluded that there were two origins of replication, located near the opposite ends of the unique portion of the genome (in the vicinity of genes *2* and *32*, respectively; Fig. 8).

E. Regulation of Transcription in Virus Development

1. *The Temporal Sequence*

The genome of phage SPO1 provides a relatively complex temporal sequence of gene expression (Gage and Geiduschek, 1967, 1971a,b). It is clear that the genetic regulation which generates this program operates primarily at the level of transcription. Three major phases of gene expression—early, middle, and late—are readily distinguished. Early gene expression starts within a minute of infection, but the activities of different early genes are turned off at different times: Some are repressed within 3–4 min (at 37°C), whereas others continue to be active until about the onset of DNA replication. These genes are designated *e* and *em*, respectively. Subsequent viral gene expression is positively controlled by proteins that must be made during the infectious cycle. *Middle* genes are first activated about 4 in after infection. Some of the middle genes are again repressed at about the time of starting DNA replication, and others continue to be active until the end of the viral eclipse period (i.e., until there is, on the average, one complete assembled virion per infected cell). These middle genes are designated *m* and m_1l, respectively. The third, *late* phase of gene expression involves genes that are turned on asynchronously. Two classes of late genes have been distinguished on the basis of the time of first appearance of their conjugate transcripts and on the basis of the transcriptional properties of certain mutants, which will be discussed below. These two classes of genes have been designated m_2l and *l*, respectively. The m_2l transcripts first appear about the time that viral DNA replication starts (10 min after infection at 37°C). The *l* transcripts first appear a few minutes later. It should be noted at the outset that this classification, which distinguishes two classes of late genes, is not generally agreed upon (see also Talkington and Pero, 1977).

Analysis of protein synthesis in phage SPO1-infected cells yields a sequence that can be accommodated reasonably well in this scheme (Parker, 1979; Heintz and Shub, 1981), although the late proteins evidently appear in a more staggered temporal sequence than the late mRNA. It is, however, important to state several qualifications: (1) The resolution of virus-coded proteins in published work has been limited by the use of one-dimensional gels; (2) residual host protein synthesis further limits the ability to detect less abundant viral proteins, and the available methods for eliminating the host genome's activity introduce their own complications (see below); (3) the resolution of RNA synthesis has been limited by analytical methods yielding only global categories; (4) in order to connect protein and RNA synthesis precisely, one must know something about translational efficiency and messenger lifetime. What is clearly required here is a much more detailed analysis based on a precise mapping of proteins and their conjugate transcripts. On balance, however, the currently available analyses of protein and RNA synthesis give a generally consistent picture of the sequence of viral gene action during the infectious cycle and thus do not point to any elements of posttranscriptional regulation.*

The problem of continuing host protein synthesis during phage infection can be overcome in two ways: *either* by infecting *B. subtilis* minicells which then make only phage-coded proteins (Reeve and Cornett, 1975), or by UV-irradiating the host bacteria before infection (Heintz and Shub, 1981). Neither method permits progeny phage to be produced, and each distorts gene expression. Not a single abundant SPO1 structural protein is synthesized in minicells (Reeve *et al.*, 1978). Apparently, *E. coli* and

* Posttranscriptional regulation of certain early proteins has been postulated on the basis of experiments with minicells (Reeve *et al.*, 1978), but there are alternative interpretations of the data.

B. subtilis minicells also lack "regulatory fidelity" in the sense that they change quantitative relations in the production of different virus-coded proteins and, more importantly, that they do not invariably obey the same regulatory rules as complete, chromosome-bearing cells (Mertens *et al.*, 1979; Magazin *et al.*, 1978). Much is ultimately to be learned from these relationships between normal cells and minicells but, conversely, the interpretation of minicell experiments should be approached with caution. Ultraviolet-irradiated *B. subtilis* do not replicate SPO1 DNA and, although less obviously defective in late protein synthesis than minicells, fail to make normal quantities of a small number of late proteins.

Be that as it may, infection with phage SPO1 wild type and with mutants that are blocked at different stages of the viral gene expression sequence permits the identification of early, middle, and late proteins. About 20 early proteins have been found, with an aggregate molecular weight of approximately 500,000, implying a coding capacity of about 13.5 kbp of DNA (Reeve *et al.*, 1978; Heintz and Shub, 1981). Heintz, and Shub (1981) detected 9 middle proteins with an aggregate molecular weight of approximately 400,000, whereas Reeve and co-workers assigned 19 proteins, with an aggregate molecular weight of 970,000, to this class. Both may well represent underestimates of the complexity of middle genes. Heintz and Shub resolved 16 late proteins with an aggregate molecular weight in excess of 700,000. (It should be recalled, from Section III,A, that 53 polypeptide chains have thus far been identified in the virion.)

The sequence of RNA and protein synthesis during phage SP82 infection has also been analyzed (Lawrie *et al.*, 1976, 1978; Hiatt and Whiteley, 1978). In general the observations on SPO1 and SP82 are similar. The temporal classification scheme constructed for SP82 differs in certain details from that constructed for SPO1, but we shall not dwell on these differences.

At least some features of the temporal sequence of gene expression during phage multiplication also depend on the particular conditions of infection. Parker (1979) found that he could introduce very considerable variations in time sequence, particularly for late proteins, when he raised the concentration of infected cells above certain limits. The timing of certain aspects of early and middle transcription might also, for example, depend on multiplicity of infection, but this remains to be investigated.

2. *Regulatory Genes*

The incomplete genetic analysis of phage SPO1 has thus far identified four genes whose products are required for the activation of middle and late genes. Gene products (gp) 33 and 34 are required for late (m_2l and l) transcription, and gp 28 is required for middle transcription (m and m_1l; Fujita *et al.*, 1971). Crude extracts of SPO1 phage-infected cells can be programmed with mature DNA to execute positively regulated transcription (Grau *et al.*, 1971). RNA polymerase bearing *de novo* synthesized subunits can be purified from SPO1- and SP82-infected extracts and shown to execute positively regulated transcription (Duffy and Geiduschek, 1973, 1975; Spiegelman and Whiteley, 1974a,b; Lawrie *et al.*, 1975; Fox and Pero, 1974; Pero *et al.*, 1975a,b; Swanton *et al.*, 1975; Shub *et al.*, 1979). Indeed, gp 28, 33, and 34 are RNA polymerase-binding proteins (Fox and Pero, 1974; Fox *et al.*, 1976; Fox, 1976) which change the DNA-binding and RNA chain initiation properties of the *B. subtilis* RNA polymerase core (Duffy and Geiduschek, 1976; Tjian & Pero, 1976; Talkington and Pero, 1978). RNA polymerase bearing gp 28 selectively transcribes SPO1 middle genes (Duffy *et al.*, 1975; Duffy and Geiduschek, 1977) and selectively binds to fragments of SPO1 DNA containing middle genes (Talkington and Pero, 1978). RNA polymerase bearing gp 33 and 34 selectively binds to and transcribes a set of SPO1 DNA fragments, several of which are associated with late genes (Talkington and Pero, 1978). This clearly signifies that the phage-specific modifications of the host's RNA polymerase determine changes in transcription specificity by changing DNA binding.

Similar experiments have been done with RNA polymerase isolated from phage SP82-infected bacteria. RNA polymerase isolated from uninfected bacteria and from bacteria taken during the middle and late phases of viral transcription binds to restriction fragments of SP82 DNA that are associated with early, middle, and late transcription (Spiegelman and Whiteley, 1978).

Quite recently, gp 27 has also been shown to be required for late transcription (unpublished experiments referred to by Heintz and Shub, 1981; Greene *et al.*, 1982). It is known that the 20-kDalton gp 27 is required for DNA replication. This dualism reminds one of the T4 gp 45 that plays a role in replication and transcription (Wu *et al.*, 1975). The 24-kDalton T4 gp 45 is known to interact with RNA polymerase (Coppo *et al.*, 1975) and with a complex of other T4 replication proteins (Liu *et al.*, 1979). T4 gp 45 is associated with a DNA-dependent ATPase (for review, see Liu *et al.*, 1979). In contrast, essentially nothing is known so far about the biochemical properties of SPO1 gp 27.* However,

* It is worth noting that Glassberg and collaborators (1977b) have isolated temperature-sensitive mutants in SPO1 gene *27* that are not absolutely defective in DNA replication. Since these mutants are defective in phage production, they might contain an altered gp 27 which at least partly preserves its replication functions but lacks its transcription functions.

it is likely that gp 27 is, in fact, a replication protein: Middle transcription appears to be normal in the absence of gp 27, so the replication-defective phenotype of the gene-*27* mutant is not a priori likely to be due to pleiotropy. In contrast, the effect of gp 28 on DNA replication is likely to be indirect, because of its regulation of the synthesis of replication proteins. Thus, the products of the closely linked genes *27* and *28* apparently affect viral gene expression in different ways.

In comparison with the smaller ϕ29 phages, we have here a more differentiated temporal sequence of gene expression in which several phage-coded proteins act as positive regulators of transcription—for ϕ29 only one, the gene-*4* product, has thus far been implicated.

3. *Mapping Viral Transcripts*

Before taking the discussion of transcriptional regulation further, we shall review what is known about the physical location of viral transcripts. Consider, first, the results of Southern blot hybridization with RNA labeled at various times after infection of *B. subtilis* by wild-type and regulatory mutant phages (Talkington and Pero, 1977). Such an analysis provides a general picture of the transcriptional program. The distribution of hybridized radioactive RNA among the bands of the Southern blot represents the relative transcriptional activity of the conjugate segments of the genome, subject to certain practical limitations. First, the Southern analysis averages transcription; low transcriptional activity of an entire large fragment and high transcriptional activity of only a small section of this fragment give similar signals. Second, continual activity of a large DNA fragment throughout infection is not distinguished in a simple Southern blot analysis from compensatory turning on and turning off of different subsegments of this fragment. Third, equally effective transfer of all DNA fragments of the digest must be ensured (Wahl *et al.*, 1979). Nevertheless, a qualitative picture readily emerges (Fig. 9). Early transcription predominantly comes from the redundant extremities of the genome, but several restriction fragments located between 84 and 99 kbp and between 115 and 132 kbp hybridize weakly with early RNA, as does one fragment covering the 29–33 kbp region. These very weakly transcribed segments include the region of the early (*e*) regulatory gene *28* (located near 90 kbp).

A group of *e* genes must be located in the region between 2 and 6 kbp, because the transcription of this part of the terminal redundancy clearly is repressed after just a few minutes of infection. Middle and late transcripts come from different regions of the DNA. Part of the middle RNA hybridizes very strongly with certain restriction fragments that hybridize weakly with early RNA (e.g., in the region of 84–89 and

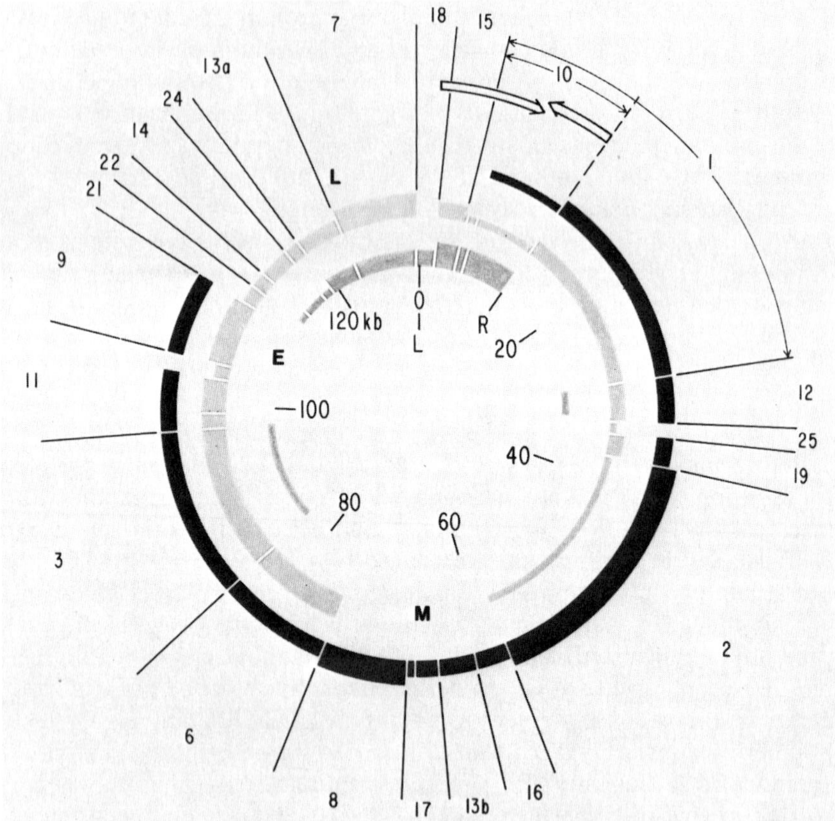

Fig. 9. Transcription map of phage SPO1. Southern hybridization of RNA from infected bacteria to *Eco*RI* digests of SPO1 DNA provided the data for this analysis (Talkington and Pero, 1977). The inner, middle, and outer rings represent early, middle, and late transcription, respectively. In each ring, the thickness of the shaded band represents the relative hybridization of radioactive RNA with different restriction fragments of DNA. Hybridization is *not* corrected for the size of the DNA fragment and thus represents the aggregate transcription rates of DNA fragments of *unequal* size. Within this limitation, low, intermediate, and high levels of transcription are qualitatively distinguished by the three different bandwidths. The polarity of early transcription in the 1.4–12.4 kbp region is also shown (Brennan *et al.*, 1981). The *Eco*RI fragments are numbered according to Pero *et al.* (1979).

123–132 kbp) and may therefore contain *em* genes. Some restriction fragments hybridize with middle RNA but not early RNA (e.g., in the region of 75–82 and 99–112 kbp) and must contain *m* or $m_l l$ genes. Similarly, some of the late RNA hybridizes with restriction fragments that are apparently not transcribed during the early or middle periods (e.g., in the region of 58–75 kbp), whereas much late RNA hybridizes with restriction fragments that are relatively weakly transcribed during the middle phase of transcription (e.g., in the region of 37–58 kbp). Some restriction fragments apparently hybridize with RNA labeled at all stages of viral development (e.g., in the region of 84–98 kbp). Of course, this does not prove that there is continued activity of individual units of transcription.*

By and large, the Southern analysis with RNA from nonpermissive cells infected with regulatory mutant phage or in the presence of protein synthesis inhibitors confirms what the temporal analysis shows: RNA extracted from cells infected with gene-*28* mutant phage or infected with wild-type phage in the presence of chloramphenicol is early RNA; cells infected with gene-*33* or -*34* mutants make middle but not late RNA. Two intriguing complications also appear.

1. Cells infected with a gene-*28* mutant or with wild-type phage in the presence of chloramphenicol ultimately make RNA that hybridizes weakly with a DNA restriction fragment (*Eco*RI-12; 29–33 kbp) normally not transcribed until the middle period of transcription. Possibly, a small part of transcription that is exclusively mediated by host components (and in that more fundamental functional sense is early transcription) is sufficiently delayed that it appears together with positively regulated middle RNA in the normal sequence of viral gene expression.†
2. RNA samples from cells infected with a gene-*33* and a gene-*34* mutant have slightly different Southern blot patterns, suggesting that the absence of these two RNA polymerase-binding proteins might generate distinguishable defects.§ The temporal program of SP82 transcription (Lawrie *et al.,* 1976) has also been reanalyzed by Southern blot hybridization (Lawrie *et al.,* 1977; Whiteley, 1981) with similar findings: During the first minutes of infection, transcription predominantly comes from the terminally redundant segments of the genome.

More precise transcriptional maps are under construction but are still far from complete. Detailed mapping of early *in vitro* synthesized SPO1

* However, continuously generated SP82 transcripts have been detected by hybridization–competition (Lawrie *et al.,* 1976).

† A similar kind of complication also arises in analyzing phage T4 early and middle gene expression. Because early and middle T4 transcription units overlap so extensively, and because the genetic control of T4 transcription is relatively complex (indeed, not yet completely worked out), it has proven very difficult to distinguish time-delayed early from middle T4 gene expression.

§ Another parallel can be seen here with phage T4: Genes *33* and *55* of T4 code for RNA polymerase-binding proteins (Horvitz, 1973; Ratner, 1974) which are required for late gene expression. However, mutations in genes *33* and *55* apparently do not equally affect T4 late transcription (unpublished experiments referred to in Rabussay and Geiduschek, 1977).

transcripts and polypeptides has been undertaken (Brennan *et al.*, 1981; Chelm *et al.*, 1981; Heintz and Shub, personal communication; Romeo *et al.*, 1981). As already mentioned, the early transcripts come predominantly from the terminal redundancy. There, the transcriptional pattern is quite remarkable. Thirteen promoters are distributed over approximately 10 kbp of DNA in a convergent array of overlapping transcription units (Fig. 10). Eleven of these promoters are very strong, though not equally so; P_E1 is weaker, and its neighbor, P_E1', is quite weak. Transcripts from these promoters converge on strong termination sites located near 8.2 kbp. Subsidiary, weaker termination sites generate a complex set of overlapping *in vitro* transcripts, some of which are up to 7 kb long. *In vivo* transcripts of comparable length have never been found (Imamoto, 1970; Reeve *et al.*, 1978). This obviously implies one or both of the following: (1) that mRNA is posttranscriptionally processed and (2) that termination of transcription is more effective *in vivo* than it is *in vitro* and possibly utilizes additional sites not detected with pure RNA polymerase holoenzyme.

What might be the function of such a very high promoter concentration in the terminal redundancy? Both termini of the SPO1 genome are likely to become available for transcription very soon after infection. One is

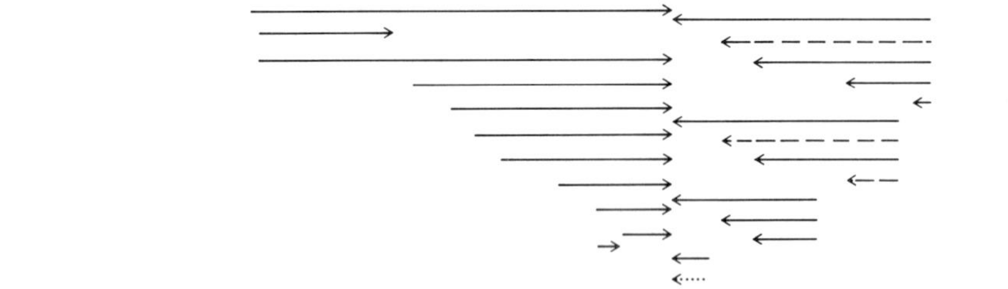

Fig. 10. Map of the major SPO1 early transcription units. The locations of promoters are shown above the line, and sites for terminating transcription *in vitro* are shown below the line. Termination at sites H1, L1.5, L2, and L2.5 is only partly effective; termination at H2 and L3 occurs only at low concentrations of ribonucleoside triphosphates. The collection of transcripts generated by these signals is shown in the bottom part of the figure. The dashed lines designate minor transcripts whose existence is inferred but which have not been specifically identified. The dotted line indicates a transcript whose termini are not precisely mapped. Data taken from Brennan *et al.* (1981).

drawn to this conclusion by knowing that the SP82 genome injects relatively rapidly, taking only 1.4 min to inject extremal markers on the genome at 33°C; at 37°C injection probably takes about 1 min (McAllister, 1970). The high density of promoters in the terminal redundancy would allow these transcription units to compete very effectively for transcriptional activity in the newly infected cell and to reach their transcriptional steady state very rapidly (as we have already pointed out, transcription of the host genome is not irreversibly shut off after infection with these phages). When fully loaded with RNA polymerase molecules along its entire terminal redundancy, a single genome might sop up 250–500 RNA polymerase molecules.

The early (*in vitro*) transcripts from the SP82 terminal redundancy have been mapped (Panganiban and Whiteley, 1981) in sufficient detail to allow one to see the similarity of organization of SPO1 and SP82 transcription units. In SP82, as in SPO1, there is convergent transcription to a bidirectional termination region.

Weaker *B. subtilis* RNA polymerase binding sites and promoters are located in other regions of the SPO1 genome (Chelm *et al.*, 1981). In the presence of limiting quantities of RNA polymerase, these segments of the genome are weakly transcribed *in vitro* unless the strong-promoter-laden terminal segments of the DNA are removed. A detailed analysis of the 91–113 kbp region has located three such binding sites (Fig. 11). *Bacillus subtilis* RNA polymerase also binds at two additional sites in this region but does not readily initiate RNA chains there. The relationship of the three early transcription initiation sites near 91, 99, and 113 kbp to *in vivo* transcription has not yet been established by high-resolution mapping but might be reconciled as follows: If the sites near 91 and 99 kbp were used for transcription *in vivo,* they would primarily yield RNA that hybridizes with the *Eco*RI-3 restriction fragment; the latter shows weak early transcription activity (Fig. 9). The promoter near 113 kbp yields transcripts *in vitro* that run off at (i.e., extend beyond) the right-hand end of Fig. 11 and might mainly hybridize with *Eco*RI-22 and possibly (clockwise in Fig. 9) beyond.* Hybridization–competition experiments with *in vivo* and *in vitro* RNA synthesized from the early initiation site near 113 kbp show that this segment does, indeed, generate *em* RNA *in vivo,* which first appears after other early transcripts. It is probably significant that RNA associated with a weak early promoter should first appear some time after the start of early viral transcription. This is what one would expect if there were strong competition for RNA polymerase molecules by SPO1 promoters at the outset of the infection.

* However, on this basis, low-level hybridization of *in vivo* early RNA with *Eco*RI-11 and -*21* might have been expected but has not been reported (Fig. 9).

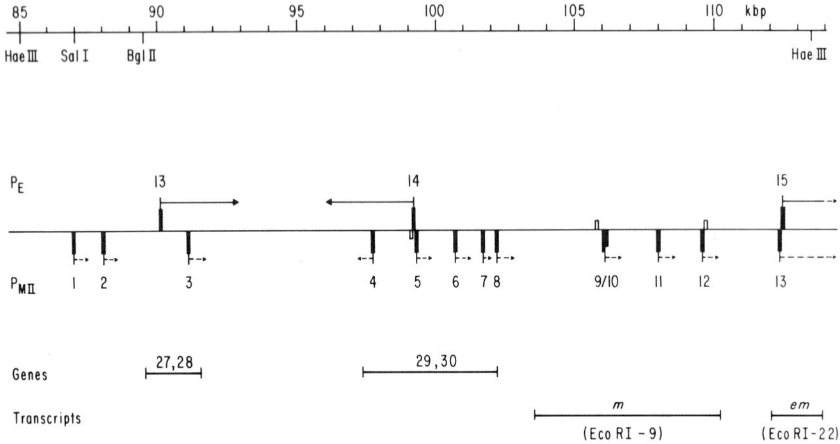

Fig. 11. Map of early and middle transcription in the 88–114 kbp region of the SPO1 genome. Sites at which *B. subtilis* RNA polymerase holoenzyme (Eσ) binds to SPO1 DNA are shown above the line. Initiation of RNA synthesis occurs at sites indicated by closed boxes. *Bacillus subtilis* RNA polymerase also binds at the sites indicated by open boxes (near 106 and 110 kbp), but synthesis of RNA is not efficiently initiated there. Sites at which E.gp 28 RNA polymerase, containing gp 28 in place of σ, binds to SPO1 DNA and initiates RNA synthesis are shown below the line. Polarities of transcription are indicated by arrows. Hybridization–competition experiments referred to in the text show that the segments designated at the bottom right of the figure yield *em* and *m* transcripts *in vivo*. The locations of genes *28–30* are from Fig. 8. Restriction sites are from Fig. 7. *Eco*RI restriction fragments are numbered according to Pero *et al.* (1979; see also Fig. 9). Data from Chelm *et al.* (1981).

Transcription of the 87–114 kbp segment of the SPO1 genome increases greatly during the middle phase of phage development (Fig. 9) and, in fact, 13 middle promoters have been located thus far (Fig. 11). Two of these middle promoters are very close to early promoters (at 99 and 113 kbp). Obviously, one sees here a way of generating *em* transcripts: They come from segments of the genome serviced by early *and* by middle promoters.

The aggregate transcriptional strength of these middle promoters greatly exceeds that of the early promoters, generating *em* transcripts whose abundance increases with middle transcription. Indeed, the early hybridization–competition analysis pointed to the existence of this kind of *em* RNA (Gage and Geiduschek, 1971b).

The other way of generating *em* transcripts would be to have two kinds of early transcription units, some of which are shut off after (or less completely than) others. It is not inconceivable that this latter kind of *em* transcription unit exists. Some bacterial holoenzyme must persist during the middle phase of transcription, because some host chromosome transcription also persists at this time.

In the region of 106–111 kbp there is a cluster of initiation sites for *in vitro* middle transcription, but none for *in vitro* early transcription. Hybridization–competition analysis between *in vitro* RNA from this region and *in vivo* RNA confirms the validity of these promoter assignments by showing that the *in vivo* RNA corresponds to the *m* class.

We remarked earlier on the density of early promoters in the heavily transcribed terminal redundancy and now see that the density of initiation sites in the heavily transcribed middle gene region is also high. Five middle promoters have been identified and sequenced in a 7.6-kbp region near 30 kbp (Talkington and Pero, 1979; Lee *et al.,* 1980b). It appears that the middle promoters, like the early promoters, will turn out to be very numerous.

No work on detailed mapping of late transcription units has been published at the time of completing this chapter.

4. *Early and Middle Promoters*

Two early and five middle SPO1 promoters have now been sequenced (Talkington and Pero, 1979; Lee *et al.,* 1980b) and show interesting properties (Fig. 5B).

To judge by their locations on restriction fragments, one of the early promoters is P_E5 and the other must be P_E2, P_E3, or P_E4. Both conform to the Pribnow box and the −35 region consensus sequences of 54 *E. coli* promoters (Siebenlist, 1979; Siebenlist *et al.,* 1980), as do the 3 early ϕ29 promoters shown in Fig. 5. Indeed, these two and all the other early SPO1 promoters in the terminal redundancy are recognized by *E. coli* RNA polymerase (Brennan, personal communication). The RNA synthesized from promoter P_E5 should be *e* RNA to judge by the hybridization map shown in Fig. 9. We have already presented data on the DNA sequences at 3 of the ϕ29 early promoters, and the simlarities are striking.

Five middle promoters, located in the 29–37 kbp region have been sequenced. To judge by their locations, these must be associated with *m* or m_1l transcription. These 5 promoters have common sequences in the region of the Pribnow box and the −35 region. In the Pribnow box region, the consensus sequences of the middle SPO1 promoters and of *E. coli* promoters are identical in three of the latter's six nucleotides, including the invariant T (hmU) at the 3′-end. However, the SPO1 middle promoters have a heavy bias to hmU in the DNA-coding strand. Further upstream, the 5 SPO1 middle promoters have the consensus sequence UxAgGaGAxxA, again coinciding with the *E. coli* −35 region consensus sequence at three of the latter's six nucleotides. The SPO1 middle promoters deviate significantly in substituting Ag for the highly conserved TT of the *E. coli* consensus sequence. (The significance of the capital and lowercase letters is indicated in the legend for Fig. 5B.) Not a single one of the 54 sequenced *E. coli* promoters has a purine at both of these positions, and only one has even a single purine at this position. The middle promoter −35 region consensus sequence is striking in its complementarity to the 3′-end of *B. subtilis* 16 S RNA (Fig. 5B). Such complementarity is characteristic of the ribosome-binding site of bacterial mRNA (Shine and Dalgarno, 1974; reviewed by Steitz, 1979). Ribosomal proteins interact with this segment of mRNA–rRNA duplex. In a sense, therefore, part of the interaction of the phage-modified E.gp 28 with its recognition site in DNA resembles an interaction of cellular proteins with their recognition sites in RNA.

The hmU for T replacement of SPO1 DNA is essential for middle promoter activity but not for early promoter activity *in vitro* (Lee *et al.*, 1980a). This probably reflects a specific hmU-binding property of the gene-*28* protein either alone or as part of its RNA polymerase complex. On should also bear in mind that substituting hmU for thymine lowers the melting temperature of DNA. For DNA with 43 mol% of GC base pairs (the average composition of SPO1 DNA) the difference in melting temperature is 10°C in neutral salt solution (Kallan *et al.*, 1962). All the SPO1 promoters arepoor in GC base pairs in the entire region between the initiating nucleotide and the Pribnow box and should melt readily.

Both the resemblances and the differences between early and middle promoters deserve attention. It is surely significant that the *spacing* of the conserved sequences is retained in early and middle promoters. This most probably reflects the constraint that the structure of the RNA polymerase enzyme core places on the location of recognition sites in DNA. It is also remarkable to see middle promoters distinguished from early promoters in so many features: After all, replacement of the σ subunit by gp 28 alone suffices to change the selectivity of RNA polymerase from one group of promoters to the other. There is evidence that the *E. coli* σ subunit is closely associated with DNA in the vicinity of the initiating nucleotide (Simpson, 1979). Does gp 28 bind to the comparable segment of middle SPO1 promoters and yet generate a change in polymerase–DNA interaction three entire helical turns upstream? It would be interesting to reconstruct hybrids between early and middle promoters (cf. Stefano *et al.*, 1980), so that one could test directly for the role these sequence features play in determining the selectivity of SPO1 promoters.

5. *Effects of Viral DNA Replication on Transcription*

We have seen that the sequence of early, middle, and late SPO1 transcription is primarily due to successive modification of the *B. subtilis* RNA polymerase by the products of three phage SPO1 regulatory genes which change the binding initiation characteristics of the enzyme, selecting different conjugate promoters with different nucleotide sequences. The selectivity of these enzymes can be demonstrated with mature double-stranded DNA extracted from virions. It appears, then, that late and middle SPO1 promoters do not absolutely require activation by DNA structural modification, as T4 promoters are thought to (Rabussay and Geiduschek, 1977). Nevertheless, SPO1 DNA replication strongly affects phage transcription.

The primary effect of phage DNA replication (and of replication proteins) is on the partition of transcription activity between bacterial and phage genes in the infected cell (Fig. 12). If DNA replication is blocked by mutations in gene *22* (Fujita *et al.*, 1971), *31* (Sarachu *et al.*, 1978), or *27* (Greene, unpublished observations), viral transcription is greatly decreased starting about the time of onset of late transcription. Novobiocin, a DNA gyrase inhibitor that completely blocks SPO1 DNA replication also considerably reduces phage transcription in relation to total RNA synthesis late in infection (Sarachu *et al.*, 1980). The synthesis of virion proteins is, as expected, decreased (Parker, unpublished observations; personal communication), and many fewer subassemblies of virions such as phage tails are seen when lysates of these abortively infected cells are examined in the electron microscope. A replication-defective mutant of SP82 makes about one-quarter as much of the major phage antigen as the wild type (Stewart *et al.*, 1972).

Undoubtedly, these effects of DNA replication are, in part, due simply to limitations of gene dosage. For example, the residual viral transcription in replication-blocked cells increases with mutliplicity of infection (Sarachu *et al.*, 1978). But this cannot be all there is to this replication dependence: Viral transcription actually decreases from its prereplication level, a decrease that is accentuated rather than relieved by blocking the transition to late transcription through the introduction of a mutation in gene *33* or *34* (Fujita *et al.*, 1971).

There is conflicting evidence on the question of whether blocking viral DNA replication selectively affects the transcription of different late RNA species. Gage and Geiduschek (1971b) concluded that the delayed-appearance late (*l*) RNA was selectively depleted relative to m_2l RNA when replication is blocked by a mutation in gene *22* (see also Fujita *et al.*, 1971). Talkington and Pero (1977) found no sign of this in a

Southern analysis with labeled RNA from bacteria infected with another mutation in the same gene. On the other hand, two late proteins whose properties might correspond with the postulated *l* class of RNA have been found (Heintz and Shub, 1981). They first appear after the other late proteins, and their rate of synthesis, relative to that of other late proteins, is decreased in the absence of DNA replication. Sarachu and co-workers have reported selective blocking of *l* RNA synthesis in the presence of novobiocin (in addition to the already recounted failure to shut off *e* transcription; Sarachu *et al.*, 1980) and have described a general failure of late transcription in cells infected with a gene-*31* mutant (Sarachu *et al.*, 1978).

So the information on this subject seems, at present, to be incomplete and contradictory. The underlying theoretical issue, of course, is whether other regulatory components or interactions, in addition to the products

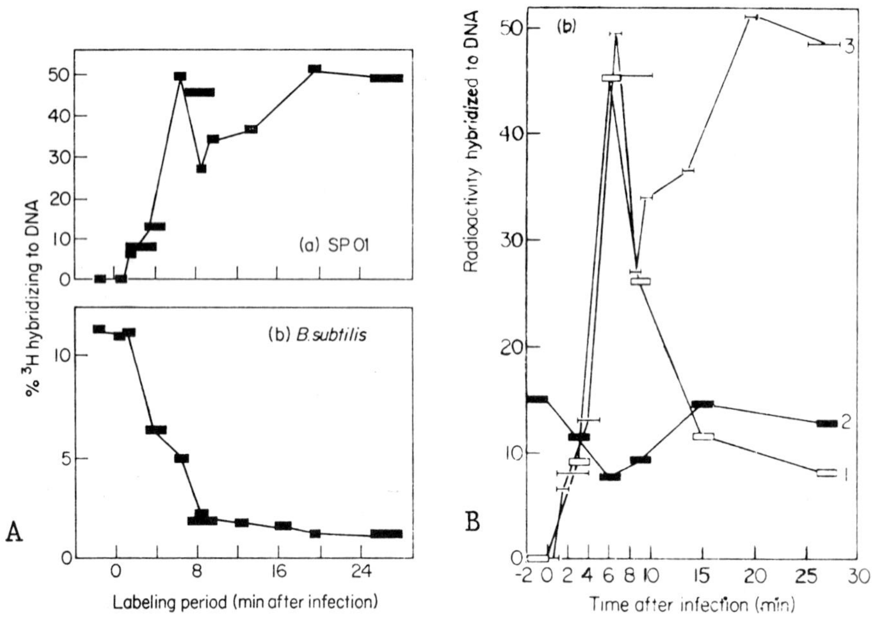

Fig. 12. DNA replication and viral transcription. Relative rate of phage transcription at various times after infection of *B. subtilis* 168M (su^-, nonpermissive for *sus* mutant phage) infected with wild-type and mutant phage SPO1. RNA labeled during the intervals shown was hybridized with denatured SPO1 or *B. subtilis* DNA. Infection was with (A) wild-type phage (at 37°C) (from Gage and Geiduschek, 1971a): (top) phage transcription; (botton) host transcription. (The absolute hybridization efficiency was not the same for phage and bacterial DNA.) (B) SPO1 replication-defective phage *susF30* (gene *22;* Fujita *et al.*, 1971): curve 1, phage transcription; curve 2, host transcription; curve 3, phage transcription after infection with wild-type virus.

of the four already known regulatory genes, determine the progression of viral gene expression during development.

6. *Repression of Viral Transcription during Development*

Certain *e* and *m* genes are repressed at characteristic times of the normal phage SP82 and SPO1 developments (Section III,E,1). The mechanism is currently not understood. Of course, one plausible and simple notion is that repression of these genes might simply be due to conversion of the conjugate RNA polymerase to another form: *e* transcripts would cease to be made when the conjugate host enzyme, Eσ, is converted to the phage-modified E.gp 28; *m* transcripts would cease to be made when the latter is converted to the late phage-modified E.gp 33.gp 34. The observation that at least some DNA segments yielding *em* transcripts have early and middle promoters (Section III,E,3) is consistent with this idea. Nevertheless, on the basis of relatively little evidence, we guess that, at least with respect to *e* genes, repression involves other components and interactions as well (or entirely).

With regard to repression of *e* genes, the pertinent observations are the following: (1) repression of *e* transcription occurs at the *start* of middle transcription. It is now known that this repression involves segments of the genome that are served by some of the strongest early promoters (Section III,E,3). If these promoters continued to compete effectively for the unmodified Eσ, one would expect to see them remaining active until almost all the Eσ had been converted, that is, until some time after the start of middle transcription. (2) Repression of *e* transcription can be uncoupled from the onset of *m* transcription when bacteria are irradiated before infection with a sufficient UV dose to reduce bacterial protein synthesis greatly (Heintz and Shub, 1981) or when they are treated with the DNA gyrase inhibitor novobiocin (Gellert *et al.*, 1977; Sarachu *et al.*, 1980). For instance, early and middle gene expression is turned on normally in UV-irradiated bacteria, but repression of early transcription is delayed.

There is even less information about repression of *m* genes than about *e* genes. However, it may be pertinent to point out that, during the normal course of SPO1 infection, total viral transcription decreases at about the time at which middle genes are turning off, slightly before the onset of m_2l late transcription (Fig. 12A). It seems that, here also, one may be looking at a repression event occurring independently of the middle → late polymerase conversion event.

A protein that selectively inhibits transcription of *hmU*-containing SPO1 DNA has been isolated from phage-infected bacteria (Wilson and

Geiduschek, 1969; Shub and Johnson, 1975). This protein, named TF1, has interesting properties. It is a small (11.5-kdalton) basic, phage-coded DNA-binding protein which reaches high concentrations (greater than 10^5 promoters per cell) during infection and selectively binds to hmU-containing DNA (Johnson and Geiduschek, 1972, 1977). *In vitro,* this protein prevents the formation of promoter complexes between bacterial RNA polymerase and SPO1 DNA and can even displace RNA polymerase from preformed open promoter complexes (Geiduschek *et al.*, 1977). The TF1 protein is partly homologous with the *E. coli* HU protein (Maziarz and Heinrickson, unpublished; Mende *et al.*, 1978). It may thus be a viral chromatin-specific structural analog of HU thought to be part of a nucleosome corelike complex in prokaryotic chromatin (cf. Griffith, 1976; Rouvière-Yaniv *et al.*, 1979; Rouvière-Yaniv and Kjeldgaard, 1979), although it is not packaged into virions (Johnson and Geiduschek, 1972). Whether TF1 plays any role in the repression of viral transcription *in vivo* is not known.

Acknowledgments

We are grateful to the following colleagues for permission to refer to their unpublished work or for sending us copies of articles in advance of publication: F. Eiserling, R. Heinrickson, M. Salas, D. Shub, J. Sogo, C. Stewart, and H. Whiteley. We also thank B. Chelm, F. Eiserling, J. Pène, and C. Stewart for their comments on the text. Our own research on this subject has been supported by grants from the National Institute of General Medical Sciences, the National Science Foundation, and the American Cancer Society.

References

Alegria, A. H., Kahan, F. M., and Marmur, J. (1968). *Biochemistry* **7,** 3179–3186.

Anderson, D. L., and Mosharrafa, E. T. (1968). *J. Virol.* **2,** 1185–1190.

Anderson, D. L., and Reilly, B. E. (1974). *J. Virol.* **13,** 211–221.

Anderson, D. L., and Reilley, B. E. (1976). "Microbiology, 1976" (D. Schlessinger, ed.), pp. 254–274. American Society for Microbiology, Washington, D.C.

Anderson, D. L., Hickman, D. D., and Reilly, B. E. (1966). *J. Bacteriol.* **91,** 2081–2089.

Aposhian, H. W., and Tremblay, G. Y. (1966). *J. Biol. Chem.* **241,** 5095–5101.

Arwert, F., and Venema, G. (1974). *J. Virol.* **13,** 584–589.

Bradley, D. E. (1965). *J. Gen. Microbiol.* **41,** 233–241.

Brennan, S. M., Chelm, B. K., Romeo, J. M., and Geiduschek, E. P. (1981). *Virology* **111,** 604–628.

Camacho, A., Moreno, F., Carrascosa, J. L., Viñuela, E., and Salas, M. (1974). *Eur. J. Biochem.* **47,** 199–205.

Carrascosa, J. L., Viñuela, E., and Salas, M. (1973). *Virology* **56,** 291–299.

Carrascosa, J. L., Camacho, A., Viñuela, E., and Salas, M. (1974). *FEBS Lett.* **44,** 317–321.

Carrascosa, J. L., Camacho, A., Moreno, F., Jimenez, F., Mellado, R. P., Viñuela, E., and Salas, M. (1976). *Eur. J. Biochem.* **66,** 229–241.
Chelm, B. K., Romeo, J. M., Brennan, S. M., and Geiduschek, E. P. (1981). *Virology* **112,** 572–588.
Cocito, C. (1974). *J. Virol.* **14,** 1482–1493.
Coppo, A., Manzi, A., and Pulitzer, J. F. (1975). *J. Mol. Biol.* **96,** 601–624.
Cregg, J. M., and Stewart, C. R. (1978a). *Virology* **85,** 601–605.
Cregg, J. M., and Stewart, C. R. (1978b). *Virology* **86,** 530–541.
Davison, B. L., Leighton, T., and Rabinowitz, J. C. (1979). *J. Biol. Chem.* **254,** 9220–9226.
Davison, B. L., Murray, C. L., and Rabinowitz, J. C. (1980). *J. Biol. Chem.* **255,** 8819–8830.
Dickel, C. D., Burtis, K. C., and Doi, R. H. (1980). *Biochem. Biophys. Res. Commun.* **95,** 1789–1795.
Duffy, J. J., and Geiduschek, E. P. (1973). *FEBS Lett.* **34,** 172–174.
Duffy, J. J., and Geiduschek, E. P. (1975). *J. Biol. Chem.* **250,** 4530–4541.
Duffy, J. J., and Geiduschek, E. P. (1976). *Cell* **8,** 595–604.
Duffy, J. J., and Geiduschek, E. P. (1977). *Nature (London)* **270,** 28–32.
Duffy, J. J., Petrusek, R. L., and Geiduschek, E. P. (1975). *Proc. Nat. Acad. Sci. USA* **72,** 2366–2370.
Eiserling, F. A. (1979). *In* "Comprehensive Virology" (H. Fraenkel-Conrat and R. R. Wagner, eds.), Vol. 13, pp. 543–580. Plenum Press, New York.
Eiserling, F. A., and Beard, C. (1972). Unpublished observations.
Eiserling, F. A., and Romig, W. R. (1962). *J. Ultrastruct. Res.* **6,** 540–546.
Escarmis, C., and Salas, M. (1981). *Proc. Nat. Acad. Sci. USA* **78,** 1446–1450.
Fox, T. D. (1976). *Nature (London)* **262,** 748–753.
Fox, T. D., and Pero, J. (1974). *Proc. Nat. Acad. Sci. USA* **71,** 2761–2765.
Fox, T. D., Losick, R., and Pero, J. (1976). *J. Mol. Biol.* **101,** 427–433.
Fujita, D. J. (1971). Thesis, Univ. of Chicago.
Fujita, D. J., Ohlsson-Wilhelm, B. M., and Geiduschek, E. P. (1971). *J. Mol. Biol.* **57,** 301–317.
Gage, L. P., and Fujita, D. J. (1969). *J. Bacteriol.* **98,** 96–103.
Gage, L. P., and Geiduschek, E. P. (1967). *J. Mol. Biol.* **30,** 435–440.
Gage, L. P., and Geiduschek, E. P. (1971a). *J. Mol. Biol.* **57,** 279–300.
Gage, L. P., and Geiduschek, E. P. (1971b). *Virology* **44,** 200–210.
Geiduschek, E. P., Armelin, M. C. S., Petrusek, R., Beard, C., Duffy, J. J., and Johnson, G. G. (1977). *J. Mol. Biol.* **118,** 825–842.
Gellert, M., O'Dea, M. H., Itoh, M., and Tomizawa, J. (1977). *Proc. Nat. Acad. Sci. USA* **74,** 4772–4776.
Glassberg, J., Franck, M., and Stewart, C. R. (1977a). *Virology* **78,** 433–441.
Glassberg, J., Slomiany, R. A., and Stewart, C. R. (1977b). *J. Virol.* **21,** 54–60.
Grau, O., Ohlsson-Wilhelm, B. M., and Geiduschek, E. P. (1971). *Cold Spring Harbor Symp. Quant. Biol.* **35,** 221–226.
Green, D. M., and Laman, D. (1972). *J. Virol.* **9,** 1033–1046.
Green, J. R., Chelm, B. K., and Geiduschek, E. P. (1982). *J. Virol.,* in press.
Griffith, J. D. (1976). *Proc. Nat. Acad. Sci. USA* **73,** 563–567.
Hagen, E. W., Reilly, B. E., Tosi, M. E., and Anderson, D. L. (1976). *J. Virol.* **19,** 501–517.
Harding, N. E., and Ito, J. (1976). *Virology* **73,** 389–401.
Harding, N. E., and Ito, J. (1980). *Virology* **104,** 323–338.
Hawley, L. A., Reilly, B. E., Hagen, E. W., and Anderson, D. L. (1973). *J. Virol.* **12,** 1149–1159.

Heintz, N., and Shub, D. (1981). *J. Virol.*, submitted.
Hempill, H. E., and Whiteley, H. R. (1975). *Bacteriol. Rev.* **39,** 257–315.
Hermoso, J. M., and Salas, M. (1980). *Proc. Nat. Acad. Sci. USA* **77,** 6425–6428.
Hiatt, W. R., and Whiteley, H. R. (1978). *J. Virol.* **25,** 616–629.
Hinkle, D. C., and Chamberlin, M. J. (1972). *J. Mol. Biol.* **70,** 157–185.
Hirokawa, H. (1972). *Proc. Nat. Acad. Sci. USA* **69,** 1555–1559.
Holland, M., and Whiteley, H. R. (1973). *Proc. Nat. Acad. Sci. USA* **70,** 2234–2237.
Horvitz, H. R. (1973). *Nature (London) New Biol.* **244,** 137–140.
Ikeda, Y., Saito, H., Miura, K., Takagi, J., and Aoki, H. (1965). *J. Gen. Appl. Microbiol.* **11,** 181–190.
Imamoto, F. (1970). Unpublished observations.
Inciarte, M. R., Viñuela, E., and Salas, M. (1976a). *Eur. J. Biochem.* **71,** 77–83.
Inciarte, M. R., Lazaro, J. M., Salas, M., and Viñuela, E. (1976b). *Virology* **74,** 314–322.
Inciarte, M. R., Salas, M., and Sogo, J. M. (1980). *J. Virol.* **34,** 187–199.
Ito, J. (1978). *J. Virol.* **28,** 895–904.
Ito, J., and Kawamura, F. (1976). *In* "Microbiology, 1976" (F. E. Young *et al.*, eds.), pp. 367–379. American Society for Microbiology, Washington, D.C.
Ito, J., and Roberts, R. J. (1979). *Gene* **5,** 1–7.
Ito, J., Meinke, W., Hathaway, G., and Spizizen, J. (1973). *Virology* **56,** 110–122.
Ito, J., Kawamura, F., and Yanofsky, S. (1976). *Virology* **70,** 37–51.
Ito, J., Harding, N. E., and Saigo, K. (1979). *Cold Spring Harb. Symp. Quant. Biol.* **43,** 525–536.
Ivarie, R. D., and Pène, J. J. (1973). *Virology* **52,** 351–362.
Johnson, G. G., and Geiduschek, E. P. (1972). *J. Biol. Chem.* **247,** 3571–3578.
Johnson, G. G., and Geiduschek, E. P. (1977). *Biochemistry* **16,** 1473–1485.
Kahan, E. (1966). *Virology* **30,** 650–660.
Kahan, E. (1971). *Virology* **46,** 634–637.
Kallan, R. G., Simon, M., and Marmur, J. (1962). *J. Mol. Biol.* **5,** 248–250.
Kawamura, F., and Ito, J. (1974). *Virology* **62,** 414–425.
Kawamura, F., and Ito, J. (1975). *In* "Spores VI" (P. Gerhardt, R. N. Costilow, and H. L. Sadoff, eds.), pp. 231–240. American Society for Microbiology, Washington, D.C.
Kawamura, F., and Ito, J. (1977a). *J. Virol.* **23,** 562–577.
Kawamura, F., and Ito, J. (1977b). *J. Gen. Appl. Microbiol.* **23,** 231–248.
Kawamura, F., and Ito, J. (1977c). *Virology* **83,** 233–245.
Kelly, T. J., Jr., and Thomas, C. A., Jr. (1969). *J. Mol. Biol.* **44,** 459–476.
Kornberg, A. (1974). "DNA Synthesis." Freeman, San Francisco, California.
Lavi, U., and Marcus, M. (1972). *Virology* **49,** 668–674.
Lawrie, J. M., and Whiteley, H. R. (1977). *Gene (Amsterdam)* **2,** 233–250.
Lawrie, J. M., Spiegelman, G. B., and Whiteley, H. R. (1975). *J. Virol.* **15,** 1286–1288.
Lawrie, J. M., Spiegelman, G. B., and Whiteley, H. R. (1976). *J. Virol.* **19,** 359–373.
Lawrie, J. M., Spiegelman, G. B., and Whiteley, H. R. (1977). *Gene (Amsterdam)* **2,** 251–262.
Lawrie, J. M., Downard, J. S., and Whiteley, H. R. (1978). *J. Virol.* **27,** 725–737.
Lee, G., Hannett, N. M., Korman, A., and Pero, J. (1980a). *J. Mol. Biol.* **139,** 407–422.
Lee, G., Talkington, C., Hannett, N., Jolly, J., and Pero, J. (1980b). Abstract, EMBO Workshop on Protein-DNA interaction in bacteriophages, Salamanaca, Spain, June 30–July 4, 1980.
Levner, M. H., and Cozzarelli, N. R. (1972). *Virology* **48,** 402–416.

Lica, L., and Ray, D. S. (1976). *J. Virol.* **18,** 80–84.
Liu, C. C., Burke, R. L., Hibner, U., Barry, J., and Alberts, B. (1979). *Cold Spring Harbor Symp. Quant. Biol.* **43,** 469–487.
Loskutoff, D. J., and Pène, J. J. (1973). *J. Virol.* **11,** 87–97.
Loskutoff, D. J., Pène, J. J., and Andrews, D. P. (1973). *J. Virol.* **11,** 78–86.
Magazin, M., Reeve, J. N., Maynard-Smith, S., and Symonds, N. (1978). *FEMS Microbiol. Lett.* **4,** 5–9.
McAllister, W. T. (1970). *J. Virol.* **5,** 194–198.
McGuire, J. C., Gilpatrick, M. W., and Pène, J. J. (1977). *Virology* **78,** 234–240.
McGuire, J. C., Pène, J. J., and Barrow-Carraway, J. (1974). *J. Virol.* **13,** 690–698.
Mellado, R. P., Moreno, F., Viñuela, E., Salas, M., Reilly, B. E., and Anderson, D. L. (1976). *J. Virol.* **19,** 495–500.
Mende, L., Timm, B., and Subramanian, A. R. (1978). *FEBS Lett.* **96,** 395–398.
Mendez, E., Ramirez, G., Salas, M., and Viñuela, E. (1971). *Virology* **45,** 567–576.
Mertens, G., Amann, E., and Reeve, J. N. (1979). *Mol. Gen. Genet.* **172,** 271–279.
Mosharrafa, E. T., Schachtele, C. F., Reilly, B. E., and Anderson, D. L. (1970). *J. Virol.* **6,** 855–864.
Nishihara, M., Friedman, N., and Aposhian, H. V. (1969). *J. Virol.* **3,** 164–170.
Okubo, S., Strauss, B., and Stodolsky, M. (1964). *Virology* **24,** 552–562.
Okubo, S., Yanagida, T., Fujita, D. J., and Ohlsson-Wilhelm, B. M. (1972). *Biken J.* **15,** 81–97.
Ortin, J., Viñuela, E., Salas, M., and Vasquez, C. (1971). *Nature (London) New Biol.* **234,** 275–277.
Panganiban, N., and Whiteley, H. R. (1981). *J. Virol.* **37,** 372–382.
Parker, M. (1979). Ph.D. Thesis, UCLA.
Parker, M., and Eiserling, M. (1981). Manuscript in preparation.
Pène, J. J., Murr, P. C., and Barrow-Carraway, J. (1973). *J. Virol.* **12,** 61–67.
Pero, J., Nelson, J., and Fox, T. D. (1975a). *Proc. Nat. Acad. Sci. USA* **72,** 1589–1593.
Pero, J., R. Tjian, J. Nelson, & R. Losick (1975b). *Nature (London)* **257,** 248–251.
Pero, J., Hannett, N. M., and Talkington, C. (1979). *J. Virol.* **31,** 156–171.
Price, A. R., Dunham, L. F., and Walker, R. L. (1972). *J. Virol.* **10,** 1240–1241.
Rabussay, D., and Geiduschek, E. P. (1977). *In* "Comprehensive Virology" (H. Fraenkel-Conrat and R. Wagner, eds.), Vol. 8, pp. 1–196. Plenum Press, New York.
Ratner, D. (1974). *J. Mol. Biol.* **88,** 373–383.
Reeve, J. N., and Cornett, J. B. (1975). *J. Virol.* **15,** 1308–1316.
Reeve, J. N., Mertens, G., and Amann, E. (1978). *J. Mol. Biol.* **120,** 183–207.
Reilly, B. E. (1965). Ph.D. Thesis. Western Reserve Univ., Cleveland, Ohio.
Reilly, B. E. (1976). *In* "Microbiology, 1976" (F. E. Young *et al.*, eds.), pp. 228–237. American Society for Microbiology, Washington, D.C.
Reilly, B. E., and Spizizen, J. (1965). *J. Bact.* **89,** 782–790.
Reilly, B. E., Nelson, R. A., and Anderson, D. L. (1977). *J. Virol.* **24,** 363–377.
Rekosh, D. M. K., Russel, W. C., Bellett, A. J. D., and Robinson, A. J. (1977). *Cell* **11,** 283–295.
Revie, D., Tseng, B. Y., Grafstrom, R. H., and Goulian, M. (1979). *Proc. Nat. Acad. Sci. USA* **76,** 5539–5543.
Rima, B. K., and Steensma, H. Y. (1971). *Antonie van Leeuwenhoek* **37,** 425–434.
Romeo, J. M., Brennan, S. M., Chelm, B. K., and Geiduschek, E. P. (1981). *Virology* **111,** 588–603.
Roscoe, D. H. (1969). *Virology* **38,** 527–537.

Roscoe, D. H., and Tucker, R. G. (1964). *Biochem. Biophys. Res. Commun.* **16,** 106–110.
Roscoe, D. H., and Tucker, R. G. (1966). *Virology* **29,** 157–166.
Rosenberg, M., and Court, D. (1979). *Ann. Rev. Genet.* **13,** 19–53.
Rouvière-Yaniv, J., and Kjeldgaard, N. O. (1979). *FEBS Lett.* **106,** 297–300.
Rouvière-Yaniv, J., Yaniv, M., and Germond, J.-E. (1979). *Cell* **17,** 265–274.
Rubio, V., Salas, M., Viñuela, E., Usobiaga, P., Saiz, J. L., and Llopis, J. F. (1974). *Virology* **57,** 112–121.
Salas, M., Mellado, R. P., Viñuela, E., and Sogo, J. M. (1978). *J. Mol. Biol.* **119,** 269–291.
Sarachu, A. N., Añon, M.-C., and Grau, O. (1978). *J. Virol.* **27,** 483–489.
Sarachu, A. N., Alonso, J. C., and Grau, O. (1980). *Virology* **105,** 13–19.
Schachtele, C. F., DeSain, C. V., Hawley, L. A., and Anderson, D. L. (1972). *J. Virol.* **10,** 1170–1178.
Schachtele, C. F., DeSain, C. V., and Anderson, D. L. (1973). *J. Virol.* **11,** 9–16.
Shimizu, N., Miura, K., and Aoki, H. (1970). *J. Biochem.* **68,** 277–286.
Shine, J., and Dalgarno, L. (1974). *Proc. Nat. Acad. Sci. USA* **71,** 1342–1346.
Shub, D. (1966). Ph.D. Thesis, MIT, Cambridge, Massachusetts.
Shub, D. A., and Johnson, G. G. (1975). *Mol. Gen. Genet.* **137,** 161–169.
Shub, D. A., Swanton, M., and Smith, D. H. (1979). *Mol. Gen. Genet.* **172,** 193–198.
Siebenlist, U. (1979). *Nucleic Acids Res.* **6,** 1895–1907.
Siebenlist, U., Simpson, R. B., and Gilbert, W. (1980). *Cell* **20,** 269–281.
Simpson, R. B. (1979). *Cell* **18,** 277–285.
Sogo, J. M., Inciarte, M. R., Corral, J., Viñuela, E., and Salas, M. (1979). *J. Mol. Biol.* **127,** 411–436.
Spiegelman, G. B., and Whiteley, H. R. (1974a). *J. Biol. Chem.* **249,** 1476–1482.
Spiegelman, G. B., and Whiteley, H. R. (1974b). *J. Biol. Chem.* **249,** 1483–1489.
Spiegelman, G. B., and Whiteley, H. R. (1978). *Biochem. Biophys. Res. Commun.* **81,** 1058–1065.
Stahly, D. P., and Ito, J. (1981). *Mol. Gen. Genet.* **182,** 180–182.
Stefano, J. E., Ackerson, J. W., and Gralla, J. D. (1980). *Nucleic Acids Res.* **8,** 2709–2723.
Steitz, J. A. (1979). *In* "Biological Regulation and Development" (R. Goldberger, ed.), pp. 219–277. Plenum Press, New York.
Stewart, C. R., and Franck, M. (1981). *J. Virol.* **38,** 1081–1083.
Stewart, C. R., Click, B., and Tole, M. F. (1972). *Virology* **50,** 653–663.
Sugino, A., Peebles, C. L., Kreuzer, K. N., and Cozzarelli, N. R. (1977). *Proc. Natl. Acad. Sci. USA* **74,** 4767–4771.
Swanton, M., Smith, D. H., and Shub, D. A. (1975). *Proc. Nat. Acad. Sci. USA* **72,** 4886–4890.
Takagi, J., and Hirokawa, H. (1979). *J. Gen. Appl. Microbiol.* **25,** 315–321.
Talavera, A., Salas, M., and Viñuela, E. (1972). *Eur. J. Biochem.* **31,** 367–371.
Talkington, C., and Pero, J. (1977). *Virology* **83,** 365–379.
Talkington, C., and Pero, J. (1978). *Proc. Nat. Acad. Sci. USA* **75,** 1185–1189.
Talkington, C., and Pero, J. (1979). *Proc. Nat. Acad. Sci. USA* **76,** 5465–5469.
Tjian, R., and Pero, J. (1976). *Nature (London)* **262,** 753–757.
Tosi, M., and Anderson, D. L. (1973). *J. Virol.* **12,** 1548–1559.
Tosi, M. E., Reilly, B. E., and Anderson, D. L. (1975). *J. Virol.* **16,** 1282–1295.
Truffaut, N., Revet, B., and Soulie, M. (1970). *Eur. J. Biochem.* **15,** 391–400.
Viñuela, E., Camacho, A., Jimenez, F., Carrascosa, J. L., Ramirez, G., and Salas, M. (1976). *Proc. R. Soc. London* **276,** 29–35.
Wahl, G. M., Stern, M., and Stark, G. R. (1979). *Proc. Natl. Acad. Sci. USA* **76,** 3683–3687.
Watson, J. D. (1972). *Nature (London) New Biol.* **239,** 197–201.

Whiteley, H. R., Spiegelman, G. B., Lawrie, J. M., and Hiatt, W. R. (1976). *In* "RNA Polymerase" (R. Losick and M. Chamberlin, eds.), pp. 587–600. Cold Spring Harbor Laboratory, Cold Spring Harbor, New York.
Wilson, D. L., and Gage, L. P. (1971). *J. Mol. Biol.* **57,** 297–300.
Wilson, D. L., and Geiduschek, E. P. (1969). *Proc. Natl. Acad. Sci. USA* **62,** 514–520.
Winnacker, E. L. (1978). *Cell* **14,** 761–773.
Wood, W. B., and King, J. (1979). *In* "Comprehensive Virology" (H. F. Fraenkel-Conrat and R. R. Wagner, eds.), Vol. 13, pp. 581–634. Plenum Press, New York.
Wu, R., and Geiduschek, E. P. (1975). *J. Mol. Biol.* **96,** 513–538.
Yanofsky, S., Kawamura, F., and Ito, J. (1976). *Nature* (*London*) **259,** 60–63.
Yamada, Y., Whitaker, P. A., and Nakada, D. (1974). *J. Mol. Biol.* **89,** 293–303.
Yehle, C. O. (1978). *J. Virol.* **27,** 776–783.
Yehle, C. O., and Ganesan, A. T. (1972). *J. Virol.* **9,** 263–272.
Yoshikawa, H., and Ito, J. (1981a). *Proc. Natl. Acad. Sci. USA* **78,** 2596–2600.
Yoshikawa, H., and Ito, J. (1981b). Submitted.
Yoshikawa, H., Friedmann, T., and Ito, J. (1981). *Proc. Natl. Acad. Sci. USA* **78,** 1336–1340.
Young, E., and Sinsheimer, R. (1964). *J. Mol. Biol.* **10,** 562–564.
Young, F. E. (1967). *Proc. Natl. Acad. Sci. USA* **58,** 2377–2384.

8

Temperate Bacteriophages of *Bacillus subtilis*

LARS RUTBERG

I. Introduction

Infection of sensitive bacteria with a temperate phage can have either of two consequences. The phage may enter a lytic path with autonomous replication of phage DNA and production of all the components required for the making of an infectious phage particle. The end result will be lysis of the infected bacteria, with the liberation of a burst of phage particles. Alternatively, the lytic functions of the phage may be repressed, and the bacteria will survive the infection and continue to divide. Most of the survivors will acquire the potential to produce phage of the original

The Molecular Biology
of the Bacilli

ISBN 0-12-222701-8

infecting type, and this property will be transmitted to essentially all the progeny of these cells, a phenomenon called lysogeny. Lysogeny is very common and probably occurs in most bacterial species. Most of our concepts about temperate phage and lysogeny are derived from studies on a few coliphages, in particular λ (Hershey, 1971; Echols, 1979), but also P2 (Bertani and Bertani, 1971) and Mu (Bukhari, 1976).

In lysogenic bacteria the phage genome (DNA) is integrated into the host genome as a prophage. In λ and in P2, integration occurs by a site-specific reciprocal recombination, catalyzed by a phage-coded integration protein, between a phage and a bacterial attachment site. Normally, λ uses only one site on the *Escherichia coli* chromosome for integration. P2 generally integrates at one preferred site but can integrate at a limited number of other sites. In contrast, phage Mu can integrate literally anywhere into the *E. coli* chromosome by a mechanism not fully known. In both λ and P2, prophage DNA is permuted relative to mature DNA, which reflects the fact that the phage attachment site is located internally in mature DNA. Before integration, DNA of these two phages (and many others) is circularized by way of the complementary single-stranded (cohesive) ends of mature DNA. During establishment and maintenance of lysogeny, phage lytic functions are repressed by a phage immunity repressor, which also makes the lysogen immune to infection with homoimmune phage. In the supernatant of a liquid culture of lysogenic bacteria some infectious phage particles are generally found. These particles are produced by "spontaneously" induced lysogens, where the system for repression of phage lytic functions has broken down, with subsequent lytic development of the resident prophage. Early in induction the prophage generally is excised from its site on the bacterial chromosome. In many lysogenic systems essentially all the bacteria can be induced by agents that interfere with DNA metabolism or by mutations that decrease the stability of the immunity repressor. Lysogenic bacteria thus acquire two new properties, namely, immunity to superinfection with homoimmune phage, and the potential to produce phage upon induction. Not too rarely temperate phage also carry genes which provide the lysogenic bacterium with other new properties seemingly unrelated to lysogeny per se, a phenomenon called lysogenic conversion (Barksdale and Arden, 1974). Outside the Enterobacteriaceae, the most intensively studied species from the point of view of molecular biology is *Bacillus subtilis*. A number of temperate phages active on *B. subtilis* have been described, but in comparison with the detailed knowledge available for coliphages, our knowledge of temperate phages of other bacteria is indeed limited. In the following discussion I will try to review what is known about some of the temperate *B. subtilis* phages. Rather than speculating on mechanisms about which too few facts are available, I will try to concentrate on summarizing what is known and, where possible, point out some problems which may be worthwhile to pursue in greater experimental detail. A more general background on the phages of *B. subtilis* can be found in the excellent review by Hemphill and Whiteley (1975).

II. Major Groups of Temperate *Bacillus subtilis* Phages

Recently, Dean *et al.* (1978a) proposed an ordering of nine temperate *B. subtilis* phages into the following four groups on the basis of cross-reactivity with neutralizing antibody, immunity, host range, and adsorption site on *B. subtilis*. Group I: ϕ105, ρ6, ρ10, ρ14; group II: SP02; group III: ϕ3T, ρ11, and SPβ; group IV: SP16. This last phage is interesting, especially in view of its broad host range, but little information has been published regarding SP16 (Thorne and Mele, 1974) and it will not be further dealt with here. Some properties of phages from each of groups I, II, and III are summarized in Table I.

A. Group I

The molecular weight of the DNA from all four phages in this group is about 26×10^6 as determined from the contour length of mature DNA

TABLE I Properties of Some Temperate *B. subtilis* Phages

Group	Phage	Size (nm) Head	Size (nm) Tail	Molecular weight of DNA ($\times 10^6$)	DNA base composition (% G + C)	Bacterial attachment site	Other phages in group
I	ϕ105	52 × 52	10 × 220[a]	25–26[a–c]	43.5[a]	*pheA–leuB*[d]	ρ6, ρ10, ρ14
II	SP02	50 × 50	10 × 180[b]	26[b,c]	43[f]	*rpsE–lin*[g]	None
III	SPβ	72 × 82	12 × 358[h]	62[h]	31[h]	*ilvA–kauA*[i]	ϕ3T, ρ11

[a] Birdsell *et al.* (1969).
[b] Boice *et al.* (1969).
[c] Chow *et al.* (1972).
[d] Rutberg (1969).
[e] Dean *et al.* (1978b).
[f] Romig (1968).
[g] Iselburg *et al.* (1969); Smith and Smith (1973).
[h] Warner *et al.* (1977).
[i] Zahler *et al.* (1977).

in electron micrographs (Rudinsky and Dean, 1979). Electron microscopy of heteroduplexes between mature φ105 DNA and DNA from the other phages in Group I, respectively, indicate at least 80% base pair homology. A region of nonhomology was found in the central parts of heteroduplexes φ105–ρ6 and φ105–ρ14. In several φ105–ρ6 heteroduplexes split ends were observed. It would be of interest to map the bacterial attachment site and determine the structure of the prophage DNA of ρ6, particularly in view of the fact that the φ105 attachment site is located at or very close to the cohesive ends (Armentrout and Rutberg, 1970; Chow and Davidson, 1973a). Maps of restriction endonuclease fragments of mature DNA from φ105, ρ10, and ρ14 also reveal the close relationship among these phages (Perkins *et al.*, 1978). Reportedly, the phages in group I recombine in genetic crosses (Perkins *et al.*, 1978). Bacteria lysogenic for any of the group I phages fail to plate any of the other phages in this group. Antibody prepared against any of the group I phages neutralizes all other phages in the group and also cross-reacts with group II phage SP02 (Boice *et al.*, 1969; Dean *et al.*, 1978a).

B. Group II

The only phage in this group is SP02, which was isolated from soil by Okubo (Okubo and Romig, 1965). SP02 is morphologically similar to φ105 (Birdsell *et al.*, 1969; Boice *et al.*, 1969) and contains about the same amount of DNA (Table I). Heteroduplexes between SP02 and φ105 DNA have revealed a 14% stretch of partial homology (Chow *et al.*, 1972). Since the two phages are serologically related, it has been proposed that this region contains gene(s) coding for parts of the phage tail. SP02 does not complement any known φ105 gene, and the two phages do not recombine in genetic crosses (Rutberg *et al.*, 1972). Also, the restriction fragment map of SP02 is distinct from that of φ105; e.g., φ105 DNA contains eight *Eco*RI sites, whereas SP02 has four sites (Yoneda *et al.*, 1979; Perkins *et al.*, 1978).

C. Group III

The genomes of the three phages of this group, φ3T, ρ11, and SPβ, are considerably larger than those of the phages of groups I and II. There is some uncertainty regarding the molecular weights of φ3T and ρ11 DNA, however. For φ3T, Dean *et al.* (1976c) calculated a value of about 64×10^6 from the summed weights of 22 *Eco*RI restriction endonuclease fragments. Ehrlich *et al.* (1976), however, found 30 *Eco*RI fragments with a summed molecular weight of about 72×10^6, whereas electron microscopy of φ3T DNA by the same authors gave a molecular weight of about 83 × 10^6. Finally, Cregg and Ito (1979), using different restriction endonucleases, arrived at a molecular weight of φ3T DNA of 77–80 × 10^6. Cregg and Ito (1979) report that deletion mutants can be isolated from φ3T and, more important, phage produced by the induction of lysogenic bacteria derived from single colonies is heterogeneous with respect to the restriction endonuclease fragment patterns of its DNA.

Dean *et al.* (1976c) found 20 fragments after *Eco*RI digestion of ρ11 DNA with a summed molecular weight of about 60 × 10^6. Sixteen of the ρ11 and φ3T *Eco*RI fragments comigrated on agarose gel electrophoresis. Recently Kawamura *et al.* (1979b) reported that *Eco*RI cleaved ρ11 DNA into 30 fragments with a summed molecular weight of about 80 × 10^6. The reason for these differing values for the molecular weight of ρ11 and φ3T DNA is unclear. Both ρ11 and φ3T convert thymine auxotrophs to prototrophy (Dean *et al.*, 1976c; Tucker, 1969).

SPβ is heteroimmune to both ρ11 and φ3T, but the three phages show serological cross-reactivity (Warner *et al.*, 1976). SPβ DNA has a molecular weight of 62 × 10^6 as determined by band sedimentation in alkaline solution. The G + C content is about 31%, which is slightly lower than the 34% reported for ρ11 and φ3T (Dean *et al.*, 1976c). The bacterial attachment sites for SPβ (Zahler *et al.*, 1977) and φ3T (Williams and Young, 1977) map close together but are not identical (Henner and Hoch, 1980).

III. φ105 and Related Phages

φ105 is probably the most studied of the temperate *B. subtilis* phages. It was isolated from soil by B. E. Reilly. The purified phage particle has an icosohedral head with a diameter of about 52 nm and a flexible, noncontractile tail about 220 nm long (Birdsell *et al.*, 1969; Boice *et al.*, 1969). To the tip of the tail is attached a baseplate with six appendages which probably serve in adsorption to sensitive bacteria. Mature DNA is nonpermuted, has cohesive ends, and has a molecular weight of about 26 × 10^6. The G + C content is 43.5% as determined by buoyant density and by melting (Table I). A genetic map for φ105 has been constructed by two- and three-factor crosses using about 40 temperature-sensitive (*ts*) or suppressor-sensitive (*sus*) mutants (Armentrout and Rutberg, 1970). Eleven complementation groups essential for lytic growth have been identified. One of these, gene *K*, is required for replication of φ105 DNA. This is probably true also for gene *J* (Flock, 1978). The functions of the other essential genes are unknown. A number of mutants with altered ability to lysogenize have been described (Armentrout and Rutberg, 1970; Garro, 1973a), and they seem to represent several complementation groups (Scher *et al.*, 1978; Flock, 1977). A mutation, *cts23*, that renders φ105 heat-inducible has been isolated. It probably defines the structural gene for the φ105 immunity repressor (Armentrout and Rutberg, 1971). The *cts23* mutation maps close to gene *J* (there is no gene *I*). Genes involved in lysogenization are generally not required for lytic growth of the phage. A series of φ105 deletion mutants that give turbid or clear plaques have been isolated (Flock, 1977). The deletions cover a region about 55–70% from the left end of the DNA molecule (Fig. 1). Mutants with deletions covering the 55–65% region are not impaired in lysogenization, whereas deletions extending into the 65–70% region give a clear-plaque phenotype. Phage carrying these deletions are unable to lysogenize and do not complement the *cts23* mutation.

Transfection has also been used to construct a genetic map for φ105 (Armentrout and Rutberg, 1970; Rutberg and Armentrout, 1970). In these experiments the frequency of corescue of a series of marker pairs compared to rescue of single markers was measured as a function of the recombinational distance of the markers in two-factor crosses. The corescue frequency using mature DNA was proportional to this distance for marker pairs separated by less than 10% recombination. Similar experiments were also performed using half-molecules of mature DNA obtained by mechanical shearing. A shear-sensitive center was found roughly in the middle of the genetic map, indicating that the DNA was nonpermuted. This has been confirmed by characterization of restriction endonuclease fragments of mature DNA (Scher *et al.*, 1977, Perkins *et al.*, 1978) and by marker rescue from mature DNA after limited digestion with exonuclease III of *E. coli* (Armentrout *et al.*, 1971). Corescue of marker pairs in transfection was also used to determine the gene order in prophage DNA (Armentrout and Rutberg, 1970). In these experiments the corescue frequency was proportional to the recombinational distance for all marker pairs tested. The prophage map derived from these experiments was colinear with the mature DNA map, in agreement with more limited data obtained previously using PBS1 transduction (Rutberg, 1969) and transformation (Peterson and Rutberg, 1969).

Prophage φ105 integrates linearly into the bacterial chromosome at a (unique?) attachment site between *pheA* and *ilvB, ilvC* (Rutberg, 1969; Dean *et al.*, 1976a). In an electron microscopic study of heteroduplexes between mature φ105 DNA and DNA from φ105 lysogenic bacteria, no circular structures of φ105 monomeric length were found, although such structures were found in similar experiments using SP02 (Chow and Davidson, 1973a). This negative result is consistent with colinearity between mature and prophage DNA and supports the idea that prophage integration occurs at or very close to the cohesive ends.

Prophage φ105 is induced by a number of agents that interfere with DNA metabolism (Birdsell *et al.*, 1969; Arwert and Rutberg, 1974b). After heat induction of *B. subtilis* (φ105*cts23*) replication is initiated at the resident prophage before it is excised from the bacterial chromosome (Armentrout and Rutberg, 1971). Replication does not stop at the prophage–bacterium junction but proceeds bidirectionally into the host chromosome (Armentrout and Rutberg, 1971; Rutberg, 1973), and linkage of prophage and host markers as measured by transformation is preserved after several rounds of replication after prophage induction. Replication seems to be a prerequisite for excision of φ105 prophage, since the prophage remains integrated after the induction of su^- bacteria lysogenic for the DNA-negative mutant φ105*susK7* (Armentrout and Rutberg, 1971), in contrast to the excision after induction of DNA-negative mutants of phage SP02 (Arwert and Rutberg, 1974c). Initiation of replication at the φ105 prophage occurs also in a lysogenic initiation-negative *ts* mutant of *B. subtilis* induced under nonpermissive conditions (Rutberg, 1973). What advantage (if any) prophage φ105 enjoys by remaining covalently linked to bacterial DNA during (at least) the first rounds of replication after induction is unknown. Also, the mechanism by which host and prophage DNA ultimately become separated after induction is uncertain. Association of phage DNA with host DNA is not seen after lytic infection with φ105 (Rutberg, unpublished experiments). Runoff DNA synthesis has been observed also after induction of prophage λ (Imae and Fukasana, 1970). Our knowledge of the mode of replication of φ105 DNA is limited. During replication a "high-frequency transfecting" DNA species appears (Rutberg and Rutberg, 1970), which has been identified as concatemeric DNA (Flock, 1978). There is evidence from electron microscopy of replicating φ105 DNA that it may replicate at some stage as a rolling circle (Flock, 1978). Unlike that of most other *B. subtilis* phages (Brown, 1970), replication of φ105 is inhibited by 6-(*p*-hydroxyphenylazo)uracil (Rutberg *et al.*, 1972) (HpUra) an inhibitor of *B. subtilis* DNA polymerase III (pol III) (Gass *et al.*, 1973; Mackenzie *et al.*, 1973). In bacteria carrying a resistant pol III, φ105 replicates in the presence of the drug (Rowley and Brown, 1977). In superinfected φ105-lysogenic *B. subtilis* the superinfecting genome is degraded (Flock and Rutberg, 1977).

Mature φ105 DNA has eight *Eco*RI recognition sites. In agarose gel electrophoresis of *Eco*RI-digested DNA, 10 fragments are detected which have been labeled A–J (Perkins *et al.*, 1978; Scher *et al.*, 1978). Fragments C and D represent the right and the left ends, respectively. The largest fragment, A, with a molecular weight of about 11×10^6, is a composite of fragments C and D associated via the cohesive ends. Fragment A disappears when DNA is treated with single-strand-specific S1 endonuclease, whereas fragments C and D disappear after treatment with ligase with a corresponding increase in fragment A (Scher *et al.*, 1977). These experiments clearly establish the presence of cohesive ends in mature φ105 DNA. Analysis of partial *Eco*RI digests has established the order of the fragments to be D, I, E, J, G, B, H, F, C (Perkins *et al.*, 1978). This order has been correlated with the genetic map by using purified restriction fragments in transfection marker rescue with a set of *ts* and *sus* mutants and some clear-plaque and deletion mutants (Scher *et al.*, 1978). The results of these and other experiments are summarized in Fig. 1, where the genetic and physical maps of mature φ105 DNA are compared. The two maps are pleasingly congruent. The bar labeled "Del." in Fig. 1 represents the nonessential region defined by deletions. In this region are located genes involved in the establishment and maintenance of lysogeny (Flock, 1977; Scher *et al.*, 1978). *Eco*RI fragment F has recently been cloned using the plasmid vector pUB110 (Cully and Garro, 1980) or pHV14 (Uhlen *et al.*, 1980). Bacteria carrying either of these hybrid plasmids are immune to infection with φ105, most probably

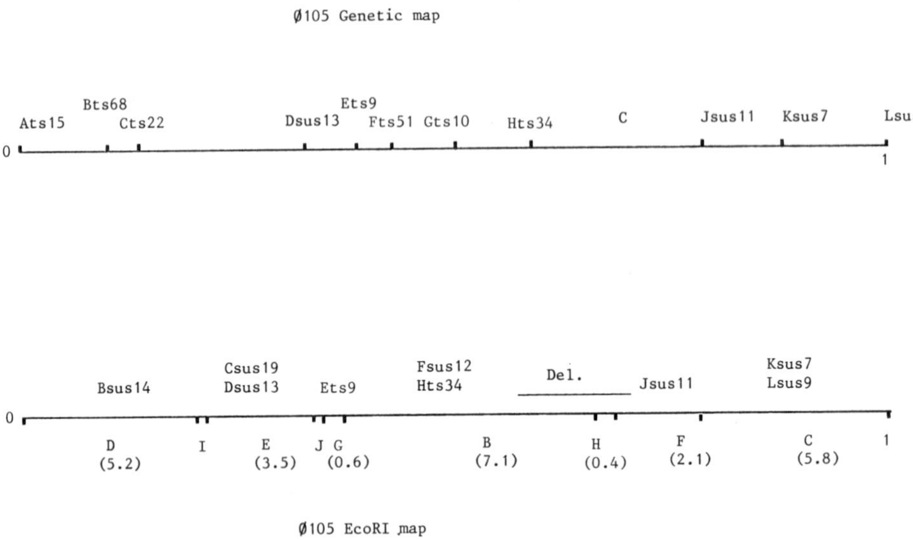

Fig. 1. Genetic map: The distances between the respective mutations are drawn proportional to recombinational distances. *Eco*RI map: The symbols above the line indicate genes carried on the respective fragments as identified by marker rescue in transfection. "Del." indicates a nonessential region which can be deleted. The capital letters below the line indicate respective *Eco*RI fragments. The numbers within parentheses give the size of the respective fragments in megadaltons.

because of the production of an immunity repressor. These experiments thus place the gene for the φ105 immunity repressor in the rightmost part of the nonessential region, in good agreement with heteroduplex analysis of deletion mutants (Flock, 1977). The genetic map of φ105 is far from saturated, and little is known about the function of most genes. Clearly, many more mutants must be isolated and characterized genetically and physiologically before a more in-depth comparison can be made between φ105 and, for example, the temperate phages of *E. coli.*

The frequency of genetic transformation is decreased in *B. subtilis* lysogenic for either phage φ105 or SP02 (Yasbin *et al.,* 1973; Garro and Law, 1974), whereas transduction with PBS1 or SP10 is not impaired (Peterson and Rutberg, 1969). Garro (1973a) found that a clear-plaque mutant φ105c4 could lysogenize at a very low frequency Prophage φ105c4 was not inducible with, for example, mitomycin C, and bacteria lysogenic for this mutant showed transformation frequencies comparable to those of nonlysogenic bacteria. This result suggested that induction of prophage in competent bacteria might cause the reduction in transformation frequency in wild-type lysogenic cells. This suggestion was supported by the finding that *B. subtilis* lysogenic for various induction-negative (Ind^-) mutants of φ105 or SP02, respectively, gave a normal frequency of transformation for auxotrophic markers (Garro and Law, 1974). Induction of prophage in competent lysogenic bacteria was also demonstrated as an increase in free phage after the addition of DNA (Garro, 1973b; Yasbin *et al.,* 1975a). Induction of prophage φ105 by DNA was dependent on the source of DNA (Garro, 1973b). Homologous DNA (*B. subtilis*) was an efficient inducer, whereas heterologous DNA (*B. licheniformis*) did not induce. Thus, the capacity of the donor DNA to recombine with the recipient genome seems essential for induction of prophage φ105 in competent lysogens. Surprisingly, DNA from φ105c4 lysogens was a poor inducer compared to DNA from wild-type lysogens. The reason for this is not understood.

In contrast to transformation, transfection with DNA from, for example, the virulent phage SP01 or φ29, is unaffected in φ105 or SP02 lysogenic bacteria (Yasbin *et al.,* 1973). Actually, the presence of prophage φ105 in competent bacteria may enhance transfection with SP02 DNA (Yasbin *et al.,* 1975b). The loss of transformants due to prophage induction in competent bacteria can be prevented by inhibiting protein synthesis for a few hours after the uptake of DNA before plating on selective media (Yasbin *et al.,* 1975a).

Phages ρ6, ρ10, and ρ14. The three ρ phages of the φ105 group were isolated by J. A. Hoch. They are homoimmune to φ105, and DNA of all four phages show extensive homology (Section II, A). Mature DNA has cohesive ends (Scher *et al.*, 1977; Rudinski and Dean, 1979). Interest in these phages has concentrated on their potential use as cloning vehicles in *B. subtilis* (Dean *et al.*, 1978b). Analysis of restriction endonuclease fragments of DNA from these phages has revealed that ϕ105 and ρ14 have single *Sal*I sites (Perkins *et al.*, 1978). In ϕ105 the *Sal*I site is located 0.9 Mdalton from the right end, whereas in ρ14 the *Sal*I site is located about 7.5 Mdaltons from the right end and is not located in an essential gene. Dean and co-workers (Kroyer and Dean, 1980; Kroyer *et al.*, 1980) isolated four deletion mutants of ρ14 which were analyzed with restriction endonucleases and by electron microscopy of heteroduplexes. The size of the deletions ranged from 2.2 to 3.3 kb. All deletions were within one contiguous segment 55–70% from one end of mature DNA, demonstrating that about 15% of ρ14 DNA does not code for any functions essential for lytic growth. One deletion mutant, ϕ*do7*, could lysogenize *B. subtilis*, and the ϕ*do7* lysogens were inducible. The deletion in ϕ*do7* was located at 55–63% from one end of mature DNA. The other three deletion mutants could not lysogenize but were immunity-sensitive. The respective deletions extended beyond the 63% end of the ϕ*do7* deletion, which places the immunity region of ρ14 at 63–70% from one end of mature DNA. The physical location of the immunity region of ρ14 is thus very similar to that of ϕ105 (Flock, 1977). The single *Sal*I site was preserved in all deletion mutants and was located about 400 bases from the right end of deletion *doc39* (64.2–70.4%). It has been shown that DNA can be inserted into the *Sal*I site of ϕ*do7* (Dean *et al.*, 1979). The resulting hybrid phage gives slightly turbid plaques, and it is presumably able to lysogenize. It is possible that expression (if any) of the cloned fragment is controlled by the immunity system. Phage ϕ*do7* may prove useful for gene amplification and complementation studies in *B. subtilis*.

IV. Phage SP02

Phage SP02 was isolated from soil by Okubo (Okubo and Romig, 1965). It is morphologically similar to ϕ105 and crossreacts serologically with this phage, but the two phages show little DNA homology and appear not to be related genetically (Section II, A and B).

Yasunaka and co-workers (1970) have constructed a linear genetic map of SP02 from two-factor crosses between a set of suppressor-sensitive mutants. The map contains 16 essential genes and a locus called *cI*, proposed to represent the structural gene for the SP02 immunity repressor. The genes are labeled *A–P*. Cistrons *J* through *M* have a DNA-negative phenotype under nonpermissive conditions. The *cI* locus maps between genes *J* and *K*. Another series of conditionally lethal SP02 mutants was isolated and characterized by Farrell (1970). The genetic map constructed by Farrell showed clustering of genes involved in DNA synthesis, in tail synthesis, and in head-related functions, respectively. Unfortunately, Farrell's mutants have not been compared with those of Yasunaka and co-workers. However, in both studies, gene(s) involved in lysogeny were found among genes coding for early functions, i.e., genes whose products are needed for phage DNA synthesis.

Prophage SP02 integrates at a (unique?) site between a cluster of genes for ribosomal proteins and a set of rRNA genes (Inselburg *et al.*, 1969; Smith and Smith, 1973). From electron microscopy the bacterial SP02 attachment site has been located about 6000 bases from the rDNA loci (Chow and Davidson, 1973b). Prophage SP02 is induced by treating lysogenic bacteria with agents that interfere with DNA metabolism (Boice, 1969; Arwert and Rutberg, 1974b). Prophage SP02 is induced at high temperature in bacteria carrying the *tsi-23* mutation, a condition that also induces the defective phage PBSX (Siegel and Marmur, 1969). In a study on heteroduplexes between mature SP02 DNA and DNA from SP02 lysogenic bacteria, Chow and Davidson (1973a) found circular, double-stranded structures of SP02 monomeric length. These structures were positively identified as containing SP02 DNA by the use of SP02 deletion mutants. From the position of the single-stranded loop in heteroduplexes formed between wild-type prophage DNA and mature DNA from deletion mutants, the SP02 attachment site was located 40% from one end of mature DNA. Thus, the SP02 prophage is permuted relative to mature DNA. Marker rescue of pairs of markers in transfection with prophage or mature SP02 DNA, respectively, has also been used to show that the prophage genetic map is permuted relative to the vegetative map (Arwert *et al.*, 1976). In these experiments, the phage attachment site was mapped between genes *I* and *J*, which is about 0.4 fractional lengths from one end of the genetic map of Yasunaka and co-workers (1970).

Excision of prophage SP02 after induction has been studied under conditions nonpermissive for both phage and bacterial DNA synthesis (Arwert and Rutberg, 1974c; Arwert *et al.*, 1976). After induction, DNA was extracted and centrifuged in sucrose gradients. In DNA isolated after induction, phage DNA was found (by transfection marker rescue) in the gradients at a position corresponding to monomeric mature DNA, indicating excision of the prophage. An increased frequency of cured bacteria was also found among the survivors after induction. Excision of the prophage was not seen when protein synthesis was blocked with chloramphenicol during induction. When suppressor-negative *B. subtilis* lysogenic for the DNA-negative mutant SP02*susL244* was induced, the excised prophage could be recovered as covalently closed circular (CCC) DNA (Arwert *et al.*, 1976). The CCC DNA was not infectious in transfection of bacteria carrying the $su^{+}3$ suppressor, but it was active in marker rescue.

Unlike ϕ105, phage SP02 can replicate in the presence of the pol III inhibitor HpUra (Section III). Infection of a DNA pol I- deficient mutant of *B. subtilis* with SP02 is followed by up to 100-fold increased DNA polymerase activity in partially purified extracts of infected cells (Rutberg and Armentrout, 1972). *In vivo* SP02-induced DNA polymerase can replicate SP02 DNA selectively (Rutberg, unpublished results). No increased DNA polymerase activity was found after infection with the DNA-negative mutant SP02*susL244*, whereas an increase was found after infection with the DNA-negative mutant SP02*susJ385*, indicating that gene *L* may code for an SP02 DNA polymerase. SP02 gene *L* has recently been cloned from partial *Hin*dIII digests in *B. subtilis* using the plasmid pC194 (Flock and Rutberg, unpublished experiments). Kolenbrander *et al.* (1972) have studied the synthesis of macromolecules in SP02-infected *B. subtilis*. Their results indicate a temporal control of synthesis of different classes of SP02 mRNA. During the later part of the latent period host DNA synthesis was strongly decreased. No breakdown of host DNA was found during infection and, consequently, radioactive thymidine incorporated into host DNA before infection was not detected in progeny phage.

Yoneda *et al.* (1979) have characterized and ordered the fragments obtained after treating mature SP02 DNA with six different restriction endonucleases. The fragments were arranged in a circular physical map. Single restriction sites were found for *Bgl*II, *Sal*I, and *Sma*I. In *Eco*RI-treated DNA a 1.62-Mdalton fragment disappeared upon heating, and instead two bands of 0.78 and 0.75 Mdalton appeared, indicating that these bands represent cohesive end fragments. Treatment with T4 ligase prevented heat dissociation of these fragments, supporting the suggestion that they contain cohesive ends. Deletion mutants of SP02 have been isolated following repeated cycles of heat–EDTA treatment (Graham *et al.*, 1979). Out of 30 turbid plaque formers that survived five cycles of treatment, 21 had restriction endonuclease fragment patterns identical to those of wild-type phage. Of the remaining 9 phages 2 could establish stable lysogeny. Both these phages had suffered a 2.3-Mdalton deletion which included the single *Bgl*II site, which is located 64% from one end of mature SP02 DNA (Yoneda *et al.*, 1979). The phenotypes of the two deletion mutants were different, however, indicating that the two deletions may be nonidentical. The presence of a single *Bgl*II site in what appears to be a genetically silent region may be useful for developing SP02 as a cloning vehicle for *B. subtilis*.

V. Phages φ3T, ρ11, and SPβ

Phage φ3T was isolated from garden compost by Tucker (1969). It was originally called φ3 but was later renamed φ3T to distinguish it from a previously isolated virulent phage called φ3. The interest in φ3T stems mainly from the finding that it can grow in thymine auxotrophs and that φ3T prophage converts thymine auxotrophs of *B. subtilis,* which lack thymidylate synthetase, to prototrophy (Tucker, 1969). In bacteria infected with a clear-plaque mutant, φ3Tc, the thymidylate synthetase activity rises to about 10 times the level found in wild-type *B. subtilis*. In thymine auxotrophs lysogenic for φ3T, thymidylate synthetase activity is comparable to that found in wild-type bacteria. There is no evidence for the presence of a defective *thy*-transducing particle in φ3T lysates. Essentially all plaque formers in a lysate seem able to convert thymine auxotrophs to prototrophy. Among 1364 φ3T lysogenic clones of a *B. subtilis* thymine auxotroph, Tucker (1969) found 46 that were not converted to prototrophy by the prophage. DNA extracted from purified φ3T or φ3Tc particles can transform thymine auxotrophs to prototrophy with at least the same frequency as DNA from wild-type bacteria. The integration site for the φ3T thymidylate synthetase gene, *thyP,* has been mapped by PBS1 transduction at a site between *thyA* and *thyB,* close to the bacterial attachment site for φ3T (Williams and Young, 1977). Thymidine auxotrophs transformed to prototrophy with φ3T mature DNA do not carry the complete phage genome. In contrast, about 50% of the Thy$^+$ transformants obtained by transformation with φ3T prophage DNA are lysogenic for φ3T. In thymineless bacteria converted to prototrophy by lysogenization with φ3T, the frequency of phage-producing bacteria was close to 100% at a multiplicity of infection of 1, but the frequency decreased with decreasing multiplicity (Williams and Young, 1977). Restriction endonuclease cleavage fragments of mature φ3T DNA can transform thymine auxotrophs to prototrophy. The *thyP* gene has been located on specific fragments (Ehrlich *et al.,* 1976; Graham *et al.,* 1977).

Cregg and Ito (1979) have produced a detailed physical map of φ3T DNA by ordering the fragments generated by cleaving with nine different restriction endonucleases. To my knowledge there is no published work on the genetics of φ3T. *Eco*RI-generated fragments of φ3T have been cloned in *E. coli* using the plasmids pMB9 and pSC101 (Ehrlich *et al.*, 1976; Duncan *et al.*, 1977). Some hybrid plasmids containing the φ3T *thyP* gene could convert *E. coli* thymine auxotrophs to prototrophy. The *thyP*-containing fragment remained with the plasmid in *E. coli*. DNA from *thyP*-carrying hybrid plasmids could transform thymineless *B. subtilis* to prototrophy.

Phage ρ11 was isolated by J. A. Hoch. Basic information about this phage has been published by Dean *et al.* (1976c). Phage ρ11 is very similar to φ3T morphologically, in DNA content, in G + C content, and in *Eco*RI restriction endonuclease fragment pattern. The two phages are homoimmune (Section II, C). Like φ3T, phage ρ11 converts thymine auxotrophs to prototrophy. ρ11 has recently been used by Kawamura *et al.* (1979a) for *in vitro* construction of a specialized transducing phage (Section VI).

Phage SPβ is probably identical to a phage originally found by electron microscopic examination of culture supernatants of *B. subtilis* 168M (Eiserling, 1964). Most derivatives of this strain are lysogenic for SPβ, but Warner *et al.* (1977) happened to observe that turbid plaques were formed when supernatants from strain-168 cultures were plated on a *B. subtilis* 168 derivative that carried the $su^{+}3$ suppressor gene (Georgopoulos, 1969). Several other derivatives of strain 168 that also carry the $su^{+}3$ gene do not plate SPβ but produce the phage. It seems that the SPβ-sensitive strain described by Warner *et al.* (1977) was cured of its SPβ prophage by chance. Whether the $su^{+}3$ gene increases the curing frequency of SPβ prophage is not known, but at least one other *B. subtilis* 168 derivative carrying this gene is also sensitive to SPβ (Dean, D. H. as cited by Warner *et al.*, 1977).

SPβ prophage is inducible with mitomycin C, but only a small fraction of the bacteria produced active phage and the resulting lysates have low titers. Considerably higher titers are obtained by infection with a clear-plaque mutant (Warner *et al.*, 1977) or by heat induction of bacteria lysogenic for a mutant, SPβ*c2*, which probably produces a temperature-sensitive immunity repressor (Rosenthal *et al.*, 1979). SPβ is heteroimmune to φ3T and ρ11 but is serologically related to these phages (Dean *et al.*, 1978a) (Section II, C). Zahler *et al.* (1977) have mapped the SPβ bacterial attachment site between *ilvA* and *kauA*, close to the terminus of the *B. subtilis* chromosome. In the same study, specialized transduction of *kauA* and of *citK* by SPβ was demonstrated. Transduction by SPβ is treated in detail by Zahler in Chapter 9. S. Zahler (personal communication) has isolated an integration-negative (int^{-}) mutant of SPβ. In a mixed infection with Int^{+} phage and the Int^{-} mutant SPβ*c2int5*, the mutant phage integrates at its normal bacterial attachment site. The phage yield obtained after induction of Int^- lysogens is reduced about 10,000-fold. After single infection with SPβ*c2int5,* rare lysogens can be isolated. Many of these have the prophage integrated outside the normal attachment site. Lysates obtained after the induction of such lysogens contain SPβ transducing particles which can transduce markers adjacent to the abnormal attachment site (Rosenthal *et al.,* 1980; S. A. Zahler, personal commun.).

Recently, a bacterial mutant lacking a functional SPβ attachment site has been isolated (S. A. Zahler, personal commun.). This mutant and the SPβ*c2int5* mutant can probably be used to generate specialized transducing phage for most parts of the *B. subtilis* chromosome.

SPβ has not been reported to convert thymine auxotrophs to prototrophy. However, it does confer some new properties on lysogenic bacteria besides immunity and the ability to produce SPβ phage. Bacteria lysogenic for SPβ (or a related phage called Z) produce a bacteriocin called betacin (H. E. Hemphill, personal commun.). SPβ lysogens are immune to betacin. Hemphill has isolated mutants of SPβ that do not induce production of betacin but still give immunity to the bacteriocin. Mutations in a bacterial gene can also give resistance to betacin. Betacin is not made in some SPβ lysogenic asporogenous mutants of *B. subtilis.*

SPβ lysogens do not support growth of the virulent phage φ1m. Mutations in SPβ or in the host *spoA* gene, or growth of the lysogenic bacteria at high temperature, makes the bacteria permissive for growth of φ1m (Rettenmeier *et al.,* 1979). The effect of SPβ on the growth of φ1m may be due to some alteration of the bacterial cytoplasmic membrane.

SPβ DNA is not cleaved by restriction endonuclease *Bsu* (GG CC) or *Hpa*II (C CGG) (T. A. Trautner *et.al.,* 1980). Trautner and co-workers (1980) have shown that SPβ codes for two methyltransferases that methylate cytosine at the *Bsu* site and the *Hpa*II site, respectively. They have also isolated mutants of SPβ that lack the *Bsu* site-specific methyltransferase. DNA from these mutants, grown on a nonmodifying or nonrestricting background, is cleaved by *Bsu.* This DNA contains low levels of methylcytosine compared to DNA from wild-type phage. SPβ is thus a self-modifying phage with respect to the *Bsu* site. The SPβ methyltransferases are not identical to the chromosomally determined enzyme (T. A. Trautner *et al.,* 1980). φ3T DNA is also resistant to *Bsu* restriction endonuclease, and this phage is also possibly self-modifying.

The fact that SPβ codes for a *Bsu* site-specific methyltransferase may explain the observation (Arwert and Rutberg, 1974a) that phage SP02 produced by mitomycin C induction of SP02 lysogenic bacteria or by infection of mitomycin C-treated *B. subtilis* is less sensitive to *Bsu* restriction *in vivo* compared to phage produced by lytic infection of untreated bacteria.

VI. Transduction with Temperate *Bacillus subtilis* phage

Phages PBS1, SP10, SPP1, and to some extent SP15 have been used for generalized transduction in *B. subtilis*. None of these is a true temperate phage, although the genomes of SP10 and PBS1 can exist in a carrier state in their hosts, a phenomenon sometimes referred to as pseudolysogeny (Hemphill and Whiteley, 1975). In the present context I will not deal with the above phages but only discuss some aspects of transduction with the temperate phages φ105 and ρ11. Transduction with SPβ is treated in detail by Zahler in Chapter 9.

Shapiro *et al.* (1974) have reported that φ105 lysates produced by induction of lysogenic *B. subtilis* contain particles that can transduce markers adjacent to the φ105 bacterial attachment site. The frequency of transduction decreased strongly with increasing distance between the marker tested and the prophage. Attempts to produce high-frequency transducing lysates were unsuccessful. The transducing activity of φ105 lysates was not DNase-sensitive, and the activity banded in CsCl density gradients at the position of infectious phage. It is unclear whether the transducing activity is sensitive to anti-φ105 antibody. The frequency of transduction is roughly proportional to input phage (Dean *et al.*, 1976b). An analysis of Leu^+ transductants showed that these segregated Leu^- cells, most of which were still immune to φ105. Attempts to obtain higher transduction frequencies by using *sus* mutants of φ105 were unsuccessful. Thus φ105 seems capable of specialized transduction of markers close to the bacterial attachment site. The frequency of transduction is low and is influenced by the presence of φ105 prophage in the recipient. Transducing particles cannot be demonstrated in all lysates from φ105 lysogenic strains including the ones used by Shapiro and co-workers (Flock and Rutberg, unpublished experiments).

The φ105 attachment site is located at or close to the cohesive ends of mature phage DNA (Section III). As φ105 deletion mutants with a decreased amount of DNA can easily be isolated, it is likely that DNA maturation normally depends on the presence of the cohesive ends as has been shown, for example, for λ, although λ can also package DNA by a "head-full" mechanism (Sternberg and Weisberg, 1975). Cohesive ends are also required for replication of φ105 DNA (Flock, 1978). The simplest model for generation of φ105 specialized transducing particles is to assume that a phage–bacterium DNA hybrid molecule is formed by imprecise excision.

This leads to loss of the right or the left cohesive end. Such DNA would be unable to replicate even in the presence of helper phage and probably could not integrate as a prophage. Lysogenic recipients would provide a region of essentially complete homology with the phage–bacterium hybrid DNA, which might facilitate integration by some general recombinational mechanism (Shapiro *et al.*, 1974). One obvious experimental approach in studying φ105 transduction is to introduce suitable genetic markers into the prophage of the donor and recipient bacteria, respectively, and to analyze the distribution of these markers in the transductants. Such experiments might be difficult to interpret, however, if the high multiplicities of infection used are essential for obtaining a sufficiently high transduction frequency.

Kawamura *et al.* (1979a) have recently developed a technique for *in vitro* construction of transducing *B. subtilis* temperate phage. Bacterial DNA was extracted from purified PBSX particles (Section VII). The DNA was cleaved with *Eco*RI restriction endonuclease and mixed with similarly cleaved ρ11 DNA. The mixture was incubated with T4 ligase and then used to transform ρ11 lysogenic histidine or lysine auxotrophs to prototrophy. The transformants obtained were pooled, and ρ11 lysates were produced by mitomycin C induction. These lysates were then used to transduce nonlysogenic histidine or lysine auxotrophs to prototrophy. All transductants obtained were immune to ρ11. Lysates were prepared from several transduced clones and tested for transducing activity. The lysates fell into two classes. One type contained no plaque-forming units (PFUs) but gave about 10^5–10^6 transductants/ml lysate. The other type contained plaque-forming particles and, in addition, gave about 10^{-2} histidine or lysine transductants per PFU. DNA from one plaque-forming transducing phage, ρ11hisA, was isolated and the ρ11 *Eco*RI fragments flanking the inserted bacterial *hisA*$^+$-carrying fragment were isolated. These "linker" fragments should prove useful for the cloning of DNA in infectious ρ11 phage. Kawamura *et al.* (1979b) have shown that ρ11 contains at least 5 Mdaltons of nonessential DNA distributed in two regions of the phage genome.

A similar approach was used to construct *metB*-specialized transducing φ105 particles, called φ105*dmetB* (Ijima *et al.*, 1980). These particles were defective, but transducing lysates could be obtained by providing a φ105 helper phage. φ105*dmetB* particles are less dense than wild-type phage and can be purified from the helper phage by banding in CsCl density gradients. Analysis of φ105*dmetB* DNA showed that the bacterial fragment carrying *metB* had replaced the phage *Eco*RI fragments E and

G (Fig. 1). φ105 DNA is not cleaved by *Bam*HI restriction endonuclease. Interestingly, the cloned *metB* fragment contains one *Bam*HI site. Cleavage with *Bam*HI does not inactivate *metB* transforming activity. The site, in φ105*dmetB*, might prove useful for cloning of DNA.

VII. Defective Phages

Defective phages are found in many bacterial species. The principal method for identifying defective phages is electron microscopy. Defective phages may contain some or all of the attributes of an infectious phage particle such as DNA-containing head and a tail. Unlike a complete phage, however, a defective phage by definition is unable to infect and multiply in any known host bacterium, although it often adsorbs to and kills a restricted range of bacteria. Defective phages are often inducible, and induction is the only means of obtaining sufficient quantities of phage for, for example, chemical characterization. Bacteria are generally immune to the killing action of the defective phage they carry.

All *B. subtilis* strains seem to carry one or more defective phages (see Garro and Marmur, 1970 for a review). One of these, PBSX or PBSH, has been studied in some detail. PBSX is induced by mitomycin C or ultraviolet irradiation (UV). About 2–3 hr after the induction of exponentially growing *B. subtilis* 168 the bacteria lyse and liberate a burst of 100–150 PBSX particles (Okamoto *et al.*, 1968b). PBSX particles have a hexagonal head with a diameter of about 41 nm and a rigid tail about 200×20 nm. The particles adsorb to and kill *B. subtilis* strain W23 but do not inject their DNA. They do not adsorb to and do not kill strain 168. Following adsorption to W23, macromolecular synthesis is rapidly inhibited by some unknown mechanism. The PBSX head contains double-stranded DNA with a molecular weight of about 8.3×10^6. After induction of PBSX a 22 S DNA begins to accumulate in the bacteria as a result of specific cleavage of the bacterial chromosome. The 22 S DNA is later packaged into PBSX heads. Cleavage of bacterial DNA and packaging occur also when net DNA synthesis is inhibited, but not when protein synthesis is blocked early after induction (Okamoto *et al.* 1968a). By what mechanism bacterial DNA is cleaved into pieces of 8.3×10^6 molecular weight is not known, but some phage protein (head?) seems essential for this to occur. All DNA in purified PBSX particles is of bacterial origin (Okamoto *et al.* 1968b).

Haas and Yoshikawa (1969a,b,c) studied the properties of *B. subtilis* 168 defective phage PBSH. This phage is most likely identical to PBSX. Phage particles purified from induced lysates have a DNA-containing head with a diameter of about 41 nm and a rigid, contractile tail about 200 × 17 nm. PBSH DNA is double-stranded, and its molecular weight was determined as 12×10^6, somewhat higher than the value calculated for PBSX DNA. All PBSH DNA is of bacterial origin, and DNA from PBSH (and PBSX) is active in transformation. Marker frequency analysis of DNA from PBSH particles revealed a great excess of the *ade-16* marker which is located close to the origin of replication of the *B. subtilis* chromosome (Henner and Hoch, 1980). The reason for the overrepresentation of *ade-16*-containing DNA in PBSH particles, as well as in DNA extracted from induced cells, is that induction is followed by repeated reinitiations of replication. This is most likely an effect of, for example, mitomycin C, as there is no evidence for PBSH or PBSX genes close to *ade-16* (Garro *et al.* 1970). The *ade-16* marker in DNA extracted from PBSH exhibited aberrant renaturation kinetics compared to other markers. The reason for this is unknown.

Purified PBSX particles contain at least seven structural proteins that can be resolved by sodium dodecyl sulfate–polyacrylamide gel electrophoresis (SDS–PAGE) (Thurm and Garro, 1975a). Their relative molecular weights range from 10,000 to 83,000. Five of the proteins are associated with the tail and two with the head. Six of the PBSX structural proteins could be identified in induced cells in addition to two other proteins which also may be specific for PBSX. A number of host proteins are continuously synthesized in PBSX-induced bacteria.

Bacillus subtilis mutants, called Kil^-, which do not produce particles that can kill strain W23 bacteria, have been isolated. The mutations (*xtl*) that produce this phenotype are linked by transformation to *metC* (Garro *et al.* 1970). Mitomycin C lysates of Kil^- mutants contain PBSX heads but no tails. Thurm and Garro (1975b) have succeeded in isolating additional mutations that affect PBSX. The mutants fall into three categories: defective head (*xhd*), defective tail (*xtl*), and not inducible with mitomycin C (*xin*). In total lysates of *xtl* or *xhd* mutants, all structural PBSX proteins could be identified by SDS–PAGE. However, the two head proteins were not precipitated from *xhd* lysates with anti-PBSX antibody, and the two tail proteins were not precipitated with the same antibody from *xtl* lysates. Electron microscopy of high-speed centrifugation pellets of lysates from these mutants showed, for *xhd,* tails with no heads, and for *xtl,* heads with no tails.

The *xin* mutant was less sensitive to UV than wild-type bacteria. The *xin* mutation seems specific for PBSX, since both φ105 and SP02 prophage are induced by mitomycin C in the *xin* mutant. All three mutations, *xhd, xtl,* and *xin,* map between *argC* and *metC*. Marker frequency analysis of DNA from mitomycin C-induced wild-type bacteria and the *xin* mutant showed a relative increase for markers close to the origin in both bacteria (Thurm and Garro, 1975b; Garro *et al.,* 1976). However, a similar increase in *metC* transforming activity was found in DNA from wild-type bacteria but not in DNA from the *xin* mutant. Thus reinitiation at the origin is probably an effect of mitomycin C, whereas replication of *metC* is a PBSX-specific phenomenon. The fact that all PBSX mutations isolated (although they are rather few) map at one chromosomal location supports the suggestion (Garro and Marmur, 1970) that the ancestor of PBSX was a temperate phage that lost the ability to excise and package its own DNA efficiently (Garro *et al.,* 1976).

Acknowledgments

I am deeply indebted to Dr. John Spizizen who once introduced me to the world of *Bacillus* phages. To the many colleagues who provided manuscripts and communicated results not yet published I am truly grateful. Work done in my laboratory was supported by grants from the Swedish Cancer Society.

References

Armentrout, R. W., and Rutberg, L. (1970). *J. Virol.* **6,** 760–767.
Armentrout, R. W., and Rutberg, L. (1971). *J. Virol.* **8,** 455–468.
Armentrout, R. W., Skoog, L., and Rutberg, L. (1971). *J. Virol.* **7,** 359–371.
Arwert, F., and Rutberg, L. (1974a). *Mol. Gen. Genet.* **133,** 175–177.
Arwert, F., and Rutberg, L. (1974b). *J. Virol.* **14,** 1470–1475.
Arwert, F., and Rutberg, L. (1974c). *J. Virol.* **14,** 1476–1481.
Arwert, F., Bjursell, G., and Rutberg, L. (1976). *J. Virol.* **17,** 492–502.
Barksdale, L., and Arden, S. B. (1974). *Ann. Rev. Microbiol.* **28,** 265–299.
Bertani, L. E., and Bertani, G. (1971). *Adv. Genet.* **16,** 199–237.
Birdsell, D. C., Hathaway, G. M., and Rutberg, L. (1969). *J. Virol.* **4,** 264–270.
Boice, L. B. (1969). *J. Virol.* **4,** 47–49.
Boice, L. B., Eiserling, F. A., and Romig, W. R. (1969). *Biochem. Biophys. Res. Commun.* **34,** 398–403.
Brown, N. C. (1970). *Proc. Nat. Acad. Sci. USA 67,* 1454–1461.
Bukhari, A. I. (1976). *Ann. Rev. Genet.* **10,** 389–412.
Chow, L. T., and Davidson, N. (1973a). *J. Mol. Biol.* **75,** 257–264.
Chow, L. T., and Davidson, N. (1973b). *J. Mol. Biol.* **75,** 265–279.
Chow, L. T., Boice, L. B., and Davidson, N. (1972). *J. Mol. Biol.* **68,** 391–400.
Cregg, J. M., and Ito, J. (1979). *Gene* **6,** 199–219.
Cully, D. F., and Garro, A. J. (1980). *J. Virol.* **34,** 789–791.
Dean, D. H., Arnold, M., and Halvorson, H. O. (1976a). *J. Virol.* **20,** 339–341.
Dean, D. H., Hutchison, K. W., and Halvorson, H. O. (1976b). *In* "Microbiology" (D. Schlessinger, ed.), pp. 295–302. American Society for Microbiology.
Dean, D. H., Orrego, J. C., Hutchison, K. W., and Halvorson, H. O. (1976c). *J. Virol.* **20,** 509–519.

Dean, D. H., Fort, C. L., and Hoch, J. A. (1978a). *Current Microbiol.* **1,** 213–217.
Dean, D. H., Perkins, J. B., and Zarley, C. D. (1978b). *In* "Spores VII" (G. Chambliss and J. C. Vary, eds.), pp. 144–149. American Society for Microbiology.
Dean, D. H. *et al.* (1979). *Recombinant DNA Tech. Bull.* **2,** 9–14.
Duncan, C. H., Wilson, G. A., and Young, F. E. (1977). *Gene* **1,** 153–167.
Echols, H. (1979). *In* "The Bacteria" (J. R. Sokatch and L. N. Ornston, eds.), Vol. VIII, pp. 487–516. Academic Press, New York.
Ehrlich, S. D., Bursztyn-Pettegrew, H., Stroynowski, I., and Lederberg, J. (1976). *Proc. Nat. Acad. Sci. USA* **73,** 4145–4149.
Eiserling, F. A. (1964). Doctoral Thesis, Univ. of California, Los Angeles.
Farrell, L. D. (1970). Doctoral Thesis, Univ. of California, Los Angeles.
Flock, J.-I. (1977). *Mol. Gen. Genet.* **155,** 241–247.
Flock, J.-I. (1978). Doctoral Thesis, Karolinska Inst., Stockholm, Sweden.
Flock, J.-I., and Rutberg, L. (1977). *FEMS Microbiol. Lett.* **2,** 309–312.
Garro, A. J. (1973a). *J. Virol.* **12,** 13–17.
Garro, A. J. (1973b). *J. Virol.* **12,** 18–24.
Garro, A. J., and Law, M.-F. (1974). *J. Bacteriol.* **120,** 1256–1259.
Garro, A. J., and Marmur, J. (1970). *J. Cell Physiol.* **76,** 253–264.
Garro, A. J., Leffert, H., and Marmur, J. (1970). *J. Virol.* **6,** 340–343.
Garro, A. J., Hammer, P., and Recht, B. (1976). *In* "Microbiology" (D. Schlessinger, ed.), pp. 340–349. American Society for Microbiology, Washington, D. C.
Gass, K. B., Low, R. L., and Cozzarelli, N. R. (1973). *Proc. Nat. Acad. Sci. USA* **70,** 103–107.
Georgopoulos, C. P. (1969). *J. Bacteriol.* **97,** 1397–1402.
Graham, S., Yoneda, Y., and Young, F. E. (1979). *Gene* **7,** 69–77.
Graham, S. R., Young, F. E., and Wilson, G. A. (1977). *Gene* **1,** 169–180.
Haas, M., and Yoshikawa, H. (1969a). *J. Virol.* **3,** 233–247.
Haas, M., and Yoshikawa, H. (1969b). *J. Virol.* **3,** 248–260.
Haas, M., and Yoshikawa, H. (1969c). *J. Virol.* **4,** 844–850.
Hemphill, H. E., and Whiteley, H. R. (1975). *Bacteriol. Rev.* **39,** 257–315.
Henner, D. J., and Hoch, J. A. (1980). *Microbiol. Rev.* **44,** 57–82.
Hershey, A. D. (ed.) (1971). The Bacteriophage Lambda. Cold Spring Harbor Laboratory.
Ijima, T., Kawamura, F., Saito, H., and Ikeda, Y. (1980). *Gene 9,* 115–126.
Imae, Y., and Fukasawa, T. (1970). *J. Mol. Biol.* **54,** 585–597.
Inselburg, J. W., Eremenko-Volpe, T., Greenwald, L., Meadow, W. L., and Marmur, J. (1969). *J. Virol.* **3,** 627–628.
Kawamura, F., Saito, H., and Ikeda, Y. (1979a). *Gene* **5,** 87–91.
Kawamura, F., Saito, H., Ikeda, Y., and Ito, J. (1979b). *J. Gen. Appl. Microbiol.* **25,** 223–236.
Kolenbrander, P. E., Hemphill, H. E., and Whiteley, H. R. (1972). *J. Virol.* **9,** 776–784.
Kroyer, J. M., and Dean, D. H. (1979). *Current Microbiol.* **3,** 45–50.
Kroyer, J. M., Perkins, J. B., Rudinski, M. S., and Dean, D. H. (1980). *Mol. Gen. Genet.* **177,** 511–518.
Mackenzie, J. M., Neville, M. M., Wright, G. E., and Brown, N. C. (1973). *Proc. Nat. Acad. Sci. USA 70,* 512–516.
Okamoto, K., Mudd, J. A., and Marmur, J. (1968a). *J. Mol. Biol.* **34,** 429–437.
Okamoto, K., Mudd, J. A., Mangan, J., Huang, W. M., Subbaiah, T. V., and Marmur, J. (1968b). *J. Mol. Biol.* **34,** 414–428.
Okubo, S., and Romig, W. R. (1965). *J. Mol. Biol.* **14,** 130–142.
Perkins, J. B., Zarley, C. D., and Dean, D. H. (1978). *J. Virol.* **28,** 403–407.

Peterson, A. M., and Rutberg, L. (1969). *J. Bacteriol.* **98,** 874–877.
Rettenmier, C. W., Gingell, B., and Hemphill, H. E. (1979). *Can. J. Microbiol.* **25,** 1345–1351.
Romig, W. R. (1968). *Bacteriol. Rev.* **32,** 349–357.
Rosenthal, R., Toye, P. A., Korman, R. Z., and Zahler, S. A. (1979). *Genetics* **92,** 721–739.
Rowley, S. D., and Brown, N. C. (1977). *J. Virol.* **21,** 493–496.
Rudinski, M. S., and Dean, D. H. (1979). *Virol.* **99,** 57–65.
Rutberg, L. (1969). *J. Virol.* **3,** 38–44.
Rutberg, L. (1973). *J. Virol.* **12,** 9–12.
Rutberg, L., and Armentrout, R. W. (1970). *J. Virol.* **6,** 768–771.
Rutberg, L., and Armentrout, R. W. (1972). *J. Virol.* **10,** 658–660.
Rutberg, L., and Rutberg, B. (1970). *J. Virol.* **5,** 604–608.
Ruterbg, L., Armentrout, R. W., and Jonasson, J. (1972). *J. Virol.* **9,** 732–737.
Scher, B. M., Dean, D. H., and Garro, A. J. (1977). *J. Virol.* **23,** 377–383.
Scher, B. M., Law, M. F., and Garro, A. J. (1978). *J. Virol.* **28,** 395–402.
Shapiro, J. A., Dean, D. H., and Halvorson, H. O. (1974). *Virology* **62,** 393–403.
Siegel, E. C., and Marmur, J. (1969). *J. Virol.* **4,** 610–618.
Smith, I., and Smith, H. (1973). *J. Bacteriol.* **114,** 1138–1142.
Sternberg, N., and Weisberg, R. (1975). *Nature (London)* **256,** 97–103.
Thorne, C. B., and Mele, J. (1974). *Microbiol. Genet. Bull.* **36,** 27–29.
Thurm, P., and Garro, A. J. (1975a). *J. Virol.* **16,** 179–183.
Thurm, P., and Garro, A. J. (1975b). *J. Virol.* **16,** 184–191.
Trautner, T. A., Pawlelz, B., Günthert, U., Canosi, U., Jentsch, S., and Freund, M. (1980). *Mol Gen. Genet.* **180,** 361–367.
Tucker, R. G. (1969). *J. Gen. Virol.* **4,** 489–504.
Uhlén, M., Flock, J.-I., and Philipson, L. (1980). *Plasmid* **5,** 161–169.
Warner, F. D., Kitos, G. A., Romano, M. P., and Hemphill, E. H. (1977). *Can. J. Microbiol.* **23,** 45–51.
Williams, M. T., and Young, F. E. (1977). *J. Virol.* **21,** 522–529.
Yasbin, R. E., Wilson, G. A., and Young, F. E. (1973). *J. Bacteriol.* **113,** 540–548.
Yasbin, R. E., Wilson, G. A., and Young, F. E. (1975a). *J. Bacteriol.* **121,** 296–304.
Yasbin, R. E., Wilson, G. A., and Young, F. E. (1975b). *J. Bacteriol.* **121,** 305–312.
Yasunaka, K., Tsukamoto, H., Okubo, S., and Horiuchi, T. (1970). *J. Virol.* **5,** 819–821.
Yoneda, Y., Graham, S., and Young, F. E. (1979). *Gene* **7,** 51–68.
Zahler, S. A., Korman, R. Z., Rosenthal, R., and Hemphill, H. E. (1977). *J. Bacteriol.* **129,** 556–558.

9

Specialized Transduction in *Bacillus subtilis*

STANLEY A. ZAHLER

I. Introduction

The genomes of most temperate phages are integrated within the bacterial host's chromosome during the part of the phage life cycle called lysogeny. The phage genome (called the *prophage* while integrated) then replicates as part of the bacterial chromosome. The prophage in such lysogenic cells can, upon occasion, leave the bacterial chromosome, replicate under the control of phage genes, and cause cell lysis, accompanied by the release of new phage particles. This process, called *in-*

The Molecular Biology
of the Bacilli

ISBN 0-12-222701-8

duction, may happen spontaneously with a low probability in any lysogenic cell; or the experimenter can induce an entire lysogenic culture to lyse simultaneously.

Some temperate bacteriophages, when induced from the lysogenic state, have the ability to package, within phage capsids, bacterial genes from the chromosomal regions adjacent to the prophage as well as some or all of the phage genome. These phage particles can transfer the bacterial genes to new bacteria. This process is called *specialized transduction,* and the particles that carry it out are called specialized transducing particles (TPs). The process is of interest and importance for two reasons.

First, the transductants produced in specialized transduction and their progeny are frequently diploid for the bacterial genes carried by the phage. This permits the production of heterogenotes, which allow the experimenter to study the dominance–recessiveness relations of different alleles within a gene and to determine whether closely linked mutations lie within a single cistron. Second, during the lytic part of the phage life cycle the bacterial genes carried by a transducing phage replicate under phage control. This permits amplification of the bacterial genes.

Specialized transduction was first observed with phage λ of *Escherichia coli* (Morse *et al.,* 1956; Arber *et al.,* 1957; Campbell, 1957). A summary of the aspects that seem pertinent to studies on bacillus follows. Useful reviews can be found in Campbell (1962, 1971) and Franklin (1971).

When the temperate phage λ infects a sensitive host, the linear phage DNA genome enters the cell and cyclizes into a covalently closed double-stranded circle (Fig. 1A). In some infected cells the phage enters the *lytic cycle:* phage proteins are synthesized, phage DNA is replicated, whole phage particles mature, and the cell lyses to release the new generation of phage. In other infected cells the phage enters the *lysogenic cycle.* A few phage proteins are synthesized, including the proteins called *phage repressor* and *integrase.* The phage repressor prevents the transcription of most of the phage genome. The integrase catalyzes a site-specific recombination between a particular site (PP′, Fig. 1A) on the phage DNA and a particular site (BB′, the *attachment site*) on the bacterial chromosome. The result of the recombination is insertion of the entire phage genome at the attachment site (Fig. 1B). This process is called *integration, phage insertion,* or *lysogenization.* The circular DNA structure, which is the intermediate between the linear phage structure and the linear prophage structure, is often called a *Campbell-type* circle (see Campbell, 1962).

Bacteria with the phage genome (now called the prophage) integrated in their chromosomes are said to be *lysogenic* for the phage. The phage

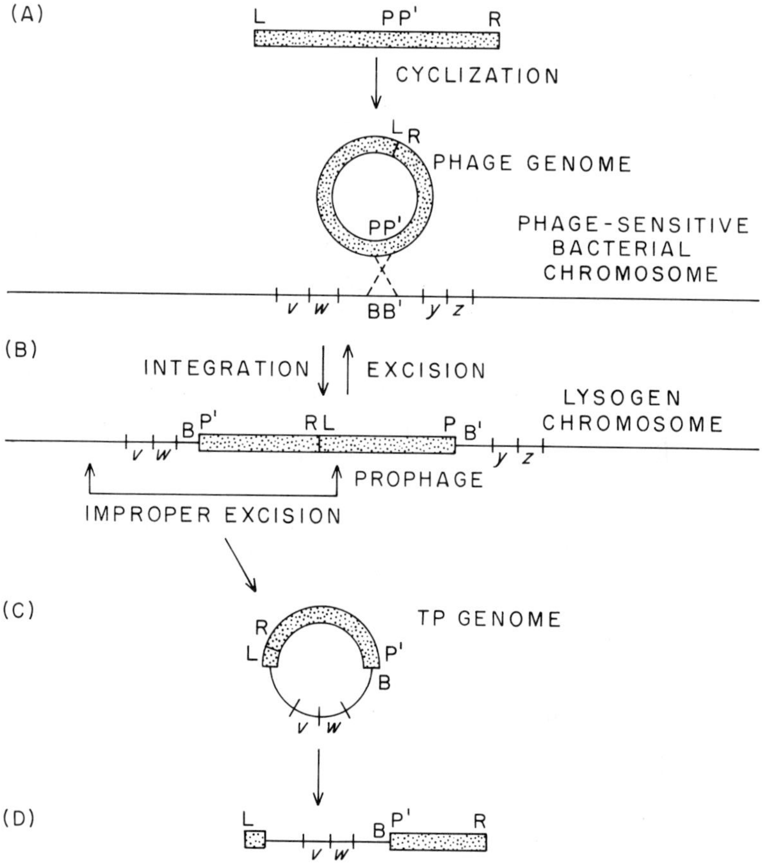

Fig. 1. Production of a defective transducing phage, modeled after phage λ. (A) Cyclization of the linear phage chromosome. (B) Integration of the phage into the normal attachment site, and normal excision. (C) Improper excision, and production of a defective transducing phage genome. (D) Linearization of the defective transducing phage genome during packaging.

repressor is synthesized by lysogenic bacteria—*lysogens*—and therefore most prophage genes are not transcribed. Lysogens cannot be lysed by phage for which they are lysogenic, because of the presence of repressor in the cells, and are said to be *immune*.

Excision of the prophage occurs when the phage repressor is inactivated. The excision of λ requires the integrase protein and a second λ-encoded protein (Gottesman and Weisberg, 1971). It results in precise reversal of the integration step (Fig. 1B). Since no repressor is present in the cells, the excised phage DNA enters the lytic cycle; it replicates, and the culture lyses, releasing new phage. The process leading from the lysogenic state to lysis and the release of phage is called *induction*. A low level of induction occurs spontaneously in cultures of lysogens. One cell in each 10^4–10^5 lyses, so that some phage particles are always present in such cultures.

Perhaps one act of λ prophage excision in 10^5 occurs improperly (Fig. 1C). Instead of reversing the act of integration precisely, an act of recombination occurs between a site within the prophage and a site on the bacterial chromosome, as indicated by the arrows below the lysogen's chromosome in Fig. 1C. It is believed that some degree of homology is present between the sites at which recombination takes place. The faulty excision may be catalyzed by bacterial or phage recombination enzymes, but not by integrase (Franklin, 1971). The result is a chimeric circle of DNA which, in the case shown in Fig. 1C, contains two bacterial genes *v* and *w*, but which lacks the phage genes that lie on the right end of the normal prophage. Because it (usually) lacks genes needed for phage growth, it is said to be *defective*. Other potential improper excisions in Fig. 1C might include only bacterial gene *w;* or they might include bacterial genes *y* or *y* and *z*, but these would lack the phage genes that lie at the left end of the prophage.

The circular genome resulting from improper excision such as that depicted in Fig. 1C is produced in a cell in which induction has occurred. Precise excision of the prophage has usually occurred from other chromosomes within the same cell. These complete phage genomes are transcribed and translated into all the products needed for new phage production. The fact that the DNA circle containing bacterial genes is defective does not prevent it from replicating or being packaged in a phage capsid, possibly using gene products from the whole helper phage genomes. When the phage particles mature in this cell, they will include some with defective phage chromosomes, provided that (1) the phage chromosome ends labeled L and R are present on the defective genome, (2) the total amount of DNA of the defective genome is not very different from that of the normal phage genome, and (3) the site of the origin of phage DNA replication is present. The improper excision diagrammed in Fig. 1C results in packaging of the defective phage chromosome depicted in Fig. 1D.

If we induce a normal lysogen and examine the lysate for TPs for genes close to the prophage, e.g., for phage particles able to transduce a v^- or w^- cell to V^+ or W^+, we may find 1 TP per 10^6 plaque-forming particles. Such a lysate is said to be a low frequency of transduction lysate. No TPs are found for genes located far from the prophage. The transduction is called *specialized transduction* to differentiate it from *generalized transduction,* in which a phage lysate can transfer any gene from one strain of bacteria to another.

When a low frequency of transduction lysate is used to transduce a v^- lysogen to V^+, many of the transductants become *double lysogens,* carrying one copy of the defective transducing phage genome and one copy of the whole prophage. When a culture of double lysogens of this sort is isolated and induced to lyse, each cell in the culture releases some normal phage particles and some TPs. Such a lysate is said to be a high frequency of transduction (HFT) lysate. In some λ HFT lysates the ratio of TP to plaque-forming units (PFU) is 1:1.

When TP genomes enter recipient cells during a transduction, two main classes of events can be detected. We will use as examples the defective transducing phage carrying v^+ and w^+ shown in Fig. 1D. The recipient culture carries the negative alleles v^- and w^-, and selection is for V^+ and W^+.

1. We may obtain *replacement* of part of the recipient's chromosome by bacterial DNA carried by the transducing phage.

2. We may see *addition* of the entire defective transducing phage genome to the recipient's chromosome. This results in a cell with two copies of the *v–w* chromosome region.

It is convenient to be able to distinguish between transductants in which a replacement of DNA has occurred and those in which the addition of the entire TP genome has occurred. The replacement cases give rise to stable V^+W^+ cells. The addition cases give rise to heterogenotes carrying both positive and negative alleles of the *v* and *w* genes. These heterogenotes are unstable because of the genetic duplication, and segregate V^-, W^-, or V^-W^- cells with fairly high frequency.

II. Specialized Transduction by Phage φ105

Shapiro *et al.* (1974) have described an aberrant type of specialized transduction in *Bacillus subtilis* carried out by the temperate phage φ105. Low frequency of transduction lysates of φ105 are produced for genes close to the prophage. Neither heterogenotes nor bacteria that can produce HFT lysates have been found.

Phage φ105 (see Chapter 8) integrates at an attachment site (*att*φ*105*) that lies between *leuB* and *pheA* (Rutberg, 1969; Dean *et al.*, 1976). When a lysogen is induced, a small fraction (less than 10^{-6}) of the phage particles produced carry *leuB, pheA,* or other nearby bacterial genes. Transduction has been demonstrated convincingly only when the recipient was lysogenic for φ105 (Shapiro *et al.*, 1974).

Phage φ105 and its close relatives are unusual in that, although they integrate at a particular attachment site on the bacterial chromosome, they do not undergo cyclization as part of the integration process. The entire linear phage DNA molecules seem to insert, still linear, at the host's attachment site.

The DNA enclosed in φ105 virions is double-stranded and probably has single-stranded cohesive ends (Chow *et al.*, 1972; Scher *et al.*, 1977). Circular molecules (not closed covalently) either are packaged in the phage heads or form spontaneously during DNA extraction. Nevertheless, it appears that φ105 prophage DNA and φ105 phage DNA are colinear, without circular permutation; Campbell-type circular intermediates do not seem to play a role either in integration or in excision and replication (Armentrout and Rutberg, 1971; Chow and Davidson, 1973).

The mechanism of φ105 integration into *att*φ*105* is not understood. Presumably there are essential sequences at the phage DNA termini that are recognized by phage-encoded enzymes to bring about the integration. The specificity for a single site on the bacterial chromosome is similar to that of phage λ, but the insertion of DNA colinear with the phage DNA is more reminiscent of the insertion of coliphage Mu or of other transposons (Bukhari *et al.*, 1977).

When bacteria lysogenic for φ105 are induced, the prophage and some nearby bacterial genes on both sides of the prophage replicate *in situ* before excision occurs (Armentrout and Rutberg, 1971; Rutberg, 1973). Later the phage DNA molecules are cut out of the newly replicated linear DNA. Presumably, improper cuts made at this stage result in the production and eventual packaging of occasional defective transducing phage genomes. The normal φ105 genomes probably replicate as linear molecules and as concatamers (Rutberg and Rutberg, 1970; Flock, 1978).

A specialized transducing phage genome produced by the improper excision of φ105 presumably lacks the special sequences at the ends of the DNA molecule that are needed for insertion into *att*φ*105*. There seems to be no mechanism available for cyclization of the genome. These characteristics may explain why specialized TPs of phage φ105 can only give replacement recombination events.

The chromosome structure of a replacement-type transductant (Fig. 2) is essentially identical to that of the donor in the transduction. (If a nonlysogen were used as recipient in Fig. 2 instead of a lysogen, replacement recombination could occur only within the region of bacterial homology; the Leu^+ transductant would remain a nonlysogen if infection by infectious φ105 particles were prevented.) The transductant's chromosome lacks the characteristic needed for the production of HFT ly-

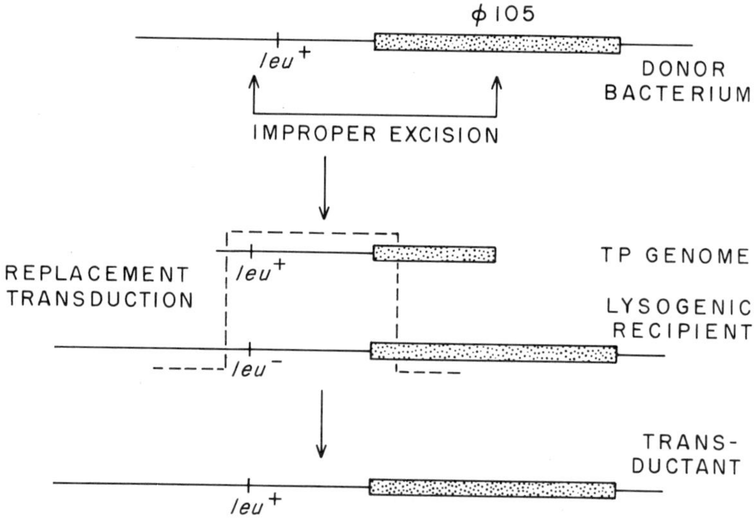

Fig. 2. Model for specialized transduction by phage φ105.

sates: double lysogeny, with one copy of a normal prophage and one copy of the prophage of a specialized transducing phage.

This line of reasoning leads to one hypothesis and one prediction: (1) phage φ105 cannot give HFT lysates of defective TPs because it cannot cyclize, and (2) its distant relative, phage SP02, may give specialized transduction and permit the production of HFT lysates, since it integrates via a Campbell-type intermediate (Chow and Davidson, 1973). Unfortunately, there are no selectable markers near the normal SP02 attachment site (Inselberg *et al.*, 1969; Smith and Smith, 1973), so this prediction cannot be tested easily.

III. Specialized Transduction by Phage SPβ

A. Introduction

In 1977 H. E. Hemphill and his colleagues at Syracuse University (Warner *et al.*, 1977) reported that strain 168 of *B. subtilis* was lysogenic for a phage which they called SPβ. [Virions of SPβ had been observed and named earlier by Eiserling (1964) during an electron microscope study of defective phage SPα (PBSX), for which strain 168 is also lysogenic.] The Syracuse University group found that a derivative of strain

168, strain su^{+3} (=CU1050), used in their laboratory had lost the SPβ prophage and become a suitable plating host for SPβ.

Zahler *et al.* (1977) located the attachment site (*attSP*β) at which SPβ integrates on the *B. subtilis* chromosome. They also observed that lysates of SPβ produced from lysogens gave low-frequency specialized transduction of nearby genes and that HFT lysates could be prepared from some of the transductants.

B. Physical and Chemical Characteristics of Phage SPβ and Its Relatives

1. *SP*β

Warner *et al.* (1977) isolated a clear-plaque mutant of SPβ that they called SPβ*c* (=SPβ*c1*). They characterized the clear-plaque mutant morphologically and chemically. We have found that the DNA of SPβ*c1* contains a deletion of about 10 kb, as determined by differences in electrophoretic mobility between restriction endonuclease fragments of SPβ and SPβ*c1* produced by the enzymes *Pst*I and *Sac*I (Fink *et al.*, 1981). Recent observations in this laboratory by P. S. Fink and M. P. Weiner suggest that the clear-plaque morphology of SPβ*c1* is caused by a mutation separable from the 10-kb deletion. *Pst*I and *Sac*I produce identical fragments from SPβ and the mutant SPβ*c2*, discussed in Section III,C,5.

The phage belongs to group B of Bradley's (1967) morphological classification scheme. Electron micrographs (Warner *et al.*, 1977) showed a phage with a head size of 76 (width) × 82 nm and a sinuous tail 12 × 357 nm. The tail ends in a hexagonal structure up to 36 nm in diameter. The buoyant density of SPβ*c1* in CsCl was reported to be 1.52 g/cm^3. P. S. Fink and J. M. Odebralski in this laboratory have found a value of 1.51 g/cm^3 for SPβ*c1* and 1.52 for SPβ*c2*.

Warner *et al.* (1977) reported that the molecular weight of the DNA of SPβ*c1* was 62 × 10^6 (94 kb). The measurement was based on band sedimentation in alkaline solution. P. S. Fink in this laboratory has estimated the size of the SPβ*c1* genome to be approximately 77 Mdaltons (118 kb), based on the electrophoretic mobilities of restriction endonuclease fragments of SPβ*c1* DNA.

The base composition of SPβ*c1* DNA was determined by Warner *et al.* (1977) to be 30.6 mol% G + C (T_m = 81.8°C; buoyant density in CsCl = 1.690 g/cm^3). The DNA of B. *subtilis* contains about 43 mol% G + C.

2. *ϕ3T, ρ11, and Z*

Warner *et al.* reported that antiserum against phage SPβ also neutralized two other temperate phages of *B. subtilis:* ϕ3T and ρ11. Hemphill later isolated a fourth temperate phage of the same serological group which he has named Z (Hemphill *et al.,* 1980).

Phages ϕ3T (Tucker, 1969) and ρ11 (Dean *et al.,* 1976) were isolated from soil by R. G. Tucker and J. A. Hoch, respectively. The phages formed plaques on strains of *B. subtilis* (NCTC3610 and 168, respectively) that were lysogenic for SPβ. (The existence of SPβ was not known when ϕ3T and ρ11 were first isolated.) Although they are related to SPβ, ϕ3T and ρ11 are obviously heteroimmune to SPβ since they form plaques on SPβ lysogens. The situation with respect to immunity is not reciprocal: SPβ cannot form plaques on lysogens of ϕ3T or ρ11 (H. E. Hemphill, personal communication; and observations in this laboratory). ϕ3T and ρ11 are homoimmune (Dean *et al.,* 1976).

Tucker (1969) has shown that phage ϕ3T carries a gene for thymidylate synthetase (*thyP*) and converts Thy^- auxotrophs of *B. subtilis* to Thy^+ when it lysogenizes them. Surprisingly, the phage gene and the enzyme it encodes are almost identical to the *B. subtilis thyA* thymidylate synthetase gene and enzyme (Duncan *et al.,* 1978; Stroynowski, 1978). Phage ρ11 also converts Thy^- hosts to Thy^+ (Dean *et al.,* 1976), and it carries a thymidylate synthetase gene very similar in sequence to that of ϕ3T (Stroynowski, 1978).

Phage Z was isolated from soil using a *B. subtilis* host that was not lysogenic for SPβ or for any of its relatives (Hemphill *et al.,* 1980). It converts Thy^- *B. subtilis* strains to Thy^+ and is homoimmune with SPβ. Like SPβ, it fails to form plaques on ϕ3T or ρ11 lysogens. Both ϕ3T and ρ11 form plaques on Z lysogens:

The virions and genomes of ϕ3T and ρ11 have been examined in several laboratories. Some pertinent characteristics are listed in Table I, with information about SPβ and Z for comparison. According to H. E. Hemphill (personal communication), temperature-sensitive (*ts*) mutants of each of the phages in the group can recombine with *ts* mutants of each of the other phages in the group to give rise to ts^+ recombinants; this suggests a fairly close relationship.

This leads to a problem first pointed out by H. E. Hemphill. Since ϕ3T and ρ11 were first isolated in lysogens of SPβ and, since they are still grown in SPβ lysogens in most laboratories, ϕ3T and ρ11 may now be "contaminated" with SPβ genetic material, picked up through recombination. Furthermore, different batches of ϕ3T or ρ11 may have different contributions of SPβ DNA in their genomes.

TABLE I Characteristics of SPβ and Some of Its Relatives

	φ3T	ρ11	SPβ*c1*	SPβ	Zeta
Virion size					
Head (nm)	57[a]	81[b]	76 × 82[c]	—	—
	74[b]	—	—	—	—
Tail (nm)	270[a]	378 × 11[b]	357 × 12[c]	—	—
DNA size (kb)	93[b]	90[b]	94[c]	128[e]	—
	117[d]	—	118[e]	—	—
	120[d]	—	—	—	—
Base composition (mol% G + C)	36[a]	34.2[b]	30.6[c]	—	—
	34.2[b]	—	—	—	—
Converts Thy^- to Thy^+	Yes[a]	Yes[b]	—	No[c]	Yes[f]
Immunity group	φ3T	φ3T	—	SPβ	SPβ
Produces betacin[f]	No	No	Yes[g]	Yes	Yes

[a] Tucker (1969).
[b] Dean *et al.* (1976).
[c] Warner *et al.* (1977).
[d] Stroynowski (1978). The value of 117 kb was determined by electrophoretic mobility of *Eco*RI endonuclease fragments. The value of 120 kb was determined from electron micrographs of phage DNA.
[e] Fink *et al.* (1981).
[f] Hemphill *et al.* (1980).
[g] Detected only in double lysogens; see Section III,C,5.

There is no very good way out of this dilemma, since the literature contains a considerable number of papers about these phages already. It would be a useful step toward standardization if strains 1L1 and 1L3 of the *Bacillus* Genetic Stock Center became the generally accepted sources of φ3T and ρ11, respectively. These strains are lysogens of CU1065, an SPβ-sensitive *trpC2* strain described in Section III,C,4. They were constructed by H. E. Hemphill.

C. Biological Characteristics

1. *Distribution*

Phage SPβ is found in *B. subtilis* strain 168 and almost all its derivatives. It is also found in strain NCTC3610, which may be the progenitor of strain 168 (Warner *et al.*, 1977). Stroynowski (1978) has presented evidence that *Bacillus coagulans* strain ATCC 7050 contains DNA homologous with that of the SPβ prophage.

The spontaneous loss of the SPβ prophage by derivatives of strain 168 has apparently occurred more than once in strains carrying the *sup-3* mutation (Warner *et al.*, 1977). No explanation for this is available. In addition, bacteria that undergo the type of spontaneous or mutagen-induced mutation called *citD* (Rutberg and Hoch, 1970; Zahler *et al.*, 1976; Rosenthal, R. *et al.*, unpublished) lack part or all of the SPβ prophage, as well as being deleted for the nearby *kauA* and *citK* genes. The *citD* deletion strains can be detected easily because their lack of *citK* activity (α-ketoglutarate dehydrogenase) makes them unable to sporulate. The sporeless morphology is distinctive on some complex media.

No host for SPβ is known other than cured strains derived from *B. subtilis* strain 168.

2. *Phage Growth*

Although *B. subtilis* strain 168 and most of its derivatives are lysogenic for SPβ and release phage particles spontaneously, the yield of phage is low. Warner *et al.* (1977) have concluded that fewer than 1% of the cells in a culture of strain 168 induced by mitomycin C release any phage. If a nonlysogen is reinfected with SPβ, the phage yield is greatly increased for both spontaneous and chemically induced release (H. E. Hemphill, 1980; Warner *et al.*, 1977). This implies that the prophage in a reinfected cell is not identical to the prophage in strain 168. This difference is not understood. It suggests the possibility that not all the phage particles released by strain 168 are identical. To guard against this, a strain of bacteria named *B. subtilis* su^{3+} (SPβ) (=CU1985) by Warner *et al.* has been used as the source of SPβ phage in the studies that follow. It was isolated from a plaque of SPβ (from strain 168) plated on strain su^{3+} (=CU1050), the original sensitive strain of Warner *et al.* (1977).

Burst sizes of SPβ and SPβ*c1* were reported as 16 (Warner *et al.*, 1977). Phage yields of 10^9/ml can be achieved by mitomycin C induction of lysogens, or by heat induction of lysogens of SPβ*c2* (a phage mutant that produces a heat-sensitive phage repressor; see below), or by lytic infection of strain CU1050 by SPβ*c1*. The latent period is 46 min for lytic infection by SPβ*c1* and 90 min for mitomycin C induction of a lysogen (Warner *et al.*, 1977). We find that heat-induced cultures of lysogens of SPβ*c2* start to lyse visibly in about 45 min.

SPβ does not replicate well at high temperatures. Plaque counts are constant when the phage is plated at temperatures between 25 and 42°C. At 45°C only about 10% of the particles form plaques. At 47°C or higher no plaques are formed, and phage is not released by induced lysogens. *Bacillus subtilis* grows well at temperatures as high as 50°C.

The plating host strain CU1050 [$=su^{+3}$ of Warner *et al.*, (1977)] gives larger and clearer plaques of SPβ than other cured derivatives of strain 168. The difference is not due entirely to the *sup-3* mutation present in

CU1050, since the transfer of *sup-3* to strain CU1064 (Section III,C,4) did not improve plaque morphology (R. H. Lipsky, personal communication). Strain CU1050 contains several mutations, known and unknown, that affect it in various ways. For example, although it carries *sup-3,* it requires methionine; its *metB5* allele is suppressible by *sup-3* in other genetic backgrounds (R. H. Lipsky, personal communication). We suspect that CU1050 carries a modifier of the *sup-3* gene product that decreases the efficiency of *sup-3* as a suppressor.

3. *Betacin*

Lysogens of SPβ produce a bacteriocin-like substance that inhibits the growth of nonlysogens. The substance has been named *betacin* (Hemphill *et al.,* 1980). Betacin is produced only during growth on agar; it has not been detected in the supernatants of broth-grown cultures.

If a plate is spread or overlaid with nonlysogenic bacteria, and a betacin-producing strain is inoculated at a point on the surface of the plate, incubation results in the appearance of a colony of betacin-producing bacteria surrounded by a clear zone of inhibition of about 2 mm. The appearance of the zone is not altered appreciably if (1) the nonlysogen is unable to adsorb SPβ because of a mutation in its *gtaB* gene, (2) the betacin-producing strain carries a defective prophage and cannot make phage particles, or (3) the incubation is carried out at 50°C, a temperature at which SPβ cannot replicate. The observations show that the zone of inhibition is not caused by SPβ virions.

A mutant of SPβ*c2* has been isolated by H. E. Hemphill that gives rise to non-betacin-producing lysogens. The mutant, SPβ*c2bet-1,* gives betacin-tolerant lysogens that seem not to be affected by betacin. We conclude that at least two genes of SPβ are involved in betacin reactions: *bet,* concerned with betacin production, and the putative gene *tol,* involved in making lysogens tolerant to betacin. Both *bet* and *tol* are expressed in SPβ lysogens.

Betacin is relatively heat-resistant. Agar containing it can be melted and recast; betacin activity is not lost. Proteases destroy the activity in molten agar held at 42°C; nucleases have no effect on it. Betacin appears to be a heat-stable protein.

Nonlysogenic bacteria can mutate to give rise to betacin-resistant clones. The bacterial *betR* gene in which such mutations occur has been mapped by transduction with phage PBS1; it lies between *metC* and *pyrD,* somewhat closer to *metC,* at about 120° on the Henner and Hoch (1980) *B. subtilis* chromosome map.

Betacin is produced by lysogens of SPβ and Z, which are also tolerant to it. It is not produced by lysogens of ϕ3T or ρ11, which are sensitive to it (Table I).

4. *Chromosomal Attachment Site*

The SPβ prophage lies on the *B. subtilis* genome between *ilvA* and *kauA* (Zahler *et al.*, 1977), at about 190° on the Henner and Hoch (1980) map. This map includes the SPβ prophage. If the prophage is 128 kb long, it makes up about 4% of the bacterial chromosome—say, from 185 to 199°.

Nonlysogens carry an unoccupied attachment site (*attSP*β) between *ilvA* and *kauA* (Fig. 3). The original SPβ-sensitive strain CU1050 carries the following linked cluster of genes: *trpC*$^+$, *metB5*, *ilvA*$^+$, *attSP*β. This entire cluster was transferred by PBS1 transduction into a lysogen that was *trpC2 ilvA3* (SPβ), with selection for *ilvA*$^+$. The transductant, strain CU1064, is *metB5 attSP*β*;* it is sensitive to SPβ. [Strain CU1065, *trpC2 attSP*β, came from the same cross; see Zahler *et al.*, (1977) for further details.]

If PBS1 transducing phage is produced in strain CU1064 (*metB5 attSP*β), it can be used to cure any derivative of strain 168 of its prophage, provided that the derivative still carries the *trpC2* marker of strain 168. Selection for Trp$^+$ results in the loss of the SPβ prophage from 50% of the transductants. Essentially all the cured transductants carry the *metB5* marker. The simplest test for curing is to transfer purified Trp$^+$ transductant colonies via toothpick to a plate overlaid with sensitive strain CU1050 and examine for betacin production after incubation.

If such a cured strain is later lysogenized, it can again be cured by PBS1 transduction using phage grown in strain CU1065 (*trpC2 attSP*β), or in strain CU1327 (*ilvD15 attSP*β) and selecting for Met$^+$; 85% of the transductants will be cured of SPβ.

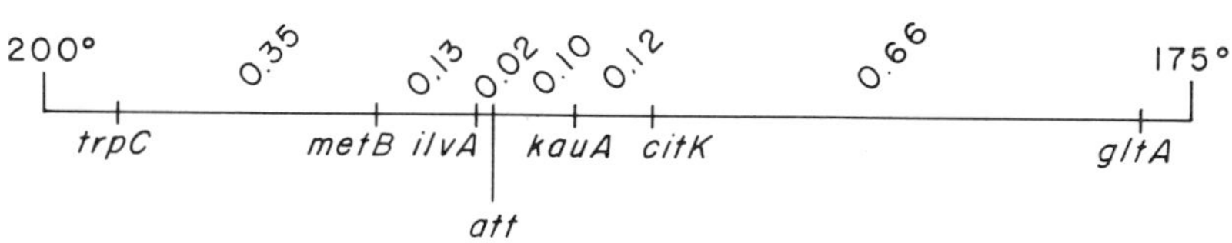

Fig. 3. Genes near *attSP*β on the *B. subtilis* chromosome. Distances are expressed as 1 minus the fraction cotransduced by phage PBS1. The *ilvD* and *thyB* genes lie between *metB* and *ilvA*. Mutations in *gltA, ilvA, ilvD, metB,* and *trpC* cause auxotrophy for glutamate or aspartate, isoleucine, isoleucine plus valine, methionine, and tryptophan, respectively. Mutations in *citK,* which encodes α-ketoglutarate dehydrogenase, cause inability to use lactate as sole carbon source and failure to sporulate. Mutations in *kauA* prevent uptake of the immediate precursors of valine and isoleucine: α-ketoisovalerate and α-ketomethylvalerate (Goldstein and Zahler, 1976). The KauA$^-$ phenotype can only be scored in strains carrying an *ilv*$^-$ mutation. Mutations in *thyB* cause auxotrophy for thymine or thymidine in strains that have mutations in *thyA*. Degree markings correspond to the map of Henner and Hoch (1980), but that map includes the SPβ prophage, *att, attSP*β.

Any of the *attSP*β phage-sensitive strains can be lysogenized by infection. Furthermore, the SPβ prophage can be cotransduced with $metB^+$ or $ilvD^+$ by phage PBS1, so that even strains unable to adsorb SPβ can be lysogenized by it.

Williams and Young (1977) reported that the prophage of ϕ3T lies in the same general region of the *B. subtilis* chromosome, between *ilvA* and *gltA*. The mapping in the Williams and Young paper may be distorted by the presence of SPβ prophage, which was present in both the donors and the recipients in the PBS1 crosses. The SPβ prophage is homologous in part to ϕ3T (Stroynowski, 1978), and recombination between ϕ3T and SPβ prophages may occur. Furthermore, ϕ3T may be able to integrate into an SPβ prophage. The existence of SPβ was not known at the time the Williams and Young work was done. J. M. Odebralski has recently located the ϕ3T prophage between *ilvA* and *kauA* in bacteria not lysogenic for SPβ (unpublished).

5. *Mutations of the* c *(Repressor) Gene*

The *c1* clear-plaque mutation of SPβ was found by H. E. Hemphill and his associates (Warner *et al.*, 1977) following *N*-methyl-*N'*-nitro-*N*-nitrosoguanidine treatment of the induced lysogen su^{3+} (SPβ) (=CU1985). SPβ*c1* is unable to lysogenize sensitive strains of *B. subtilis*. [Bacteria that survive infection by SPβ*c1* are unable to adsorb SPβ or ϕ3T and carry mutations for phage resistance linked by PBS1 transduction to *hisA1*. The mutations conferring resistance probably lie in the *gtaB* gene; F. E. Young (personal communication) has found that bacterial mutants resistant to clear-plaque mutants of ϕ3T lie in *qtaB*, at about 310° on the Henner and Hoch (1980) map.]

Eight other clear-plaque mutants of SPβ isolated by R. M. Herenstein in this laboratory failed to complement the *c1* mutation (Rosenthal *et al.*, 1979) or each other to permit lysogenization. Thus there is no evidence for more than one gene in which mutations give rise to the clear-plaque morphology. However, it is possible that most or all of these mutations contain deletions that may cover more than one *c* gene.

R. Rosenthal isolated a mutant of SPβ that produces a temperature-sensitive phage repressor (Rosenthal *et al.*, 1979). Lysogens of this phage, SPβ*c2*, can be induced to lyse by a brief heat shock (usually 5.5 min at 50°C). Since SPβ cannot replicate at 50°C, the temperature must then be lowered (usually to 37°C). The culture lyses after 45–75 min of incubation at 37°C; the more turbid a culture is, the longer it takes to lyse.

With specialized transducing phages it is possible to construct doubly lysogenic strains of *B. subtilis* that carry two copies of the *c* gene (Rosenthal *et al.*, 1979). We have examined strains that carry *c1* and *c2*, c^+ and *c2*, or two copies of either c^+ or *c2*. SPβ*c1* can form a stable double lysogen in a strain that already carries a *c2* allele; the sites at which the *c* gene product (repressor) works must still be present despite the 10-kb deletion present in SPβ*c1*. Double lysogens containing *c1* and *c2*, or two copies of *c2*, can be heat-induced, although *c2-c2* double lysogens require about 10 min of heating at 50°C instead of 5.5 min for optimum induction. Double lysogens containing c^+ and *c2* cannot be heat-induced. The results are consistent with a model that describes the product of c^+ as a heat-stable repressor, and the product of a *c2* as a heat-labile repressor. Phages carrying the *c1* allele apparently do not make any active repressor.

Strain CU1147 is isogenic with strain 168, but it carries SPβ*c2* rather than SPβ. CU1065 is an SPβ-sensitive strain isogenic with strain 168. When early logarithmic-phase cultures of these three strains grown at 37°C in tryptone broth (TB) (Rosenthal *et al.*, 1979) were subjected to heat shock (50°C, 5.5 min) and returned to 37°C, strain CU1147 lysed and released SPβ*c2;* strains CU1065 and 168 continued to grow. When 37°C cultures of the three strains were shifted to 50°C and incubated at this temperature, the growth of all three strains continued to stationary phase without lysis (P. A. Toye, unpublished experiments).

When the three strains, growing in early logarithmic phase at 37°C, were diluted, plated on tryptose–blood–agar base (TBAB) plates and then incubated at either 50 or 37°C, the colony counts at the two temperatures were essentially identical. All (100) of the colonies of strain 168 tested, grown at either temperature, were lysogenic for SPβ. The colonies of CU1065, of course, were still phage-sensitive. All the colonies of CU1147 incubated at 37°C retained the SPβ*c2* prophage, even when the medium contained anti-SPβ antiserum. However, at least 95%, and in most trials more than 99%, of the colonies of CU1147 grown for 18 hr at 50°C had lost SPβ*c2* and were SPβ-sensitive. It appears that, when the SPβ*c2* repressor is inactivated at 50°C, the prophage is excised from the bacterial chromosome. The excised phage genome apparently cannot replicate at 50°C and is diluted out, although bacterial multiplication proceeds normally.

6. *The* int5 *Mutation*

R. Rosenthal has isolated a mutant of SPβ*c2* that is defective in integration into *attSP*β (Rosenthal, R., Toye, P. A., Rosenberg, R. A., Korman, R. Z., Hemphill, H. E. and Zahler, S. A., unpublished). The mutant, SPβ*c2int5,* was isolated following treatment with *N*-methyl-*N'*-nitro-*N*-nitrosoguanidine. Fewer than one sensitive cell in 10^4 infected by SPβ*c2int5* becomes lysogenic. The mutant is similar to integration-deficient (*int*⁻) mutants of coliphage λ.

The mutant phenotype can be detected as follows. The turbid plaque centers of SPβ*c2int5* (and, as control, SPβ*c2*) are picked with toothpicks and patched onto TBAB plates containing anti-SPβ antiserum. After overnight incubation at 37°C, the patches are transferred by toothpick to plates overlaid with sensitive strain CU1050 to see if the cells in the patches produce betacin. True lysogens—patches from SPβ*c2* plaques—produce betacin. Patches from SPβ*c2int5* plaques do not produce betacin and prove to be sensitive to SPβ*c1* after growth in the presence of antiserum. The turbid plaques of SPβ*c2int5* probably contain abortively lysogenized cells in which the phage genome has not integrated. They segregate sensitive bacteria which become infected by phage released by a fraction of the cells that lyse. Antiserum prevents this reinfection. See Gingery and Echols (1967) and Zissler (1967) for similar observations in coliphage λ.

When a mixture of SPβ*c2int5* and SPβ*c1* was placed on a plate seeded with the sensitive strain CU1065, a region of partial clearing developed after incubation. From this region were isolated bacteria lysogenic for SPβ*c2int5*. We assume that the *int*⁺ function of SPβ*c1* was able to complement the SPβ*c2int5* phage; the SPβ*c2int5* prophage was inserted normally at *attSP*β. One such strain is CU1384 [*trpC2* (SPβ*c2int5*)].

Strain CU1384 [*trpC2* (SPβ*c2int5*)] produces only about 1/10,000 as many plaque-forming particles as strain CU1147 [*trpC2* (SPβ*c2*)] when it is induced by heat-shock or mitomycin C. Apparently *int*⁻ mutations of SPβ, like those of λ (Gottesman and Yarmolinsky, 1968), are deficient in excision as well as integration.

To learn if SPβ*c2int5* can integrate at sites distant from *attSP*β, we infected cells of sensitive strain CU1065 (*trpC2 attSP*β) with SPβ*c2int5* at a multiplicity of 10. After 30 min of incubation to permit adsorption, dilutions of cells were spread on TBAB plates that had previously been spread with 2.5×10^8 particles of SPβ*c2c5int5* (a clear-plaque mutant of SPβ*c2int5*) to kill sensitive cells. On plates spread with about 10^6 infected cells, a background of thin bacterial growth appeared after overnight incubation at 37°C. About 50 individual colonies of lysogens grew up through this background. Many of the lysogens were surrounded by zones of inhibition, presumably due to betacin production. Of 44 lysogens tested, all carried the SPβ*c2int5* prophage. A few, when first isolated, also released SPβ*c2c5int5* clear-plaque phages. This was probably due to the tandem integration of SPβ*c2int5* and SPβ*c2c5int5*. These strains were not stable; after a few passages on agar they released only SPβ*c2int5*. A few other of the lysogens probably contained tandem copies of two SPβ*c2int5* prophages. When heat-induced, these strains released 10^7–10^8 phage particles per milliliter instead of 10^3–10^5/ml typical of SPβ*c2int5* lysogens. Again, repeated transfers apparently resulted in the loss of one of the tandem prophages to give a lysogen with low phage yield. Of the 44 SPβ*c2int5* lysogens examined in this experiment, 10 had the prophage in (or close to) *attSP*β; 15 carried the prophage between the *citK* and *gltA* genes, at about 180° on the Hennis and Hoch (1980) chromosome map; 17 had the prophage between *purB* and *dal,* at about 40° on the map; and two had the prophage at other sites as yet unknown.

It is not surprising to find SPβ*c2int5* integrated at the normal *attSP*β. In coliphage λ it is known that the chromosomal *att* site has a 15-base pair sequence in common with the appropriate site on the phage genome (Landy and Ross, 1977), so that general recombination could result in integration. Furthermore, the *int5* mutation may retain some residual integration activity. It is more difficult to explain the relatively high frequencies of integration into the *citK–gltA* and *dal–purB* regions. It is possible that these regions contain sequences homologous to SPβ—possibly defective phages.

R. Z. Korman has carried out lysogenizations of strain CU156 (see Section III,C,7) with SPβ*c2int5*. Her method was quite different. Suspensions of the phage were spotted on the surface of TBAB plus 0.2% glucose plates overlaid with strain CU156 and incubated for 72–96 hr at room temperature (20–22°C). Cells from the turbid areas of phage growth were purified on TBAB containing anti-SPβ antiserum. Presumably because of strong selection for tolerance to betacin in the areas of phage growth on the original plates, many of the clones isolated this way were lysogenic for SPβ*c2int5*. Strain CU156 does not have a functional *attSP*β, so there is no strong clustering of prophages in that region. Furthermore, no strong clustering in the *citK–gltA* or *dal–purB* regions was observed. Only a few of the prophages in these lysogens have been located, but they seem to be distributed widely around the *B. subtilis* chromosome.

C. D. Vocke has studied the ability of CU1384 [*trpC2* (SPβ*c2int5*)] to form colonies at 50°C. When cells growing in broth logarithmically at 37°C were diluted and plated on TBAB at 50°C, only about 1 cell in 5000 formed colonies. About one-half of the survivors were morphologically nonsporulating; all of these proved to carry deletions extending from within the SPβ*c2int5* prophage, through *citK* (Fig. 3). Three of about 50 studied had longer deletions, extending from the prophage through the *gltA* gene. If the Henner and Hoch (1980) map is correct, these three strains have lost the chromosomal replication terminus. None of the survivors carried deletions extending to the left of the prophage through the *ilvA* or *ilvD* genes. All 50 Cit^+ survivors tested were still lysogenic for SPβ*c2int5*.

C. D. Vocke constructed a strain carrying the unlinked *ilvBΔ1* mutation as well as the SPβ*c2int5* prophage. Using this strain she could test for the *kauA* gene, which lies between the prophage and *citK* (Fig. 3). Although all the Cit^- survivors of this strain at 50°C were $KauA^-$ and $CitK^-$ as expected, Vocke could not find any $KauA^-$ survivors among the $CitK^+$ survivors (50 tested). Short deletions, extending from the prophage through *kauA* but ending before *citK*, must be quite rare if they occur.

The nonsporulating survivors carry deletions that correspond to the *citD* mutations described by Rutberg and Hoch (1970) and Zahler *et al.* (1976). None of them released phage. Of 12 tested by Vocke, 2 were immune to SPβ*c1* and produced betacin; 10 were sensitive to SPβ*c1* and did not produce betacin.

The inability of most cells of strain CU1384 to form colonies at 50°C must be related to the inability of SPβ*c2int5* to excise properly when its repressor is inactivated at the high temperature. We suspect that the SPβ prophage has a gene [corresponding to the *kil* gene of coliphage λ (Greer, 1975)] that kills lysogens when it is derepressed. We do not know if the inactivation of the *c2* repressor also derepresses phage genes that are responsible for the high frequency of deletions found in these experiments, or if the deletions occur spontaneously and are selected in the experiments; we suspect the former.

7. Bacillus subtilis *Strain CU156*

R. Z. Korman in this laboratory screened a number of *citD* deletion strains (see Rutberg and Hoch, 1970; Zahler *et al.*, 1976) in the Cornell University stock collection to see if any had lost the SPβ attachment site *attSP*β. One strain, CU156, acts as though it does not have a functional *attSP*β. Its genotype is *trpC2 ilvA3 citD2*. Like other *citD* strains, CU156 does not release SPβ, and it lacks *kauA* and *citK* activity: it is unable to take up the α-keto analogs of isoleucine and valine; it cannot use lactate as sole carbon source, nor can it sporulate. It does not produce betacin. Unlike most *citD* strains in our collection, it is sensitive to SPβ*c1* in a cross-patch test. Isolated plaques of SPβ or of SPβ*c1* cannot be seen on a lawn of CU156, but cleared areas are seen if a drop containing 10^3 phage or more is placed on the surface of an overlaid plate. The plating agar used for CU156 and other Cit^- bacteria is TBAB containing 0.2% glucose.

A simple way to isolate new lysogens of SPβ*c2* carrying the prophage at sites distant from the normal attachment site has been developed by

R. Z. Korman and C. J. Mackey. Drops of phage suspension containing 10^5–10^7 SPβ*c2* or SPβ*c2int5* particles were placed on the surface of TBAB plus 0.2% glucose plates overlaid with CU156. The plates were incubated at room temperature (20–22°C) for 3 or 4 days. Bacteria from turbid areas of phage growth were picked and purified on TBAB plus 0.2% glucose plus anti-SPβ antiserum. Presumably because of selection for tolerance to betacin, most of the bacteria isolated in this way were lysogens. Phage yields from these lysogens were low, usually 10^3–10^5 phages/ml, perhaps because the DNA substrate needed for the efficient precise excision of the phage genome is not present.

D. Specialized Transduction of Genes near *attSP*β

1. *Introduction*

Zahler *et al.* (1977) reported that SPβ lysates contained defective specialized TPs for two genes that lie to the right of the prophage, *kauA* and *citK*. High frequency of transduction lysates could be prepared from some of the transductants.

Assays of TPs were carried out by adding filter-sterilized phage preparations to late logarithmic-phase recipient cells in antibiotic medium no. 3 or other complex media, allowing adsorption to occur for 20 min, centrifuging, resuspending the cells in buffer, and spreading samples on selective media. Assays of TPs were usually higher (up to threefold) if the recipient cells were lysogens. Transducing particles could also be detected by placing a drop of lysate on a selective agar plate on which recipient cells had been spread, but only 5–10% of the expected number of transductant colonies appeared after incubation.

One particular transducing phage that carried *kauA*$^+$ and *citK*$^+$ was named SPβd*citK*-1; see Fig. 4A for its probable origin. Rosenthal *et al.* (1979) gave an extended description of a derivative of this defective transducing phage. They replaced the c^+ allele of the defective phage with the *c2* allele by recombination and studied the resulting SPβ*c2*d*citK*-1. The findings are summarized in Fig. 4. When nonlysogens were transduced, the addition-type heterogenotic transductants had the defective prophage integrated into the region of bacterial DNA homology. The phage genes lay between the two copies of the bacterial genes.

Double lysogens that carried the SPβ*c2* and SPβ*c2*d*citK*-1 prophages gave rise to HFT lysates that contained from 1 to 10 TPs per 1000 PFU. No HFT lysates with TP/PFU ratios above about 1:100 have been found for any marker.

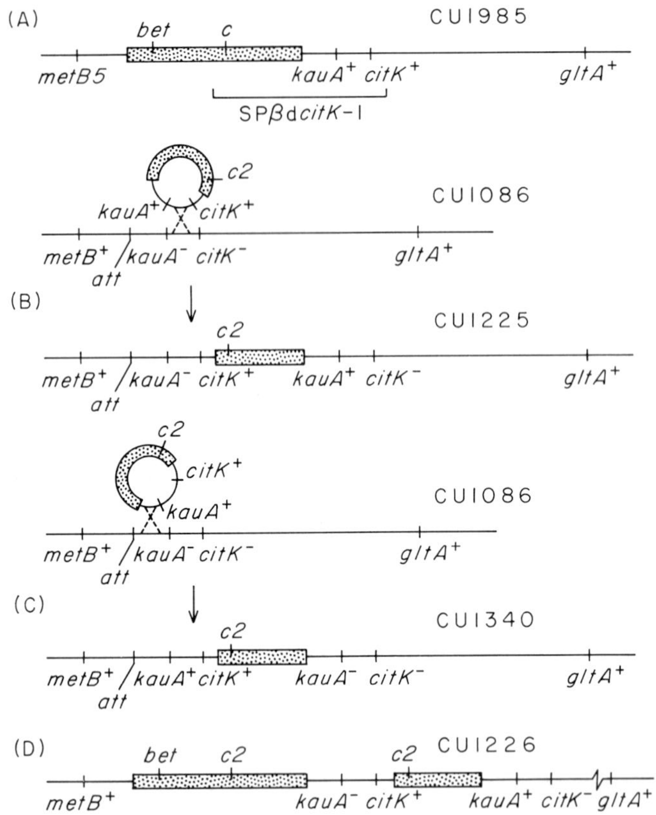

Fig. 4. Production of and transductions with SPβ*c2*d*citK*-1. (A) The origin of SPβd*citK*-1, through improper excision of SPβ from strain CU1985. (B) Insertion of SPβ*c2*d*citK*-1 between the *kauA*$^-$ and *citK*$^-$ markers of strain CU1086, producing the single defective lysogen CU1225. The *c2* allele replaced c^+ by recombination; see Rosenthal *et al.* (1979). Note that *att* (= *attSP*β) is unoccupied. (C) Insertion of SPβ*c2*d*citK*-1 between the *attSP*β and *kauA*$^-$ markers of strain CU1086, producing the single defective lysogen CU1340. Note that *attSP*β is unoccupied. (D) The double lysogen CU1226, produced by superinfecting strain CU1225 with SPβ*c2*. Strains CU1086, CU1225, CU1226, and CU1340 also carry the *ilvB*Δ*1* marker, which permits scoring of *kauA*.

It is possible to isolate defective specialized transducing phages that carry the *kauA* gene but not *citK;* they are named SPβ*c2*d*kauA* (J. M. Odebralski, unpublished experiments).

The order of known bacterial genes on the left side of *attSP*β is *metB, ilvD, thyB, ilvA, attSP*β. Fink *et al.* (1980) have isolated a series of defective transducing phages that carry *ilvD, thyB, ilvA,* or *thyB, ilvA,* or only *ilvA*. Single lysogens carrying any of these phages produce betacin, unlike SPβ*c2*d*citK* and SPβ*c2*d*kauA* lysogens. Presumably the *bet* gene of SPβ lies near the left end of the prophage (Fig. 4A).

P. S. Fink has mapped the gene order of single lysogens carrying SPβ*c2*d*thyB*-1, which includes the *thyB, ilvA* segment. Like SPβ*c2*d*citK*-1, SPβ*c2*d*thyB*-1 can apparently integrate anywhere within the region of the bacterial chromosome that is homologous to the defective phage. In each case studied, the single lysogen (as in the case of CU1225 and CU1340, Fig. 4B and C) still possessed an unoccupied *attSP*β. In our experience, defective SPβ TPs do not integrate into the attachment sites of nonlysogenic bacteria (Fink and Zahler, 1980).

An HFT lysate can be produced from double lysogens carrying the prophage of SPβ*c2* and one of the defective transducing phages of the *ilvA, thyB, ilvD* group. Such lysates contain fewer TPs than lysates of *citK* and *kauA;* the TP/PFU ratio is seldom greater than 1:10,000.

2. *Replacement Transductions*

When an HFT lysate containing SPβ*c2*d*citK*-1 is diluted to prevent double infections and used to transduce a nonlysogen to $CitK^+$ $KauA^+$, about one-half of the transductants are the result of replacement events. These transductants are still sensitive to SPβ*c1;* they do not segregate $CitK^-$ clones at detectable rates, nor can the $citK^-$ or $kauA^-$ alleles be detected in them if they are used as donors in PBS1 transductions. It seems clear that the $citK^-$ and $kauA^-$ alleles of the recipient strain have been replaced by the positive alleles carried by the SPβ*c2*d*citK*-1 particles. Replacement-type transductions also occur when the recipients are lysogenic; about one-third of the transductants are of this type.

Lysates of SPβ*c2* from lysogens in which the prophage is in the normal attachment site carry small numbers of TPs for the *metB* and *gltA* genes (Rosenthal *et al.,* 1979). These genes lie, respectively, to the left of *ilvD* and to the right of *citK* (Fig. 4A). Many $MetB^+$ and $GltA^+$ transductants have been examined (R. Z. Korman, unpublished); all act as though they have resulted from replacement-type events. No heterogenotes or HFT-producing strains have been found.

3. *Addition Transductions of Nonlysogenic Recipients*

When an HFT lysate of SPβ*c2*d*citK*-1 was diluted and used to transduce a phage-sensitive recipient to $CitK^+$ $KauA^+$, about one-half of the transductants became lysogenic for the defective prophage. These transductants were immune to SPβ*c1* and were heterogenotic for *citK* and *kauA*.

The negative alleles of these genes could be detected in the transductants in several ways (Rosenthal *et al.*, 1979). First, when PBS1-transducing lysates were prepared from the transductants, the *kauA*$^-$ and *citK*$^-$ alleles could be found linked to the *metB*$^+$ or *gltA*$^+$ genes of the transductants (Fig. 4B and C). Second, when they were streaked or plated on TBAB, the transductants occasionally segregated CitK$^-$ sectors, which could be detected because their inability to sporulate caused an altered morphology. Third, and most striking, when the addition-type transductants were streaked or plated on TBAB and incubated overnight at 50°C, every colony either was CitK$^-$, or segregated CitK$^-$ sectors. The Cit$^-$ colonies or sectors had lost their immunity to SPβ*c1*.

It was possible to determine the order of the alleles in these addition transductants. Of three tested, the defective transducing phage had inserted between *kauA*$^-$ and *citK*$^-$ in two (Fig. 4B), and between *attSP*β and *kauA*$^-$ in one (Fig. 4C), as deduced from the order in which the positive and negative alleles lay on the chromosome. The order was determined by PBS1 transductions and was corroborated by the classes of segregants found among the colonies at 50°C. In almost every case, a defective prophage-size piece of genome was lost in the act of segregation. For example, the loss of a prophage-size segment of DNA from CU1225 (Fig. 4B) gives rise to CitK$^-$ KauA$^+$, CitK$^+$ KauA$^-$, or CitK$^-$ KauA$^-$ phage-sensitive segregants; but the loss of a similarly sized segment of DNA from CU1340 (Fig. 4C) cannot give rise to CitK$^+$ KauA$^-$ segregants. We conclude that most of the segregants seen at 50°C are produced by recombination between the homologous duplicated regions of bacterial DNA, resulting in the loss of a defective prophage-size segment of DNA that includes all the DNA of SPβ origin.

Single lysogens carrying defective prophages can be superinfected by SPβ or SPβ*c2* to give rise to double lysogens that give HFT lysates when induced. Superinfections have usually been carried out by destroying the defective lysogen's phage repressor by heat shock (5.5 min at 50°C), followed immediately by infection with SPβ*c2* at a multiplicity of infection of 3–5. We have recently found that the heat treatment is not needed; SPβ*c2* can integrate efficiently into the normal attachment sites of cells lysogenic for defective prophages.

The double lysogen CU1226, produced by the superinfection of CU1225 with SPβ*c2*, is shown in Fig. 4D. Its genotype is *ilvB*Δ*1 kauA1 citK1* (SPβ*c2*) (SPβ*c2*d*citK*-1). When CU1226 was plated on TBAB and incubated at 37 or at 50°C, occasional Cit$^-$ sectors were seen. We have not analyzed this segregation extensively. Preliminary examination suggests that the great majority of colonies that form at 50°C still consist of the heterogenotic double lysogen CU1226. The Cit$^-$ segregants isolated at 50°C (found as visible sectors in about 5% of the colonies) were about one-half lysogens of SPβ*c2* and one-half SPβ-sensitive clones. The Cit$^-$ sectors isolated at 37°C were all still lysogenic for SPβ*c2*.

We have found that 95–100% of SPβ*c2* single lysogens were cured of the prophage by overnight growth at 50°C (Rosenthal *et al.*, 1979). Essentially all SPβ*c2*d*citK*-1 singly lysogenic cells plated at 50°C gave rise to colonies that segregated Cit$^-$ sectors. Nevertheless, double lysogens such as CU1226, although they could be induced to lyse by a 10-min heat shock at 50°C, grew on plates at 50°C without much sectoring and without losing either prophage. This unexpected dosage effect of the *c2* allele has not been explained.

4. *Addition Transductions of Lysogenic Recipients*

When an HFT lysate is used to transduce a lysogen, two kinds of transductants can be distinguished easily: those that give rise to HFT lysates when they are induced, and those that do not. The former are addition-type transductants. We believe that the latter are replacement-type transductants, and we have not studied them.

There should be two classes of addition transductions into lysogenic recipients. In class B, the defective transducing phage genome integrates into the region of bacterial homology (Fig. 5A), as it always does during integration into a nonlysogen. In class P, the transducing phage genome integrates into the resident prophage of the recipient (Fig. 5B).

It is not possible to distinguish these two classes for transduced genes close to *attSP*β, but they can be distinguished if the transduced gene is far from *attSP*β. In one case, the integration of SPβ*c2int5* into the *dal–purB* region at about 40° on the Henner–Hoch (1980) map allowed us to isolate a specialized transducing phage for *dal,* SPβ*c2int5*d*dal*-1. Transduction of a Dal$^-$ lysogen to Dal$^+$ by HFT lysates of SPβ*c2int5*d*dal*-1 gave rise to addition-type transductants that produced HFT lysates when induced. PBS1 transducing lysates prepared from these transductants were tested to see whether the *dal*$^+$ gene was linked to genes near *attSP*β, e.g., *metB* or *citK*. Such a linkage is expected for P integrations, but not for B integrations (P. A. Toye, unpublished experiments). Similarly, the integration of SPβ*c2* into the *citF* gene (at about 255°) of a *B. subtilis* strain lacking a functional attachment site allowed the isolation of a defective specialized transducing phage carrying the entire *ilvBC–leuACBD* segment. When a lysogenic recipient was transduced to IlvB$^+$ by an HFT lysate of SPβ*c2*d*leuD*-1, addition-type transductants were isolated and PBS1 lysates were produced in them. In both the Dal$^+$ transductions and the IlvB$^+$ transductions, about one-half of the addition-type transductants were class P, showing linkage of *dal*$^+$ or *ilvB*$^+$ to

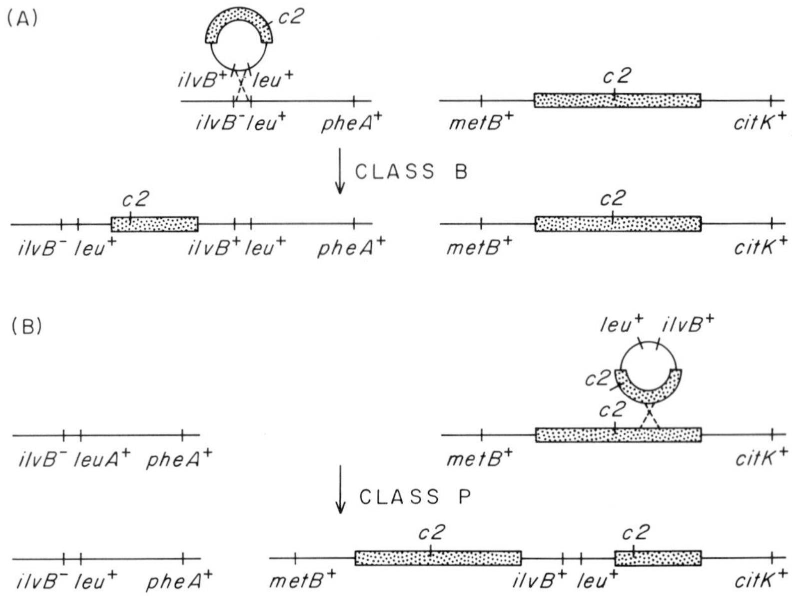

Fig. 5. Integration of SPβ*c2*d*leuD*-1 into an *ilvB*⁻ lysogen. (A) Class-B integration into the region of bacterial homology. (B) Class-P integration into the resident prophage.

metB⁺. In the case of SPβ*c2*d*leuD*-1, the linkage of *ilvB*⁺ to *pheA*⁺ (which is normally close to *ilvB* on the *Bacillus* chromosome) could also be tested; see Section III,G,3.

We believe that the specialized transduction of lysogens of SPβ will usually result in roughly equal numbers of replacement transductions, class-B addition transductions, and class-P addition transductions. Certain exceptions to this expectation have been found. One has been mentioned already (Section III,D,2): TPs that carry genes far from the prophage (e.g., *metB* and *gltA*, for SPβ in the normal attachment site) give only replacement transductions. A second exception, the defective transducing phage SPβ*c2*d*glnA*-1 (Section III,F,3), does not give class-B addition transductions.

The HFT lysates from transductants produced by class-B integrations differ in one striking way from those of class P. Class-B lysates contain many specialized TPs carrying the negative allele of the marker selected in the transduction; class-P lysates contain very few. For example, the transductant shown in Fig. 5A (class B) would release many SPβ*c2*d*leuD*-1 particles carrying the *ilvB*⁻ allele; the transductant shown in Fig. 5B (class P) would release very few.

E. Specialized Transduction of Genes near Insertion Sites of SPβ*c2int5*

The integrase-deficient phage mutant SPβ*c2int5* was described in Section III,C,6. The isolation of bacteria carrying SPβ*c2int5* at two locations distant from *attSP*β has been described (Rosenthal, R., Toye, P. A., Rosenberg, R. A., Korman, R. Z., Hemphill, H. E., and Zahler, S. A., unpublished). One site lies between *citK* and *gltA*, at about 180° on the Henner–Hoch (1980) map. The other lies between *dal* and *purB*, at 40°.

1. *SP*βc2int5 *in the Normal Attachment Site*

We have looked briefly at heat-induced lysates of strain CU1384, in which SPβ*c2int5* is integrated at the normal attachment site (see III,C,6). Although the yield of whole phage from CU1384 is only 10^3–10^4 PFU/ml, the yields of TPs for *citK* and for *ilvD* are almost as high, so that the TP/PFU ratio approaches unity. Clearly the *int5* mutation makes inaccurate excisions frequent, so that defective transducing phages can easily be isolated from SPβ*c2int5* lysogens.

2. *SP*βc2int5 *Prophage between* citK *and* gltA

J. M. Odebralski has isolated defective specialized transducing phages from strain CU1370 [*trpC2* (*citK-gltA*::SPβ*c2int5*)] that carry *citK*, or *kauA, citK*, or *ilvD, thyB, ilvA, attSP*β, *kauA, citK* (Fig. 3). The latter phage is named SPβ*c2int5*d*ilvD*-101. Single and double lysogens carrying each of these defective prophages have been constructed. The single lysogens do not produce betacin, suggesting that the *bet* gene does not lie at the left end of the prophage in CU1370. (The *bet* gene is at the left end of prophages in the normal attachment site; see Section III,D,1.) This may mean that the prophage is inserted facing in the direction opposite the prophage orientation in strain 168 (Fig. 6A and B). However, an alternative explanation is possible (Fig. 6C). The prophage may be inserted facing in the same direction as prophages in the normal attachment site, but with different end points. This possibility exists because the (presumed) circular Campbell-type intermediate was not integrated by means of the site-specific *int* function but, we assume, by nonspecific recombination functions of the phage or its host. The two orientations shown in Fig. 6B and C can be distinguished by appropriate hybridizations between SPβ DNA and the DNA of single lysogens of SPβ*c2int5*d*ilvD*-101, but these tests have not yet been carried out.

The presence of *attSP*β on SPβ*c2int5*d*ilvD*-101 makes the defective prophage a potential candidate for studying the mechanics and enzymology of SPβ integration and excision.

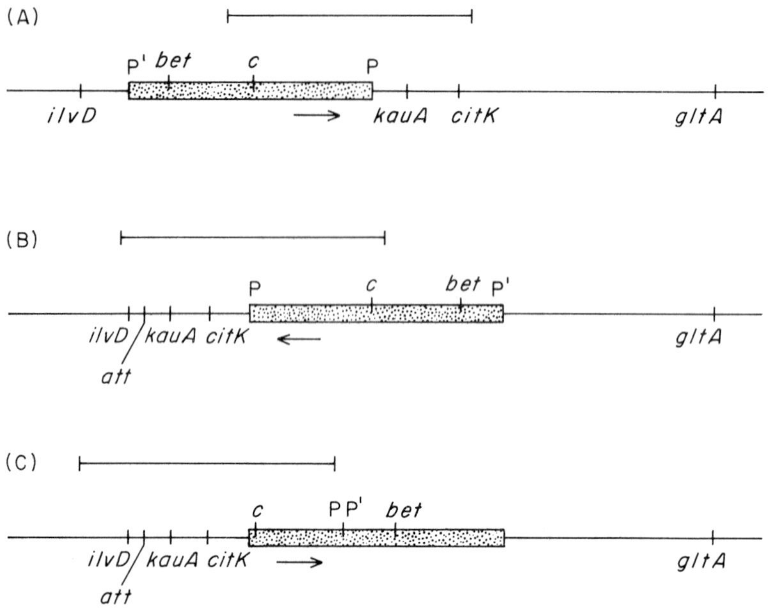

Fig. 6. Orientations of SPβ prophage explaining the origins of some defective transducing phage genomes. (A) Origin of SPβd*citK*-1 from SPβ prophage at the normal attachment site. The prophage ends are named P′ and P, as in Fig. 1, Chap. 9. The arrow indicates the prophage orientation from *c* (the repressor gene) to P. (B) One possible origin of SPβ*c2int5*d*ilvD*-101 from SPβ*c2int5* integrated at a site between *citK* and *gltA*. The prophage has the same ends as in (A), but is inserted in the opposite orientation. (C) Another possible origin of SPβ*c2int5*d*ilvD*-101 from SPβ*c2int5* integrated at a site between *citK* and *gltA*. Because the *int* gene product was not involved in integration, the prophage ends are not P and P′. Integration occurred at another point in the phage genome, and because of the Campbell-type circular intermediate in integration, the nucleotides corresponding to P and P′ are adjacent in the prophage. Note that in each case the transducing phage lacks the *bet* gene but carries the *c* gene.

3. *SP*βc2int5 *Prophage between* purB *and* dal

P. A. Toye and R. Z. Korman have studied specialized transducing phages from a prophage inserted between *dal* and *purB* at about 40° on the Henner–Hoch map (Rosenthal, R., Toye, P. A., Rosenberg, R. A., Korman, R. Z., Hemphill, H. E. and Zahler, S. A., unpublished). Specialized transducing phages for *dal* (which encodes alanine racemase) were obtained easily, and the SPβ*c2int5* prophage could be transferred by cotransduction with *dal*$^+$ by phage PBS1. Single and double lysogens were prepared. The single lysogens do not produce betacin.

Henner and Steinberg (1979) have reported the presence of an extracistronic suppressor mutation, *sup-3,* close to *dal.* The gene order in the region is *aroI, narB, sup-3, dal, purB* (Henner and Hoch, 1980). R. H. Lipsky constructed a strain called CU1968 [*metB5* (*dal-purB*::SPβ*c2int5*)]. The *metB5* mutation is suppressible by the known extracistronic suppressors *sup-3, sup-44,* and *sup-67.* He transformed CU1968 with DNA extracted from bacteria carrying each of the three suppressors, to produce strains, CU2056, CU2057, and CU2050, which carry *sup-3, sup-44,* and *sup-67,* respectively. Each is lysogenic for SPβ*c2int5* at the site near *dal,* and each still carries the *metB5* mutation (although methionine is not required because of the suppressor markers).

When CU2056 and CU2057 were heat-induced, specialized transducing phages could be found in their lysates that carried both *dal* and *sup* genes. From these, both single and double lysogens were constructed. HFT lysates from the double lysogens contain many SPβ*c2int5*d*sup3* or SPβ*c2int5*d*sup44* particles and allow us to test for the suppressibility of new auxotrophic mutations by means of a simple spot test: a drop of diluted lysate is placed on a minimal selective plate spread with the auxotroph. The appearance of prototrophic colonies means that the mutation is suppressible (Lipsky *et al.,* 1980). Difficulties with finding a transducing phage carrying *sup-67* may be technical, or may mean that the *sup-67* mutation is not located near *dal.*

F. Specialized Transduction from SPβ*c2* Prophages Integrated in *Bacillus subtilis* Strains That Lack a Functional Attachment Site

1. *Introduction*

The discovery by R. Z. Korman that strain CU156 lacks a functional *attSP*β was described in Section III,C,7. Two derivatives of CU156 were constructed by C. J. Mackey: CU1840 (*metB5 citD2*) and CU1852 (*argA2 pheA2 metB5 citD2*). They have characteristics like those of CU156 with respect to infection by SPβ: the great majority of integrations occur at sites distant from the *attSP*β region. This has permitted us to study specialized transduction of markers scattered widely on the *B. subtilis* chromosome.

We have used three techniques for finding TPs for particular genes.

1. The method of choice at present is to screen heat-induced lysates of SPβ*c2* and SPβ*c2int5* lysogens of CU156 to see which contain TPs for the marker in question. We now have a collection of more than 200 such lysogens; most of them were isolated as described in Section III,C,6

and 7. The method lacks elegance, but it works. An example is described in Section III,F,2.

2. We have heat-induced a mixed collection of lysogens of CU156 and searched in the lysate for rare TPs for the genes we wanted. This method has worked on occasion, but it has the disadvantage of not allowing us to isolate the lysogen in which the TP originated. An example is given in Section III,F,3.

3. A method we have used only once consisted of isolating lysogens in a derivative of CU156 that had two linked auxotrophic markers, one on each side of the gene for which TPs were wanted. A collection of lysogens of the derivative strain was assembled. Each lysogen was transduced by PBS1, with selection for the positive alleles of the two auxotrophic markers. Lysogens that carried SPβ*c2* prophage between the two auxotrophic markers were cured of the prophage by this transduction; this could be demonstrated by the failure of the PBS1 transductants to produce betacin. An example of this will be given in Section III,F,4.

2. *Screening a Collection of Lysogens of CU156*

J. M. Odebralski screened heat-induced lysates prepared from 44 lysogens of CU156 for TPs for glutamine synthetase. About one-half of the lysogens carried SPβ*c2* and one-half carried SPβ*c2int5*. About 20 μl of each lysate was spotted on the surface of a minimal methionine–agar plate (nine lysates per plate) that had been spread with a glutamine auxotroph, strain CU1770 [*metB5 glnA100* (SPβ*c2*)]. Two of the 44 lysates gave rise to Gln^+ colonies. One came from a lysogen of SPβ*c2*, the other from an SPβ*c2int5* lysogen.

Transductants from the SPβ*c2int5* lysogen's lysate gave HFT lysates for glutamine synthetase. Single lysogens could be prepared from these HFT lysates. The SPβ*c2int5* prophage lies between *thyA* and *glnA*, and TPs for *thyA* are also produced.

On the other hand, the SPβ*c2* lysogen produced heat-induced lysates that gave only replacement-type transductions of CU1770. We assume that the prophage in this lysogen is too far from *glnA* to give particles capable of addition-type transductions. (A similar situation is seen when we try to isolate TPs for *metB* or for *gltA* from SPβ prophage at the normal attachment site; see Section III,D,2.)

3. *Screening a Lysate Made from a Pool of Lysogens of CU156*

Plates of TBAB plus 0.2% glucose were overlaid with CU156. The plates were then inoculated with SPβ*c2* in many separate places. The method used was to wet the prongs of a 48-prong aluminum replicator (designed for inoculating microtiter plates) with a suspension containing

10^8 SPβ*c2* per milliliter. Three plates were inoculated successively. The plates were then incubated at room temperature for 4 days. During this period numerous lysogens should have been produced on the plates, with prophages located at many different regions of the bacterial chromosome. Many of the nonlysogens on the plates should have been inhibited or killed by betacin released by the lysogens. Microcolonies of the lysogens should have formed.

The overlay agar from the three plates was harvested into 500 ml of M broth (Warner *et al.,* 1977) and incubated in a 2-liter flask at 37°C for 4 hr. Then the flask was heated to 50°C for 5.5 min to induce lysis of the lysogens, cooled rapidly to 37°C, and incubated for 75 min at 37°C. Lysozyme was added to remove nonlysogens. The culture was centrifuged to remove cells and agar, and the phage particles were concentrated by adding poly(ethylene glycol) (PEG 6000; final concentration, 6%) and chilling at 4°C overnight. Phage were recovered by low-speed centrifugation and then resuspended in buffer and filtered. The 100-fold concentrated phage preparation was screened for TPs for several markers. We expected few TPs for any particular marker, so spot tests were not attempted.

One particularly interesting transducing phage found by this method carries the glutamine synthetase gene, *glnA*. It was found in a Gln^+ transductant which gave HFT lysates when induced. It is not possible to isolate single lysogens of this phage, SPβd*glnA*-1. When a Gln^- nonlysogen is transduced by an HFT lysate of SPβd*glnA*-1, only two kinds of transductants are found: replacement transductants that are phage-sensitive, and class-P transductants that probably arise from simultaneous infections by SPβ*c2* and SPβd*glnA*-1. Double lysogens are easily produced if a lysogen is transduced to Gln^+; they also are all of class P, with the defective phage integrated into (or in tandem with) the resident prophage.

The double lysogens that carry SPβ*c2* and SPβd*glnA*-1 can be fully induced by a 5.5-min heat shock, whereas most double lysogens require about 10 min. This leads us to suspect that SPβd*glnA*-1 lacks the *c* gene. However, this does not seem to offer a simple explanation for its inability to form single lysogens or class-B double lysogens. It should be pointed out that the $glnA^-$ allele can always be detected in double lysogens of SPβ*c2* and SPβd*glnA*-1; it lies at the normal *glnA* locus (160° on the chromosome), linked to *thyA*.

4. *Screening for Prophages near a Particular Gene*

C. J. Mackey searched for TPs for the *leu* genes (at about 250° on the *B. subtilis* chromosome map) by the following method. She constructed a derivative of strain CU156 that carried the following markers: *argA2, pheA2, metB5, citD2*. The strain was labeled CU1852. The *argA* and *pheA* genes are about 20% cotransduced by phage PBS1; the *leu* genes lie between them. CU1852 lacks a functional *attSP*β because of the *citD2* marker.

SPβ*c2* lysogens of CU1852 were isolated. Thirty-four of these lysogens were transduced to Arg⁺ Phe⁺ by PBS1 phage from donor strain CU1064 (*metB5 attSP*β). In 33 of the 34 cases, the Arg⁺ Phe⁺ transductants still produced betacin and thus were still lysogenic. In one of the 34 cases, none of the Arg⁺ Phe⁺ transductants produced betacin.

The SPβ*c2* prophage in this lysogen proved to lie within the *citF* gene, which codes for succinate dehydrogenase. Gene *citF* lies close to *leu*, between *argA* and *pheA*. The genotype of the lysogen, CU1881, is thus *argA2 pheA2 metB5 citD2* (*citF*::SPβ*c2*-1). Transducing particles for *leu* can easily be isolated from it. The SPβ*c2* prophage can be cotransduced with *leu* by phage PBS1 (Mackey *et al.*, 1980).

G. Genetic Analysis with Defective Transducing Phages

1. *Introduction*

The major purpose of developing a system for specialized transduction in *B. subtilis* was to allow analysis of the bacterial genome, in many of the ways that phage λ has been useful in analyzing *Escherichia coli*. The material described so far gives little insight into how *B. subtilis* works, although it does show that such genes as *ilvA*⁺, *thyB*⁺, *ilvD*⁺, *kauA*⁺, *citK*⁺, *gln*⁺, *dal*⁺, *sup-3, sup-44*, and *leu*⁺ are dominant in heterogenotes.

A more detailed analysis has been undertaken by C. J. Mackey. Her experiments with the *ilvBC–leu* region of the chromosome point the way toward using SPβ for determining cistron relationships and detecting cis-dominant mutations in *B. subtilis* (Mackey *et al.*, 1980).

2. *Specialized Transducing Phages of the* ilvBC–leu *Region*

The isolation of CU1881, carrying SPβ*c2* prophage within its *citF* gene, was described in Section III,F,4. From CU1881, C. J. Mackey isolated a transducing phage, SPβ*c2*d*leuD*-1, that carries the entire region from within *citF* past *leuD* (Fig. 7A). Single lysogens carrying this defective phage do not produce betacin, but they are immune to SPβ*c1*. An HFT lysate is produced by double lysogens. Mutants in *ilvBC* or any of the *leu* genes can be complemented by SPβ*c2*d*leuD*-1.

Mackey also isolated TPs carrying smaller amounts of the genetic material of this region. She constructed a lysogenic recipient carrying

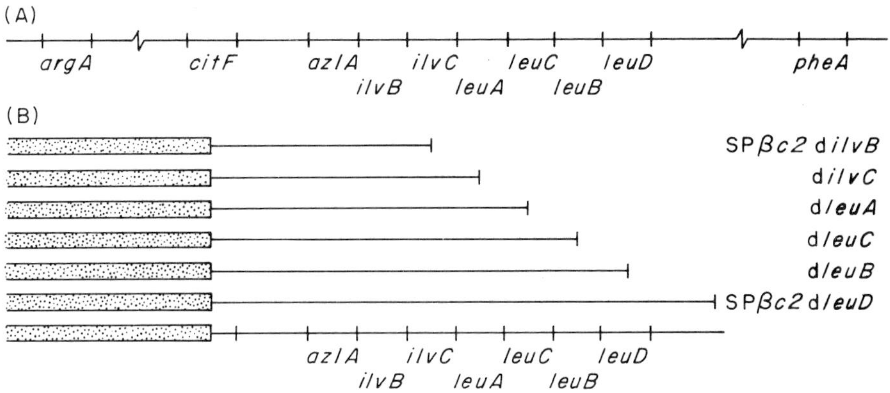

Fig. 7. The *ilvBC–leu* region of *B. subtilis.* (A) The orientation of genes in the region. Genes *argA* and *pheA* are cotransduced 20% by phage PBS1. (B) Specialized transducing phages from SPβ*c2* integrated in the *citF* gene. The phages are named according to the rightmost complete cistron they carry. Mutations in *argA, ilvBC, leuABCD,* and *pheA* cause auxotrophy for arginine, isoleucine and valine, leucine, and phenylalanine, respectively. The *citF* gene (which may include several cistrons) encodes succinate dehydrogenase; mutations (or insertion) in *citF* prevent utilization of lactate as sole carbon source, and block sporulation. Mutations in *azlA* cause constitutive overproduction of leucine, and resistance to the leucine analog 4-azaleucine.

ilvB$^-$ and *leuB*$^-$ alleles. (At that time it was thought that *leuB* was the rightmost gene of the cluster.) She isolated 100 IlvB$^+$ transductants, using an LFT lysate of CU1881 as donor. Thirty-two of the 100 were still Leu$^-$. Twenty of these 32 transductants were studied further.

Eleven of the 20 did not give HFT lysates for *ilvB*. Presumably, they were the result of replacement transductions. The other 9 transductants gave rise to HFT lysates. Two of the 9 complemented *ilvB*$^-$ mutations, but not *ilvC*$^-$, and thus carried DNA that extended into but did not include all of *ilvC;* 4 ended in *leuA;* 1 ended in *leuC;* and 2 ended in *leuB* (Fig. 7B). The distribution suggests a fairly random set of end points without obvious clustering.

When a loopful of an HFT lysate of one of these "short phages", say SPβ*c2*d*leuA*-1 (which ends in *leuC*), is spotted on various auxotrophs spread on minimal agar plates, *ilvB*$^-$, *ilvC*$^-$, and *leuA*$^-$ auxotrophs give solid regions of growth—several hundred colonies—at the site of the lysate inoculation. No colonies at all are seen with *leuB*$^-$ or *leuD*$^-$ auxotrophs. Different *leuC*$^-$ auxotrophs give different results. Some give no colonies at all; presumably their defects lie beyond the DNA carried by the TP. Other *leuC*$^-$ auxotrophs give a few colonies at the site of lysate inoculation, less than 5% of the number found with *leuA* auxotrophs. There is no difficulty in distinguishing these recombination events from the complementation found with *leuA* mutants.

3. *Distinguishing Class-B from Class-P Double Lysogens*

C. J. Mackey isolated 20 transductants from a cross in which the recipient was an *ilvB*$^-$ lysogen and the donor HFT lysate contained SPβ*c2*d*leuD*-1. Fourteen of the Ilv$^+$ transductants produced HFT lysates when induced. Six did not and were assumed to have resulted from replacement transductions. The strains that produced HFT lysates were tested to see if the defective prophage lay in the *ilvBC–leu* region (class B) or in the resident prophage (class P; see Section III,D,4).

Phage PBS1 lysates were produced from the 14 double lysogens. These phage preparations were used to transduce two different recipient strains to prototrophy: an *ilvB*$^-$ *pheA*$^-$ auxotroph and an *ilvB*$^-$ *metB*$^-$ auxotroph. Cotransduction of the prophage *ilvB*$^+$ gene with *pheA*$^+$ is diagnostic of a class-B integration. Cotransduction of the prophage *ilvB*$^+$ gene with *metB*$^+$ is diagnostic of a class-P integration, with the defective prophage near the normal attachment site. Five of the 14 addition-type transformants were of class B, 7 were of class P, and 2 showed PBS1 cotransduction of *ilvB* with both the *pheA* and the *metB* genes. We have not examined the latter class further.

4. *Transducing Particles Carrying Negative Alleles*

C. J. Mackey transduced a *leuA*$^-$ lysogenic recipient to Leu$^+$ with SPβ*c2*d*leuD*-1 (Fig. 8A). She pooled several transductants and heat-induced the mixture. We expected that some of the transductants would be of class B, and that among the particles they released would be some carrying the *leuA*$^-$ allele (Fig. 8B).

The lysate was used to transduce an *ilvB*$^-$ *leuA*$^-$ lysogen (carrying the same *leuA*$^-$ allele) to *Ilv*$^+$ (Fig. 8C). Among the transductants that gave HFT lysates for *ilvB* were some that failed to give TP for the *leuA*$^-$ allele. Their lysates were HFT for all the *ilvBC–leu* genes except *leuA*.

By analogous methods, Mackey has produced a complete set of double lysogens that give HFT lysates for all except one of the genes of the *ilvBC–leu* cluster; that is, one carries an *ilvB*$^-$ allele, one carries an *ilvC*$^-$ allele, and so on.

When this work was started, there were three *leu* genes known in *B. subtilis: leuA, leuB,* and *leuC* (Ward and Zahler, 1973a, b). Mackey tested three HFT lysates containing a *leuA*$^-$ allele, a *leuB*$^-$ allele, or a *leuC*$^-$ allele in spot tests against a number of Leu$^-$ auxotrophs from our stock collection. She was able to assign each Leu$^-$ auxotroph to one of the three genes unambiguously, except for the allele in strain CU315. That marker, *leu-117,* was complemented by each of the three lysates.

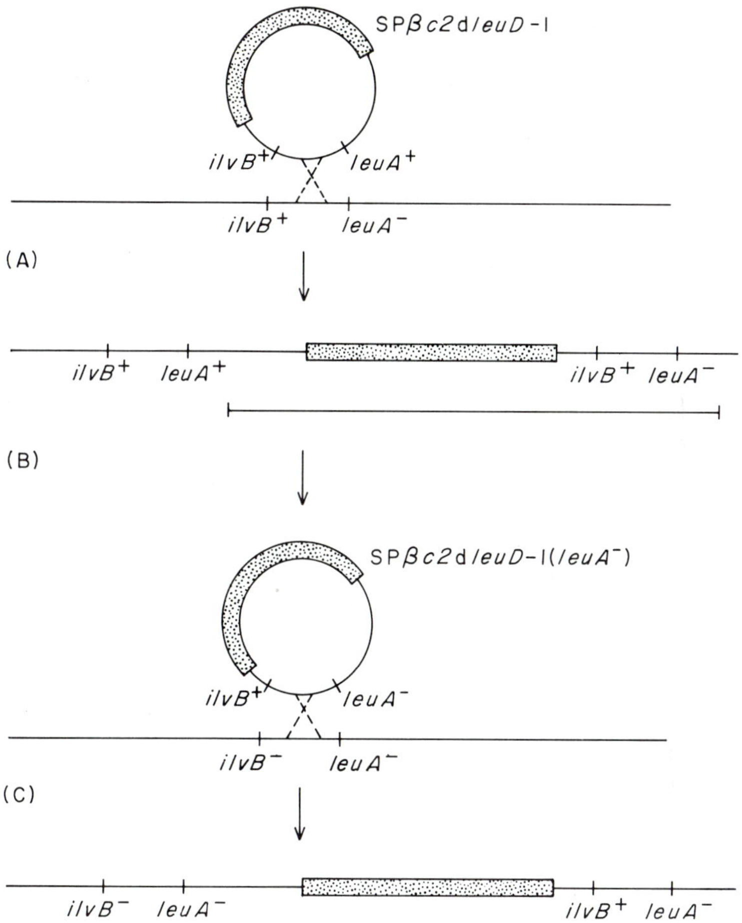

Fig. 8. Production of SPβ*c2*d*leuD*-1 carrying a *leuA*⁻ allele. (A) Transduction (class B) of a *leuA*⁻ lysogen by SPβ*c2*d*leuD*-1. (B) Excision of a TP that carries the *leuA*⁻ allele. (C) Transduction of an *ilvB*⁻*leuA*⁻ lysogen by SPβ*c2*d*leuD*-1 (*leuA*⁻). Integration of the defective phage into the region of chromosomal homology (class-B transduction) as shown, or into the resident prophage (class-P transduction), gives a double lysogen that releases SPβ*c2*d*leuD*-1 (*leuA*⁻).

When *leu-117* was mapped by transformation, it was found to lie to the right of all other *leu*⁻ mutations. It clearly defines a new cistron, *leuD*, and has been named *leuD117*. In *Salmonella typhimurium* the enzyme isopropylmalate isomerase (the product of the *B. subtilis leuB* gene) is encoded by two adjacent cistrons (Margolin, 1963). We suspect that *leuB* and *leuD* form a similar pair in *B. subtilis*.

5. *Cis–Trans Relationships*

The detailed study of how *B. subtilis* controls its biochemical reactions, and how its genome is divided into cistrons, has been hindered by our inability to construct heterogenotes easily. Specialized transduction by phage SPβ should be an important tool in solving these problems. Methods that should be useful generally are being worked out in the *ilvBC–leu* region.

If strain 168 is plated on a minimal tryptophan–agar medium containing 40 μg/ml of the leucine analog 4-azaleucine, resistant colonies appear. Two phenotypes can be distinguished: colonies that protect their neighbors from the analog and colonies that do not (Ward and Zahler, 1973b). Members of the former class have mutations that lie at a locus called *azlA* (Fig. 7A), close to *ilvB*. They protect their sensitive neighbors from 4-azaleucine by excreting leucine, which overcomes its analog's toxicity. Leucine excretion is probably due to constitutive synthesis of the products of the *ilvBC* and *leu* genes in *azlA* mutants.

We want to learn the nature of the *azlA* mutations. At least four possibilities may be considered: (1) $azlA^+$ may encode a repressor protein that prevents the transcription of *ilvBC–leu* by binding at one or more operators; $azlA^-$ mutants are constitutive for *ilvBC–leu* because they lack repressor. (2) $azlA^-$ mutations may lie in the operator of an *ilvBC–leu* operon; $azlA^-$ mutants are operator-constitutive. (3) $azlA^-$ mutations may lie in the promoter of an *ilvBC–leu* operon and cause increased constitutive transcription. (4) *ilvBC–leu* transcription may be controlled by an attenuator, and $azlA^-$ mutations prevent termination at the attenuator.

It is also possible that the $azlA^-$ mutations include more than one of these types; e.g., some may be repressor-type constitutive mutations (type 1) and others may be operator-type constitutive mutations (type 2). It is also possible that *B. subtilis* uses mechanisms of control that have not been found in gram-negative bacteria.

Of the four putative types of $azlA^-$ mutations just described, type 1—mutations in a repressor gene—should be recessive in a heterogenote of the type $azlA^-\ ilvB^+C^+\ leu^+/azlA^+\ ilvB^-C^+\ leu^+$, since the $azlA^+$ gene will make repressor that prevents the transcription of both copies of *ilvBC–leu*.

The other three types of $azlA^-$ mutations should be cis-dominant. That is, heterogenotes of the sort $azlA^-\ ilvB^+C^+\ leu^+/azlA^+\ ilvB^-C^+\ leu^+$ should be constitutive for the *ilvBC–leu* gene products and thus resistant to 4-azaleucine. Heterogenotes of the sort $azlA^-\ ilvB^+C^+\ leuA^-B^+C^+/$ $azlA^+\ ilvB^-C^+\ leuA^+B^+C^+$ are likely not to be resistant to the analog, because the constitutive operon carries a $leuA^-$ mutation, and the *leuA* gene product will be produced only from the region under *azlA*⁺ control.

C. J. Mackey has started an analysis of this system using SPβ*c2*d*leuD* phages as the carrier of one copy of the *azlA–ilvBC–leu* region. Prophages carried in the B configuration may recombine with the bacterial *azlA–ilvBC–leu* region, so only class-P clones carrying the defective prophage in (or in tandem with) the resident SPβ*c2* prophage will be used. This can be determined as described in Section III,G,3.

Mackey moved the *azlA102* mutation (Ward and Zahler, 1973b) into strain CU1886 [*argA2 pheA2* (*citF*::SPβ*c2*)] by PBS1 transduction, selecting for Phe⁺. The newly constructed strain is CU1885 [*argA2 azlA102* (*citF*::SPβ*c2*)]. From this strain she has isolated a specialized transducing phage, SPβ*c2*d*leuD*-2, that carries the *azlA102* mutation and the entire *ilvBC–leu* region. This TP will be used to construct the heterogenotes needed for the analysis. Her preliminary results suggest that *azlA102* is cis-dominant. Other *azl* mutations in this region are recessive.

6. *Precise Excision of* citF*::SP*βc2

Strain CU1886 [*argA2 pheA2* (*citF*::SPβ*c2*)] was used by C. J. Mackey to determine the stability of SPβ*c2* prophage lying within a structural gene—in this case, a gene for succinate dehydrogenase, *citF*. Because of the phage insertion CU1886 lacks a complete tricarboxylic acid cycle and is unable to grow with lactate as its major carbon source. The precise excision of SPβ*c2* regenerates the *citF* gene and permits growth on lactate.

When care was taken to prevent the selection of Cit⁺ cells during the growth of a culture of CU1886, about 1 cell in 50,000 was Cit⁺. These were all SPβ-sensitive. (The experiment was carried out in the presence of SPβ antiserum.) The number of Cit⁺ clones was not increased by plating at 50°C, suggesting that phage genes may not be responsible for the high frequency of perfect excision.

H. Prospectus

Various methods have become available for isolating TPs for any *B. subtilis* gene. In many cases these allow the production of heterogenotes to study cis–trans relationships. Relationships between alleles of sporulation or germination genes can be studied if there is a selectable marker nearby. In this context, "nearby" should be interpreted to mean that a TP can carry DNA equivalent to at least the distance between two markers that are 75% cotransduced by phage PBS1.

It is not clear whether SPβ (or its relatives, or hybrids between them) will also turn out to be useful as a tool for gene cloning. (J. M. Odebralski has shown that φ3T lysogens, in bacteria that do not carry SPβ, can give TPs for *citK*, and double lysogens of φ3T and its defective TPs give HFT lysates for *citK*.) The phage DNA is about 4% as large as the *B. subtilis* chromosome, so we cannot expect more than about 25-fold enrichment for a bacterial gene if we clone it in SPβ. R. Z. Korman has carried out preliminary experiments suggesting that it is possible to separate at least one specialized transducing phage (SPβ*c2*d*citK*-1) from whole SPβ*c2* by CsCl density gradient centrifugation; the defective phage particles have lower specific gravity. Still, the large size of the SPβ genome and our inability to achieve TP/PFU ratios that approach 1 make it likely that SPβ's main value will be for analyzing the *B. subtilis* genome and its controls, rather than for cloning its genes.

Acknowledgments

Much of the work on SPβ was carried out in the Cornell University laboratory under the direction of the author and Dr. Ruth Z. Korman. I wish to thank Dr. Korman and our co-workers. Among the people who contributed significantly to the work were three extraordinarily capable technicians: Richard Rosenthal, Patryce A. Toye, and Janice M. Odebralski; four graduate students: Pamela S. Fink, Irene Gage, Catherine J. Mackey, and Robert Lipsky; and three undergraduate students: Robert Herenstein, Richard A. Rosenberg, and Cathy Vocke. Dr. H. E. Hemphill of Syracuse University has freely exchanged bacteria, phages, and ideas with us, and I thank him. This work was supported in part by NSF Grant 78-24113 and NIH Grant GM28038.

References

Arber, W., Kellenberger, G., and Weigle, J. (1957). *Z. Allg. Pathol. Bakteriol.* **20,** 659–665.

Armentrout, R., and Rutberg, L. (1971). *J. Virol.* **8,** 455–468.

Bradley, D. E. (1967). *J. Virol.* **4,** 264–270.

Bukhari, A. I., Ljungquist, E., deBruijn, F., and Khatoon, H. (1977). *In* "DNA Insertion Elements, Plasmids, and Episomes" (A. I. Bukhari, J. A. Shapiro, and S. L. Adhya, eds.), pp. 249–261. Cold Spring Harbor Laboratory, Cold Spring Harbor, New York.

Campbell, A. (1957). *Virology* **4,** 366–384.

Campbell, A. (1962). *Adv. Genet.* **11,** 101–145.

Campbell, A. (1971). *In* "The Bacteriophage Lambda" (A. D. Hershey, ed.), pp. 13–44. Cold Spring Harbor Laboratory, Cold Spring Harbor, New York.

Chow, L. T., Boice, L., and Davidson, N. (1972). *J. Mol. Biol.* **68,** 391–400.

Dean, D. H., Arnaud, M., and Halvorson, H. O. (1976). *J. Virol.* **20,** 339–341.

Duncan, C. H., Wilson, G. A., and Young, F. E. (1978). *Proc. Nat. Acad. Sci. USA* **75,** 3664–3668.

Eiserling, F. A. (1964). Ph.D. Thesis, Univ. of California at Los Angeles.

Fink, P. S., Korman, R. Z., Odebralski, J. M., and Zahler, S. A. (1981). *Mol. Gen. Genet.* **182,** 514–515.

Fink, P. S., and Zahler, S. A. (1980). *Abstr. Amer. Soc. Microbiol. 1980,* 270.

Flock, J.-I. (1978). *Mol. Gen. Genet.* **163,** 7–16.
Franklin, N. C. (1971). *In* "The Bacteriophage Lambda" (A. D. Hershey, ed.), pp. 175–194. Cold Spring Harbor Laboratory, Cold Spring Harbor, New York.
Gingery, R., And Echols, H. (1967). *Proc. Nat. Acad. Sci. USA* **58,** 1507–1514.
Goldstein, B. J., and Zahler, S. A. (1976). *J. Bacteriol.* **127,** 667–670.
Gottesman, M. E., and Weisberg, R. A. (1971). *In* "The Bacteriophage Lambda" (A. D. Hershey, ed.), pp. 113–138. Cold Spring Harbor Laboratory, Cold Spring Harbor, New York.
Gottesman, M. E., and Yarmolinsky, M. B. (1968). *Cold Spring Harbor Symp. Quant. Biol.* **33,** 735–747.
Greer, H. (1975). *Virology* **66,** 589–604.
Hemphill, H. E. (1980). Personal communication.
Hemphill, H. E., Gage, I., Zahler, S. A., and Korman, R. Z. (1980). *Can. J. Microbiol.* **26,** 1328–1333.
Henner, D. J., and Hoch, J. A. (1980). *Microbiol. Rev.* **44,** 57–82.
Henner, D. J., and Steinberg, W. (1979). *J. Bacteriol.* **139,** 668–670.
Inselberg, J. W., Eremenko-Volpe, T., Greenwald, L., Meadow, W. L., and Marmur, J. (1969). *J. Virol.* **3,** 627–628.
Landy, A., and Ross, W. (1977). *Science* **197,** 1147–1160.
Lipsky, R. H., Rosenthal, R., and Zahler, S. A. (1981). *Abstr. Amer. Soc. Microbiol. 1981,* 129.
Mackey, C. J., Korman, R. Z., and Zahler, S. A. (1980). *Abstr. Amer. Soc. Microbiol. 1980,* 271.
Morse, M. L., Lederberg, E. M., and Lederberg, J. (1956a). *Genetics* **41,** 142–156.
Morse, M. L., Lederberg, E. M., and Lederberg, J. (1956b). *Genetics* **41,** 758–779.
Rosenthal, R., Toye, P. A., Korman, R. Z., and Zahler, S. A. (1979). *Genetics* **92,** 721–739.
Rutberg, B., and Hoch, J. A. (1970). *J. Bacteriol.* **104,** 826–833.
Rutberg, L. (1969). *J. Virol.* **3,** 38–44.
Rutberg, L. (1973). *J. Virol.* **12,** 9–12.
Rutberg, L., and Rutberg, B. (1970). *J. Virol.* **5,** 604–608.
Scher, B. M., Dean, D. H., and Garro, A. J. (1977). *J. Virol.* **23,** 377–383.
Shapiro, J. A., Dean, D. H., and Halvorson, H. O. (1974). *Virology* **62,** 393–403.
Smith, I., and Smith, H. (1973). *J. Bacteriol.* **114,** 1138–1142.
Southern, E. M. (1975). *J. Mol. Biol.* **98,** 503–517.
Stroynowski, I. T. (1978). Ph.D. thesis, Stanford Univ.
Tucker, R. G. (1969). *J. Gen. Virol.* **4,** 489–504.
Ward, J. A., and Zahler, S. A. (1973a). *J. Bacteriol.* **116,** 719–726.
Ward, J. A., and Zahler, S. A. (1973b). *J. Bacteriol.* **116,** 727–735.
Warner, F. D., Kitos, G. A., Romano, M. P., and Hemphill, H. E. (1977). *Can. J. Microbiol.* **23,** 45–51.
Williams, M. T., and Young, F. E. (1977). *J. Virol.* **21,** 522–529.
Zahler, S. A., Benjamin, L. G., Glatz, B. S., Winter, P. F., and Goldstein, B. J. (1976). *In* "Microbiology—1976" (D. Schlessinger, ed.), pp. 35–43. American Society for Microbiology, Washington, D. C.
Zahler, S. A., Korman, R. Z., Rosenthal, R., and Hemphill, H. E. (1977). *J. Bacteriol.* **129,** 556–558.
Zissler, J. (1967). *Virology* **31,** 181–189.

10

Molecular Cloning in *Bacillus subtilis*

THOMAS J. GRYCZAN

I. Introduction

The recent advances in recombinant DNA technology have been formulated using *Escherichia coli* as the host system and a variety of its endogenous plasmids and phages as molecular cloning vectors. Utilization of this system has resulted in an increased understanding of many biological phenomena in both prokaryotic and eukaryotic organisms.

The Molecular Biology
of the Bacilli

ISBN 0-12-222701-8

Since *Bacillus subtilis* is the most widely studied prokaryote other than *E. coli,* it would be advantageous to have a molecular cloning system developed for studies on the various genetic and biochemical processes of this organism. A *B. subtilis* cloning system would be a desirable asset in the study of various biological phenomena such as competence and sporulation which cannot be investigated in *E. coli.*

Bacillus subtilis has a number of useful properties that are amenable to the development of a cloning system. First, it is a nonpathogenic organism, not associated with any human disease condition. Second, it has a well-characterized chromosomal genetic map (Henner and Hoch, 1980; Chapter 1) and is associated with several endogenous plasmids (Lovett and Bramucci, 1976) and bacteriophages (Hemphill and Whiteley, 1975) which are potential cloning vehicles. Third, it possesses a high-frequency (~5%) transformation system in the competent state for exogenous DNA, making it a more suitable recipient for recombinant DNA uptake. Finally, it has a simple cell surface composition, being composed of only peptidoglycan and teichoic acid. Thus, any secretory proteins made in *B. subtilis* by recombinant DNA technology would not be contaminated by cell envelope endotoxin, which is a potential problem in *E. coli* systems. This last aspect is especially important, since *B. subtilis* is a source of many industrial, drug, and fermentation products (Erickson, 1976), and recombinant DNA technology would be an efficient way to obtain higher yields of secreted biological products.

Although *B. subtilis* is a safe organism from a medical viewpoint, one possible disadvantage is that it is a spore-forming organism and this increases the viability potential of clone-carrying host cells in the environment. In this regard, considerable research (Young *et al.*, 1977) is under way to develop asporogenic *B. subtilis* strains that would have a decreased survival capacity outside the laboratory because of the inability of an asporogenic strain to inactivate autolysins by proteases during the stationary phase of growth, resulting in cell autolysis (Brown and Young, 1970).

This chapter will summarize the recent progress in the development of plasmid and phage vectors for recombinant DNA technology in *B. subtilis,* plasmid transformation as related to cloning in this organism, and the types of clones obtained using this system.

II. Vectors

A. Plasmids Indigenous to *Bacillus subtilis* and Other *Bacillus* Species

Many *B. subtilis* strains have been shown to contain a number of endogenous plasmids (Lovett and Bramucci, 1975, 1976; Le-Hegarat and

Anagnostopoulos, 1977; Tanaka and Koshikawa, 1977; Tanaka *et al.*, 1977; Kreft *et al.*, 1978; Tanaka and Sakaguchi, 1978; Uozumi *et al.*, 1980). However, the utility of these plasmids as potential cloning vectors is limited by the observation that none of them express any useful or identifiable genetic markers (antibiotic resistances); they are cryptic plasmids. The *B. subtilis* plasmid pPL10 has been shown to code for a bacteriocin-like activity (Lovett *et al.*, 1976), but this character is not a useful selected marker in transformation experiments. Recently, a *Bacillus cereus* tetracycline resistance plasmid, pBC16, has been transformed into *B. subtilis* (Bernhard *et al.*, 1978) and has been used in the construction of a variety of molecular cloning vectors (Goebel *et al.*, 1978, 1979; Kreft *et al.*, 1978) which can be utilized in the *B. subtilis* host. In addition, one of the aforementioned cryptic plasmids has been genetically labeled with the *B. subtilis leu* gene (Tanaka and Sakaguchi, 1978), resulting in a chimeric *B. subtilis* plasmid carrying a selectable genetic marker. Further constructions and manipulations of this kind should make available an increased number of *B. subtilis* cloning plasmids.

Despite the absence of useful selectable markers on *B. subtilis* plasmids, screening of additional *B. subtilis* strains for plasmids should be pursued, since it has been observed that *B. subtilis* strains in the soil can undergo spontaneous genetic exchange (Graham and Istock, 1978, 1979). This observation indicates that *B. subtilis* may be continually acquiring new genetic information from other bacilli, or even unrelated bacteria. In this regard, *in vivo* plasmid transfer between *B. subtilis* strains (Prozorov and Glumova, 1980) and between *Staphylococcus aureus* and *B. subtilis* (Feitelson and Lederberg, 1980) has been observed. Plasmid transfer *in vivo* has also been reported between *E. coli* and *B. subtilis* (Buheitel and Klingmuller, 1979), although this finding is open to question since many of the *B. subtilis* transformants obtained are unstable for their newly acquired antibiotic resistance markers and these transformants have not been analyzed for their plasmid DNA content. *In vitro* transfer of a tetracycline resistance plasmid from a thermophilic bacillus (Bingham *et al.*, 1979) to *B. subtilis* has been observed, suggesting that plasmids from other bacilli, such as *B. thuringiensis* (Ermakova *et al.*, 1978; Stahly *et al.*, 1978; Gonzalez and Carlton, 1980), should be tested for their ability to replicate and be expressed in *B. subtilis*. Studies of this kind should further develop the library of *Bacillus*-derived plasmids available for molecular cloning.

B. *Staphylococcus aureus* Plasmids in *Bacillus subtilis*

Because of the difficulties encountered in obtaining useful endogenous *B. subtilis* plasmid cloning vehicles, attempts were made to investigate the ability of plasmids of other bacteria to replicate in *B. subtilis*. The initial experiments showed that certain tetracycline and chloramphenicol resistance plasmids of *S. aureus* were able to replicate and express their resistance markers in *B. subtilis* (Ehrlich, 1977). Subsequently, several additional *S. aureus* plasmids were transferred to *B. subtilis* (Gryczan *et al.*, 1978; Kono *et al.*, 1978; Weisblum *et al.*, 1979); these are summarized in Table I. Restriction endonuclease cleavage maps of a number of these plasmids have been constructed (Gryczan *et al.*, 1978; Lofdahl *et al.*, 1978a; Dubnau *et al.*, 1980b). These plasmids are generally stable in *B. subtilis*, except that pE194, pSA0501, and pSA2100 are temperature-sensitive and segregate at temperatures above 37°C (Iordanescu, 1975, 1976; Gryczan *et al.*, 1978, 1980b). In addition, some of these plasmids have the advantageous property of being amplifiable in the presence of hydroxyurea (Shivakumar and Dubnau, 1978a) and in certain *dna ts* replication mutants of *B. subtilis* (Shivakumar and Dubnau, 1978b). Recently, studies have begun investigating pE194-coded plasmid proteins in *B. subtilis* minicells (Shivakumar *et al.*, 1979), the nature of tetracycline resistance coded by pT127 in *B. subtilis* (Fargette *et al.*, 1979), and the pUB110 kanamycin resistance protein (Sadaie *et al.*, 1980). These and other similar studies should further advance the usefulness of these plasmids and knowledge of the resistance mechanisms conferred upon their *B. subtilis* hosts.

C. *Bacillus subtilis* Chimeric Plasmids

The *S. aureus* plasmids that replicate in *B. subtilis* have been used to construct plasmid chimeras conferring multiple antibiotic resistance

TABLE I Plasmids Transferred from *S. aureus* to *B. subtilis*

Plasmid	Antibiotic resistance	Molecular weight ($\times 10^6$)	Reference
pC194	Cm^r	2.0	Ehrlich (1977)
pC221	Cm^r	2.0	Ehrlich (1977)
pC223	Cm^r	2.0	Ehrlich (1977)
pUB112	Cm^r	2.0	Ehrlich (1977)
pT127	Tc^r	2.9	Ehrlich (1977)
pE194	Em^r	2.3	Weisblum *et al.* (1979)
pUB110	Km^r	3.0	Gryczan *et al.* (1978)
pSA0501	Sm^r	2.8	Gryczan *et al.* (1978)
pSA2100	Cm^r, Sm^r	4.8	Gryczan *et al.* (1978)
pTP4	Cm^r	2.9	Kono *et al.* (1978)
pTP5	Tc^r	2.9	Kono *et al.* (1978)

(Ehrlich, 1978; Gryczan and Dubnau, 1978), and these are listed in Table II. Restriction endonuclease cleavage maps have been constructed for these chimeras (Gryczan and Dubnau, 1978; Gryczan *et al.*, 1980b; Dubnau *et al.*, 1980b). In addition, cleavage sites have been identified in which the insertion of foreign DNA inactivates one or another of the various antibiotic resistance markers (Lofdahl *et al.*, 1978b; Gryczan *et al.*, 1980b); these are listed in Table III. The restriction sites listed that do not inactivate an antibiotic resistance marker are available for cloning, since their removal by insertion of foreign DNA or by deletion does not interfere with replication of the chimeric plasmid. Some of these restriction sites may be in essential regions in the parental plasmids (Table I), since these are not multiple replicons. Plasmid pBD64 is a spontaneous deletion derivative of pBD12 which has lost 1.3 Mdaltons of DNA from the pC194 moiety of pBD12 (Gryczan *et al.*, 1980b) but which expresses both parental antibiotic resistance markers and is more stable in *B. subtilis* than its parent plasmid. Plasmids pBD6 and pBD8 are temperature-sensitive chimeras, and the usefulness of this property in molecular cloning has been previously discussed (Dubnau *et al.*, 1980b; Gryczan *et al.*, 1980b). The *Hin*dIII fragment of pT127 that contains the Tc^r gene is incapable of self-replication. This fragment has been cloned onto the *B. subtilis* cryptic plasmid pHV400 (Niaudet and Ehrlich, 1979), resulting in the *in vitro* labeling of a cryptic plasmid with a useful genetic marker. Similar cloning of resistance marker fragments to cryptic plasmids has increased the number of useful *B. subtilis* cloning plasmids (Niaudet and Ehrlich, 1979).

TABLE II *Bacillus subtilis* Plasmid Chimeras

Plasmid	Parental plasmids	Antibiotic resistance	Molecular weight ($\times 10^6$)	Reference
pBD6	pSA0501, pUB110	Km^r, Sm^r	5.8	Gryczan and Dubnau (1978)
pBD8	pUB110, pSA2100	Km^r, Cm^r, Sm^r	6.0	Gryczan and Dubnau (1978)
pBD9	pE194, pUB110	Km^r, Em^r	5.4	Gryczan and Dubnau (1978)
pBD10	pBD8, pE194	Km^r, Cm^r, Em^r	4.4	Gryczan and Dubnau (1978)
pBD11	pBD8, pE194	Km^r, Em^r	4.4	Gryczan and Dubnau (1978)
pBD12	pUB110, pC194	Km^r, Cm^r	4.5	Gryczan and Dubnau (1978)
pBD64	pBD12	Km^r, Cm^r	3.2	Gryczan *et al.* (1980b)
pBD63	pBD6, PHV11-*Hin*dIIIB fragment	Km^r, Tc^r	7.3	Gryczan *et al.*, (1980b)
pHV11	pC194, pT127	Cm^r, Tc^r	3.5	Ehrlich (1978)

TABLE III Insertional Inactivation of *B. subtilis* Chimeric Plasmid Antibiotic Resistance Markers[a]

Plasmid	Antibiotic resistance	Single sites	Inactivated resistance marker
pBD6	Km^r, Sm^r	*Bam*HI	None
		*Bgl*II	Km^r
		*Hin*dIII	Sm^r
		*Tac*I	None
pBD8	Km^r, Cm^r, Sm^r	*Bgl*II	Km^r
		*Bam*HI	None
		*Eco*RI	Sm^r
		*Hin*dIII	Sm^r
		*Xba*I	None
pBD9	Km^r, Em^r	*Bam*HI	None
		*Bcl*I	Em^r
		*Bgl*II	Km^r
		*Eco*RI	None
		*Hpa*I	Em^r
		*Pst*I	None
		*Tac*I	None
pBD10	Km^r, Cm^r, Em^r	*Bam*HI	None
		*Bcl*I	Em^r
		*Bgl*II	Km^r
		*Hpa*I	Em^r
		*Xba*I	None
pBD11	Km^r, Em^r	*Bam*HI	None
		*Bcl*I	Em^r
		*Bgl*II	Km^r
		*Hpa*I	Em^r
		*Xba*I	None
pBD12	Km^r, Cm^r	*BamHI*	None
		*Bgl*II	Km^r
		*Eco*RI	None
		*Hin*dIII	None
		*Xba*I	None
pBD64	Km^r, Cm^r	*Bam*HI	None
		*Bgl*II	Km^r
		*Eco*RI	None
		*Xba*I	None

[a] Gryczan and Dubnau (1978); Gryczan *et al.* (1980b).

D. *Escherichia coli–Bacillus subtilis* Chimeric Plasmids

The observation that certain *S. aureus* plasmids can replicate in *B. subtilis* led to the investigation of the ability of plasmids of other bacteria to replicate in this new host. Studies of this type are important because they may shed light on the genetic barrier of gene transfer and subsequent replication and expression between gram-positive and gram-negative organisms. Replication of *E. coli* plasmids in *B. subtilis,* and vice versa, was a logical choice for investigation, since *E. coli* was the organism in which the initial cloning techniques were developed. The capacity of plasmids of one species to replicate in another would advance the potential and usefulness of molecular cloning techniques. It has been observed that certain *E. coli* plasmids are able to replicate in other gram-negative species (Bernard and Helinski, 1980). Reports of certain *E. coli* R factors being introduced into *B. subtilis* by transformation (Domarodskii *et al.*, 1976; Naumov *et al.*, 1977) are subject to question, since in the former case the transformants do not have the auxotrophic markers carried by the original recipient, and in the latter case the plasmids detected in the transformants are different from the ones used to transform them. There are currently no autonomous plasmid replicons of *E. coli* that are able to replicate in *B. subtilis*. Similarly, no autonomous *B. subtilis* plasmids have been reported to date (either endogenous or *S. aureus*-derived) that are able to replicate in *E. coli.* Other than the ability of certain *S. aureus* plasmids to replicate in *B. subtilis,* only two plasmids of another gram-positive organism have been shown to replicate in this host. First is the case of the Streptococcus faecalis tetracycline resistance plasmid pAMα1, which has been transformed into *B. subtilis* (Clewell *et al.*, 1979). Expression of the plasmid resistance determinant occurs in the new host; however, plasmids isolated from *B. subtilis* transformants were observed to have undergone a spontaneous deletion event, resulting in smaller resistance plasmids. The mechanism of this plasmid DNA rearrangement is still unknown. The second example is the *S. faecalis* erythromycin resistance plasmid pAM77, which can be transformed into *B. subtilis* and expresses erythromycin resistance (P. Tomich, Y, Yagi, and D. B. Clewell, unpublished, cited in Yagi *et al.*, 1978). In view of these findings of the inability of *E. coli* and *B. subtilis* plasmids to replicate in the alternative host, the ability of *B. subtilis–E. coli in vitro* constructed chimeric plasmids to replicate and exhibit marker expression in either host has been investigated. Plasmids of this type would be extremely useful in cloning experiments, especially if a given chimera could utilize replication genes of one parent in one host and those of the other parent in the alternative host.

A number of useful *B. subtilis–E. coli* chimeric plasmids have been constructed (Goebel *et al.*, 1978, 1979). Plasmid pHV14 was constructed by the *in vitro* ligation of pC194 and the *E. coli* plasmid pBR322 (Ehrlich *et al.*, 1978). A derivative of this plasmid, pHV33, replicates in both hosts and expresses all three antibiotic resistance markers (chloramphenicol from *B. subtilis,* and tetracycline and ampicillin from *E. coli*) in *E. coli,* but expresses only Cm^r in *B. subtilis*. This chimeric plasmid can utilize the *B. subtilis* replicon for replication in *E. coli,* since it is able to replicate in *E. coli polA ts* mutants (pBR322 requires a functional *E. coli polA* gene for replication) (Ehrlich *et al.*, 1978). Similarly, other *E. coli–B. subtilis* chimeric plasmids (Goebel *et al.*, 1978, 1979; Kreft *et al.*, 1978) show an identical pattern of antibiotic resistance expression in *E. coli,* but of limited gene expression in *B. subtilis*. In addition, these chimeras also undergo deletions and rearrangements in both hosts, but preferentially in *B. subtilis*. The resultant plasmids are more stable, can still replicate in both hosts, and can be employed as molecular cloning vectors in both bacterial hosts (Goebel *et al.*, 1979). From these studies it has been shown that the *E. coli* plasmid tetracycline resistance genes (pBR322-derived) can express in *B. subtilis* if present on chimeric plasmids. However, *B. subtilis* is unable to express other *E. coli* antibiotic resistances (β-lactamase, ampicillin, chloramphenicol, kanamycin), whereas *E. coli* is able to express some *B. subtilis*-derived antibiotic resistance markers (chloramphenicol and tetracycline). These findings indicate that these chimeric plasmids are useful for cloning in both hosts but are somewhat limited by the inability of *B. subtilis* to express some foreign (plasmid) genes. A cryptic *B. subtilis* plasmid has been cloned onto an *E. coli* antibiotic resistance plasmid (Horinouchi *et al.*, 1977), and this chimera can replicate in *E. coli* and appears to transcribe *B. subtilis* DNA.

E. Bacteriophage Vectors

Since *B. subtilis* has been shown to be a reservoir of a considerable number of bacteriophages (Hemphill and Whiteley, 1975), it is readily apparent that the potential usefulness of these phages as cloning vectors should be investigated (Young and Wilson, 1979). Restriction endonuclease cleavage maps for a number of these phages, such as SP02 (Yoneda *et al.*, 1979a), ϕ105 (Perkins *et al.*, 1978; Scher *et al.*, 1978), ϕ3T (Cregg and Ito, 1979), ϕ15 and ϕ29 (Inciarte *et al.*, 1976; Ito *et al.*, 1976), M2 (Kawamura and Ito, 1977), and ρ11 and ρ14 (Dean *et al.*, 1976; Perkins *et al.*, 1978; Cregg and Ito, 1979; Kroyer and Dean, 1979; Kroyer *et al.*, 1980), have been constructed. Further analysis has resulted in the isolation of viable deletion mutants of phages ρ11 (Kawamura *et al.*, 1979b) and ρ14 (Kroyer and Dean, 1979).

Despite the extensive phage physical maps, a number of technical problems are inherent in *B. subtilis* phage vector cloning. First, phage ρ11, which has been the most widely investigated potental cloning phage, is a large DNA molecule of 80 Mdaltons and has many multiple restriction endonuclease cleavage sites, resulting in a difficulty in isolating cloned

DNA molecules from recombinant ρ11 DNA (Kawamura *et al.*, 1979b). In addition, in *B. subtilis* no high-frequency specialized transducing phages have been found, with the exception of phages φ3T and ρ11, both of which carry thymidylate synthetase genes (Tucker, 1969; Dean *et al.*, 1976). A specialized transducing system is under development by Zahler and his collaborators (Chapter 9). A method for the construction of *B. subtilis* specialized transducing phages of ρ11 has been devised (Kawamura *et al.*, 1979a). This procedure involves digesting DNA from the temperate phage ρ11 and *B. subtilis* chromosomal DNA or from defective phage containing chromosomal DNA with a given restriction endonuclease and ligating the mixture with T4 ligase. Transformation of this ligated mixture into a *B. subtilis* auxotroph lysogenic for ρ11 resulted in transformants that produced plaque-forming or defective phages able to transduce specific markers at higher frequencies. However, the aforementioned large size of this phage DNA and the multiplicity of restriction sites on this phage make the isolation of specific DNA fragments difficult. Accordingly, this technique has been successfully applied to the construction of specialized transducing derivatives of phage φ105 (Iijima *et al.*, 1980), which has a size of 26 Mdaltons and is easier to handle in cloning experiments. A φ105 phage carrying a *metB* gene has been found to contain a *Bam*HI site within the insert DNA, and since *metB* expression is not affected by cleavage at this site, this *met* gene can be used as a selective marker after integration of foreign DNA into the *Bam*HI site of this phage.

Phage ρ14 has also been extensively analyzed as a potential phage cloning vector (Kroyer *et al.*, 1980). In addition to constructing a restriction map and isolating viable deletion mutants, Kroyer *et al.*, (1980) have shown that a *Sal*I site in the phage DNA may be available as a cloning site for the insertion of foreign DNA without inactivation of phage viability, suggesting that this phage may have some utility as a molecular cloning vehicle in *B. subtilis*.

DNA from phage φ3T has been used to construct a potential molecular cloning vehicle (Young *et al.*, 1977; Young and Wilson, 1979). In addition, the thymidylate synthetase gene (*thyP3*) of this phage has been cloned onto various *E. coli* plasmid vectors such as pMB9 and pSC101 and is expressed in both *E. coli* and *B. subtilis* (Ehrlich *et al.*, 1976, 1977; Young *et al.*, 1977). Similarly, another *B. subtilis* phage, β22, has been shown to contain a *thy* gene, and this has also been cloned onto *E. coli* plasmids and expresses in both hosts (Duncan *et al.*, 1977, 1978). The plasmid phage chimeras have been used to integrate foreign genes into the chromosome of *B. subtilis,* presumably at the phage integration site. The φ3T-labeled plasmid vectors should be useful in the study of DNA integration and expression in *B. subtilis* (Duncan *et al.*, 1978). Recently, phages φ3T (Yoneda *et al.*, 1979b) and ρ11 (Nomura *et al.*, 1980) have been employed as vectors in the cloning of the amylase gene of *Bacillus amyloliquefaciens* and *B. subtilis*, respectively, in *B. subtilis*.

III. Plasmid Transformation

A. Competent Cells

The processes involving plasmid transformation in competent *B. subtilis* cells have been extensively analyzed and reviewed (Contente and Dubnau, 1979a; Dubnau *et al.*, 1980a,b; Chapter 5). Plasmid transformation can be characterized as follows: It is first order with respect to DNA concentration; the process is inefficient, with 10^3–10^4 molecules taken up per transformation event; linearization of plasmid DNA at unique restriction endonuclease cleavage sites results in a loss of transforming ability; and nicking of plasmid DNA by DNase also inactivates plasmid transformation. Although covalent continuity may be required for plasmid transformation, superhelicity is not, since relaxation of DNA with a nicking–closing enzyme does not affect plasmid transformation efficiency (Contente and Dubnau, 1979a). Finally, plasmid transformation occurs readily in *recE* strains.

It has recently been observed (Canosi *et al.*, 1978; Mottes *et al.*, 1979) that plasmid transformation is largely due to the presence of oligomeric forms in native plasmid DNA, with covalently closed circular monomers showing little or no transformation activity. It seems possible that the uptake of oligomeric DNA allows the occurrence of *recE*-independent intramolecular annealing and repair to yield a unit length covalently closed circular plasmid molecule (Dubnau *et al.*, 1980a). It has been observed that, when a homologous resident plasmid is present in a given recipient, markers carried on linear, nicked, or monomeric plasmid DNA molecules can yield transformants (Contente and Dubnau, 1979b). This "rescue" process is *recE*-dependent and requires homology between the donor and resident plasmids. The rescue of markers carried on the nonhomologous donor moiety is dependent upon a number of factors when the resident plasmid is homologous to a specific region of the donor plasmid molecule. For example, a linearizing cleavage cannot occur within the nonhomologous region of the donor plasmid, nor can it occur at the junction between homologous and nonhomologous regions. In addition, the transforming activity of a given marker is dependent upon the distance between the aforementioned junction and a unique cleavage site in the homologous region (Contente and Dubnau, 1979b). These properties of plasmid transformation have been shown to be useful in certain *B. subtilis* molecular cloning experiments and will be described in greater detail in Section IV,B.

B. Protoplasts

As an alternative to plasmid transformation into competent cells of *B. subtilis,* the polyethylene glycol cell fusion technique has been adapted to obtain efficient plasmid transformation into protoplasts prepared from noncompetent *B. subtilis* cells (Chang and Cohen, 1979). After transformation, the protoplasts can be readily regenerated to the cellular form.

The efficiency of plasmid transformation into protoplasts is up to an order of magnitude higher than that for competent cells; values of up to 80% efficiency of input cells have been obtained with plasmid pC194 (Chang and Cohen, 1979). Subsequent studies (Gryczan *et al.,* 1980a) have shown that protoplasts can be transformed by covalently closed circular monomeric plasmid at about the same efficiency as by dimeric and oligomeric forms. Linear plasmid DNA transforms protoplasts about 1% as well as circular molecules, whereas nicked monomeric forms transform as well as covalently closed circular forms. One possible limitation of this system is that transformation of protoplasts by high-molecular-weight (concatameric) DNA may be inefficient compared to competent cell transformation (Canosi *et al.,* 1978; Chang and Cohen, 1979; Mottles *et al.,* 1979). In this regard, poor recovery of transformants using ligated DNA in protoplasts has been observed (Chang and Cohen, 1979; Gryczan *et al.,* 1980a). One possible solution to this problem is to ligate DNA under conditions that do not favor oligomer formation when being used in protoplast transformation. Another limitation of this system is that, since protoplast regeneration requires a rich synthetic medium, direct selection for clones of auxotrophic markers is difficult to perform. Currently, selection must be performed for a given drug resistance marker, with subsequent screening for insertional inactivation of another marker, or screening for the desired auxotrophic marker. Modifications or improvements in the regeneration medium may make direct selection more amenable to this system. Recently, it has been demonstrated that *Bacillus megaterium* protoplasts can be transformed with *B. subtilis* plasmids (Brown and Carlton, 1980; Vorobjeva *et al.,* 1980), suggesting that other *Bacillus* species may also have potential use in molecular cloning experiments. In addition, it has also been shown that annealed mixtures of homopolymer-tailed linear molecules of plasmid pHV14, a chimeric plasmid composed of plasmids pC194 and pBR322, and homopolymer-tailed T7 DNA fragments can transform *B. subtilis* protoplasts (Scherzinger *et al.*, 1980) at a low frequency. The resultant hybrid plasmids have been shown to undergo deletions, indicating that these plasmids are unstable in *B. subtilis*. However, this study demonstrates that annealed (gapped) DNA can transform *B. subtilis* protoplasts, which is not the case in competent cells (Contente and Dubnau, 1979b; Rapoport *et al.*, 1979).

C. Restriction and Modification

Host-controlled restriction and modification systems have been found in *B. subtilis* and have been the subject of an extensive amount of investigation (Shibata and Ando, 1974; Trautner *et al.*, 1974; Bron and Murray, 1975; Bron *et al.*, 1975; Gunthert *et al.*, 1975; Shibata and Ando, 1975; Wilson and Young, 1976; Ikawa *et al.*, 1979; Saito *et al.*, 1979). The competent strain of *B. subtilis* 168 has been reported to restrict the growth of certain bacteriophages grown on heterologous hosts, although the effect of this restriction activity is weak (Uozumi *et al.*, 1977). Another wild-type isolate of *B. subtilis* has been reported to specify another restriction system (*Bsu*) which has the same specificity as the *Hae*III enzyme (Trautner *et al.*, 1974; Bron and Murray, 1975). The genes specifying this restriction–modification system have been transferred into *B. subtilis* 168 competent cells and shown to have little or no effect on transformation by bacterial DNA (Trautner *et al.*, 1974). Similarly, the genes for another *B. subtilis* restriction–modification system similar to that of the *Pst*I enzyme have been transformed into *B. subtilis* 168 (Shibata *et al.*, 1979), but the effect of this system on transformation has yet to be determined.

Transformation of *B. subtilis* with plasmids isolated from *S. aureus* and *Bacillus cereus* was about $\frac{1}{50}$ as efficient as with those isolated from *B. subtilis* into which the plasmids have been introduced by transformation (Ehrlich, 1977; Bernhard *et al.*, 1978; Gryczan *et al.*, 1978). This finding was attributed to restriction of *S. aureus*-grown plasmid DNA in *B. subtilis*, resulting in a reduced efficiency of transformation (Ehrlich, 1977; Gryczan *et al.*, 1978). Subsequently, this difference in transformation efficiency was suggested to be due to the degree of oligomerization in plasmid DNA (Canosi *et al.*, 1978; Mottles *et al.*, 1979). Recently, the effect of restriction and modification systems on plasmid-mediated transformation in *B. subtilis* has been tested (Tanaka, 1979b). Using *B. subtilis* cryptic plasmids carrying a cloned *B. subtilis leu* gene, it was observed that, when these plasmids were isolated from a restriction–modification-deficient strain and transformed into a restricting strain that had the *leu recE4* genotype, the transformation frequency was greatly reduced. Transformation of a *rec*$^+$ strain with these plasmids, where integration of the donor plasmid DNA could occur, was not affected, either in a restriction-positive or -negative recipient. These findings suggest that plasmid-mediated transformation in *B. subtilis* is affected by host-controlled restriction and modification systems. It further implies that the use of restriction-deficient strains is an important consideration when constructing recombinant DNA molecules in *B. subtilis*. Accordingly, a restriction–modification mutant of *B. subtilis* has been isolated (Uozumi *et al.*, 1977) and shown to be a highly transformable strain for heterologous chromosomal DNA (Hoshino *et al.*, 1980). Plasmid transformation of this strain awaits further testing. Certain types of molecular cloning of foreign DNA in *B. subtilis* have not been affected by restriction systems (Keggins *et al.*, 1978b), and this may be due to a lack of specific recognition sequences on the DNA for the recipient's restriction system.

In conclusion, even a weak restriction system like that of *B. subtilis* 168 may exert a strong effect on molecular cloning in *B. subtilis,* depending on the source and type of DNA employed. In this regard, the restriction–modification systems should play an important role in the development of *B. subtilis* molecular cloning technology and must be considered in the formulation of these experiments.

IV. Cloning Techniques

A. Shotgun Cloning

The first successful demonstration of cloning of a chromosomal gene in the *B. subtilis* system was the construction of a *B. subtilis trpC2* clone using pUB110 as the cloning vector (Keggins *et al.,* 1978b). Subsequently, a great deal of difficulty was encountered in attempting to clone other bacillus chromosomal fragments by shotgun cloning experiments, although detailed experimental procedures for shotgun cloning have been presented (Lovett and Keggins, 1979). Explanations for these difficulties and methods for circumventing them will be discussed in Section IV,B. The only other published demonstrations of shotgun cloning in *B. subtilis* were the labeling of a cryptic *B. subtilis* plasmid with a *B. subtilis leu* gene (Tanaka and Sakaguchi, 1978) and the cloning of a *Bacillus licheniformis* DNA fragment into plasmid pUB110 (Gryczan and Dubnau, 1978). Despite these difficulties, methods for the specific isolation of *B.*

subtilis chromosomal restriction fragments (Harris-Warrick *et al.*, 1975; Potter *et al.*, 1977) should help give shotgun experiments a greater probability of success. In direct contrast, the cloning of plasmid fragments and construction of chimeric plasmids have been fairly straightforward (Gryczan and Dubnau, 1978; Niaudet and Ehrlich, 1979; Gryczan *et al.*, 1980b). In addition, recent reports have demonstrated the cloning of phage φ105 (Cully and Garro, 1980) and T7 (Scherzinger *et al.*, 1980) DNA fragments into plasmids pUB110 and pC194, respectively.

As an alternative to direct cloning into *B. subtilis,* a number of laboratories have cloned desired *B. subtilis* chromosomal, plasmid, or bacteriophage DNA fragments into the better characterized *E. coli* system, identified the desired fragment(s), and introduced these hybrid plasmids into *B. subtilis*. In these experiments, the *E. coli–B. subtilis* hybrid plasmids, some of which can replicate in both hosts, are of great value. Specifically, one of these plasmids (pHV33) has been used to construct a *B. subtilis* chromosomal gene bank in *E. coli* (Rapoport *et al.*, 1979; Dedonder *et al.*, 1980). Several of these clones have been shown to complement auxotrophic mutations when retransformed into *B. subtilis*. A *B. subtilis* chromosomal gene bank in *E. coli* has also been constructed using plasmid pMB9 (Hutchinson and Halvorson, 1980), although the members of this gene bank cannot replicate in *B. subtilis* directly because of the use of an *E. coli* cloning vector. These two gene banks should be useful in the identification and analysis of specific *B. subtilis* gene sequences. Recently, a *B. thuringiensis* DNA gene bank has been constructed in *E. coli* (Klier *et al.*, 1979) using the *E. coli* plasmid pBR322. In addition, other *B. subtilis* genes such as *leu* (Mahler and Halvorson, 1977; Chi *et al.*, 1978; Nagahari and Sakaguchi, 1978), *pyr* (Chi *et al.*, 1978), and *spo* (Segall and Losick, 1977; Haldenwang *et al.*, 1980; Moran *et al.*, 1980) have been cloned into *E. coli*. Specifically, one of the *spo* clones was made with an *E. coli–B. subtilis* hybrid plasmid (pMB9 plus pC194) and is able to integrate into the *B. subtilis* chromosome (Haldenwang *et al.*, 1980). A specific β-lactamase gene from *B. licheniformis* has also been cloned in *E. coli* (Brammar *et al.*, 1980; Gray and Chang, 1981). Although these examples indicate that a number of *B. subtilis* genes can be cloned and expressed in *E. coli,* it appears that certain *B. subtilis* sequences, such as specific sporulation genes, may not be expressed in *E. coli* (Section IV,C). In these cases, the detection of cloned genes will require hybridization by RNA probes or retransformation into *B. subtilis* by brute force screening methods.

Protoplasts may be used in shotgun cloning experiments, since they transform with high efficiency and oligomers are not required (Section III,B). Since protoplasts are transformed at a low frequency by annealed (gapped) DNA (Scherzinger *et al.*, 1980), this procedure may also be useful for this reason. It has been observed that *recE* protoplasts do not transform with great efficiency, resulting in a potential difficulty in cloning homologous DNA fragments (T. Gryczan, unpublished). However, chromosomal transformation by integration has been reported not to occur in protoplasts (Chang and Cohen, 1979). Thus, chromosomal integration of DNA fragments would not have a competing effect in rec^+ recipients, although clones carrying homologous chromosomal fragments may prove to be unstable in rec^+ hosts. Finally, it should be mentioned that, since high-molecular-weight (concatameric) DNA appears to transform protoplasts at low efficiency, this may place a serious limitation on the size of an insert clonable using protoplast transformation.

B. Cloning by Recombinational Rescue

As previously mentioned, difficulties have been encountered in the shotgun cloning of chromosomal fragments into plasmid vectors and transformation into *B. subtilis* recipients. One possible explanation for this problem is provided by the observation that plasmid multimeric forms are required for the transformation of plasmid-free recipients (Canosi *et al.*, 1978; Mottes *et al.*, 1979). The production of the recombinant multimeric molecules would require both vector–vector and vector–foreign DNA ligation. The probability of the occurrence of this process is extremely low. Attempts to increase the frequency of vector–foreign DNA ligation by varying DNA ratios will decrease vector–vector ligation and the yield of plasmid transformants. Plasmid transformation into a recipient carrying a homologous resident plasmid can be accomplished using linear plasmid DNA (Contente and Dubnau, 1979b). This transformation does not require oligomeric DNA and is *recE*-dependent. With this type of transformation, a marker rescue transformation system has been developed (Gryczan *et al.*, 1980a). This system has provided a way to bypass the requirement for plasmid multimers. If a linear vector molecule is ligated to a foreign DNA fragment, the resultant chimeric molecule should transform a strain carrying a resident plasmid homologous with the vector, even if the ligated molecule is monomeric and is damaged during or prior to uptake within the homologous sequence. The nonhomologous foreign DNA would be rescued by recombination with the resident plasmid. This system has been employed in shotgun cloning chromosomal DNA fragments of *B. licheniformis* into plasmid-containing recipients (Gryczan *et al.*, 1980a). This procedure has proven very efficient, yielding between 6 and 200 clones per microgram of chromosomal DNA for a given marker and is highly reproducible. In the original experiments plasmids pBD64 and pUB110 were used as donor (vector) and resident plasmids, respectively. Plasmid incompatibility plays an important role in these experiments, resulting in the eventual elimination (segregation) of one of the two plasmids. Since pUB110 and pBD64 express weak incompatibility toward each other, transformants initially contained both hybrid plasmid and resident plasmid. Because of this, insertional inactivation of Km^r on pBD64 could not be used as a screen for recombinant plasmids. Km^s clones were isolated after several growth cycles or after retransformation into plasmid-free recipients. This drawback can be alleviated by using combinations of donor and resident plasmids that exhibit strong incompatibility toward each other, such as pBD9 and pE194 (S. Contente, T. Gryczan and D. Dubnau, unpublished). Another drawback of this system is that, since marker rescue is *recE*-dependent, it is limited to cloning only heterologous DNA into *B. subtilis*. Cloning with homologous *B. subtilis* DNA into a *rec*$^+$ background would result in the integration of chromosomal DNA fragments. However, in order to clone *B. subtilis* chromosomal fragments, a parallel system for marker rescue in *B. licheniformis* with *B. subtilis* plasmids is under development. Hybrid plasmids carrying cloned *B. subtilis* DNA isolated from *B. licheniformis* could then be transformed into *recE4 B. subtilis* and analyzed.

C. Stability and Analysis of Cloned Genes

The stability of various cloned genes in the *B. subtilis* system appears to be dependent upon the source of the cloned gene, the vector employed, and the host in which the clone is present (*B. subtilis* or *E. coli*). Many of the chimeric plasmids constructed from *S. aureus*-derived plasmids are stable (Ehrlich, 1977, 1978; Gryczan and Dubnau, 1978; Niaudet and Ehrlich, 1979; Gryczan *et al.*, 1980b). However, deletions often occur in these hybrids, resulting in a loss of parental DNA (Gryczan and Dubnau, 1978). This phenomenon has also been observed in the *E. coli–B. subtilis* chimeric plasmids (Ehrlich *et al.*, 1978; Goebel *et al.*, 1978, 1979; Sgaramella *et al.*, 1978). The *Bacillus* chromosomal genes cloned by either conventional shotgun cloning (Keggins *et al.*, 1978b; Tanaka and Sakaguchi, 1978) or recombinational rescue (Gryczan *et al.*, 1980a) into *B. subtilis* appear to be reasonably stable.

Some hybrid plasmids have been used to study the expression of cloned genes. The *Bacillus pumilus trpC2* clone (Keggins *et al.*, 1978b) has been analyzed and shown to have a *Hin*dIII site in *trpC* which allows insertional inactivation of this gene, and this inactivation has a polar effect on *trpF*, another *trp* gene contained within the cloned fragment (Keggins *et al.*, 1979). Another study on the *trp* clone carrying the *trpE* gene (Kane and Hill, 1980) has yielded interesting information on gene regulation in *B. subtilis*. Complementation analysis has been used to identify genes on a cloned fragment (Gryczan *et al.*, 1980a). Using the *trp* clones, it has been shown that plasmid recombination between compatible plasmids requires the *recE4* marker (Keggins *et al.*, 1978a). A subsequent study using cloned *leu* genes has suggested that a *recE4*-independent type of plasmid recombination can also occur (Tanaka, 1979a). This apparent discrepancy may be attributed to the different type of gene sequences cloned or to the type of cloning vector employed (pUB110 in the first case and a cryptic *B. subtilis* plasmid in the second case).

A variety of *B. subtilis* genes cloned on *E. coli* or *E. coli–B. subtilis* plasmids or on lambdoid phage are able to express in *E. coli* (Horinouchi *et al.*, 1977; Mahler and Halvorson, 1977; Segall and Losick, 1977; Chi *et al.*, 1978; Duncan *et al.*, 1978; Nagahari and Sakaguchi, 1978; Rapoport *et al.*, 1979; Brammar *et al.*, 1980). When isolated clones on *B. subtilis–E. coli* plasmid chimeras were transformed into *B. subtilis*, expression often occurred (Rapoport *et al.*, 1979), but instability of plasmid chimeras was frequently observed.

Investigations into the cloning of various sporulation genes of *B. subtilis* have been initiated (Segall and Losick, 1977). Recently, a 0.4-kb *B. subtilis* DNA segment whose transcription is initated at an early stage of spore development has been cloned and amplified in *E. coli* using plasmid pMB9 (Segall and Losick, 1977). To map the genetic location of this gene, this hybrid plasmid has been ligated onto plasmid pC194 under conditions where pC194 replication is inactivated. Subsequent transformation of this plasmid into *B. subtilis* results in integration of the *spo* fragment and the plasmid DNA (both *E. coli* pMB9 and *B. subtilis* pC194) into the *B. subtilis* chromosome (Haldenwang *et al.*, 1980). Mapping of the pC194 Cm^r determinant by conventional means permitted localization of the *spo* gene on the chromosome, since it was assumed that integration of the plasmid occurred at the region of homology with the chromosome. Insertion of this cloned DNA at chromosomal regions of homology also established the physical orientation of this gene on the *B. subtilis* genetic map. Using plasmids constructed in this manner, site-directed integration of various cloned fragments into the *B. subtilis* chromsome should be useful in the orientation and mapping of other genes whose precise locations are still unknown. Recently a *Bgl*II fragment from *B. licheniformis* which complements *spo0H* mutations in *B. subtilis* has been cloned (Dubnau *et al.,* 1981). This clone was constructed using the helper cloning technique with the vector pBD64.

The analysis of cloned genes in the *B. subtilis* system has been conducted using two recently developed methods: the minicell system and the plasmid amplification system. The development of the *B. subtilis* minicell system (Reeve *et al.,* 1973) and its adaptation for the analysis of plasmid DNA (Leventhal and Chambliss, 1979; Shivakumar *et al.,* 1979) has greatly advanced the analysis of plasmid-coded proteins. The proteins specified by plasmids pUB110, pC194, pE194, and copy number mutants of pE194 (*cop-6*) have been identified (Shivakumar *et al.,* 1979) using this system. In addition, the pUB110 protein encoding kanamycin nucleotidyl transferase (Sadaie *et al.,* 1980), the pE194 protein encoding erythromycin resistance, and the pC194 chloramphenicol acetyltransferase have been identified (Shivakumar *et al.,* 1979). The map locations and mechanism of regulation of the pE194 polypeptides have been analyzed using minicells (Shivakumar *et al.,* 1980a,b). The minicell system has also been used to analyze the expression of various *E. coli* genes in *B. subtilis* when carried on *E. coli–B. subtilis* chimeric plasmids (Goebel *et al.,* 1979). Although none of the clones of chromosomal genes have yet been analyzed in *B. subtilis* minicells, eventual analysis of this kind will yield much information on the expression and regulation of these genes.

Analysis of plasmid gene-specified proteins can also be conducted using conditions of elevated gene dosage. It has been observed that certain *dna ts* replication mutants of *B. subtilis* at the nonpermissive temperature retard chromosome replication, although allowing plasmid pUB110 replication to continue (Shivakumar *et al.,* 1978b). This observation has been extended to plasmid pC194 and to the chimeric plasmid pBD9 (A. G. Shivakumar and D. Dubnau, unpublished). Utilization of this system in conjunction with minicell protein analysis should help in the further understanding of gene regulation in *B. subtilis*.

D. Examples of Cloning in *Bacillus subtilis*

A general listing of some of the genes cloned directly in *B. subtilis* is presented in Table IV. This table includes genes from the *B. subtilis* chromosome or from other sources (phages T7, φ105, *B. licheniformis*). It does *not* include cloning of genes into *E. coli* plasmids or phages, with initial isolation in *E. coli*. Only clones constructed with *B. subtilis* plasmids and phages, and in one case a *B. subtilis–E. coli* chimeric plasmid that can replicate in both hosts (pHV33), are listed.

TABLE IV Examples of Some of the Genes Cloned in *B. subtilis*

Gene[a]	Vector	Reference
trpC2	pUB110	Keggins *et al.* (1978a)
hisH2	pBD64	Gryczan *et al.* (1980a)
trpC2	pBD64	Gryczan *et al.* (1980a)
aroB2	pBD64	Gryczan *et al.* (1980a)
spo0H	pBD64	Dubnau, *et al.* (1981)
erm (macrolide–lincosamide–streptogramin B) (*B. licheniformis*)	pBD64	Docherty, *et al.* (1981)
thr	pHV33[b]	Rapoport *et al.* (1979)
leuA	pHV33[b]	Rapoport *et al.* (1979)
hisA	pHV33[b]	Rapoport *et al.* (1979)
glyB	pHV33[b]	Rapoport *et al.* (1979)
purB	pHV33[b]	Rapoport *et al.* (1979)
leu	*B. natto* (*B. subtilis*)	Tanaka and Sakaguchi (1978)
tmp	pLS103 (*B. subtilis* cryptic plasmid plus *leu* gene)	Tanaka and Kawano (1980)
ϕ105 Superinfection immunity gene	pUB110	Cully and Garro (1980)
α-Amylase		
B. amyloliquefaciens	ϕ3T	Yoneda *et al.* (1979b)
B. subtilis	ρ11	Nomura *et al.* (1979)
hisA1	ρ11	Kawamura *et al.* (1979a)
lys21	ρ11	Kawamura *et al.* (1979a)
spo077	ρ11	F. Kawamura, K. Kobayashi, and H. Saito, unpublished
metB	ϕ105	Iijima *et al.* (1980)
Phage T7 promoters	pC194	Scherzinger *et al.* (1980)

[a] Unless specifically indicated, cloned markers were from *B. subtilis*.
[b] These clones were first isolated in *E. coli* and subsequently transformed into *B. subtilis*.

V. Conclusions

The development of a molecular cloning system in *B. subtilis* has seen a steady state of advancement. In addition to the construction of more useful plasmid and phage cloning vectors, improved methods for the transformation of cloned DNA into suitable recipients and the analysis of recombinant DNA gene products have been developed. Although the *B. subtilis* system has demonstrated difficulties in the expression of *E. coli* gene sequences, the utilization of the *B. subtilis* and *E. coli* cloning systems in conjunction with each other should aid in understanding the restricted expression of foreign genes in *B. subtilis* and consequently permit the design of vectors permitting unrestricted expression.

Since many bacilli are important industrial bacterial species, the *B. subtilis* cloning system is especially important for production and analysis of bacillus gene products and possibly those of other important gram-positive organisms such as the *Streptomyces*. The importance of bacilli in the fermentation and agricultural industries is well documented (Erickson, 1976; Nickerson, 1980; Chapter 11). Further research into the development of this cloning system will hopefully make these potential uses more easily attainable.

Acknowledgments

I wish to express my gratitude to D. Dubnau for his encouragement and criticism during the course of preparing this chapter. Grateful appreciation is given to J. Hahn, G. Grandi, A. G. Shivakumar, and I. Smith for useful discussions, and to K. W. Hutchison, G. Rapoport, H. Saito, T. Tanaka, N. Ramakrishna, E. Dubnau, and I. Smith for supplying useful information prior to publication. Special appreciation is given to A. Howard, whose expert secretarial assistance was invaluable in the completion of this chapter.

References

Bernard, H. U., and Helinski, D. R. (1980). *In* "Genetic Engineering—Principles and Methods" (J. K. Setlow and A. Hollaender, eds.), Vol. 2, pp. 133–167. Plenum Press, New York.

Bernhard, K., Schrempf, H., and Goebel, W. (1978). *J. Bacteriol.* **133,** 897–903.

Bingham, A. H. A., Bruton, C. J., and Atkinson, T. (1979). *J. Gen. Microbiol.* **114,** 401–408.

Brammar, W. J., Muir, S., and McMorris, A. (1980). *Mol. Gen. Genet.* **178,** 217–224.

Bron, S., and Murray, K. (1975). *Mol. Gen. Genet.* **143,** 25–33.

Bron, S., Murray, K., and Trautner, T. A. (1975). *Mol. Gen. Genet.* **143,** 13–23.

Brown, B. J., and Carlton, B. C. (1980). *J. Bacteriol.* **142,** 508–512.

Brown, W. C., and Young, F. E. (1970). *Biochem. Biophys. Res. Commun.* **38,** 564–568.

Buheitel, G., and Klingmuller, W. (1979). *Eur. J. Appl. Microbiol. Biotechnol.* **7,** 79–84.

Canosi, U., Morelli, G., and Trautner, T. A. (1978). *Mol. Gen. Genet.* **166,** 259–267.

Chang, S., and Cohen, S. N. (1979). *Mol. Gen. Genet.* **168,** 111–115.

Chi, N.-Y. W., Ehrlich, S. D., and Lederberg, J. (1978). *J. Bacteriol.* **133,** 816–821.

Clewell, D. B., Yagi, Y., and Tomich, P. (1979). *In* "Microbial Drug Resistance" (S. Mitsuhashi, ed.), pp. 23–32. University Park Press, Baltimore, Maryland.

Contente, S., and Dubnau, D. (1979a). *Mol. Gen. Genet.* **167,** 251–258.

Contente, S., and Dubnau, D. (1979b). *Plasmid* **2,** 555–571.

Cregg, J. M., and Ito, J. (1979). *Gene* **6,** 199–219.

Cully, D. F., and Garro, A. J. (1980). *J. Virol.* **34,** 789–791.

Dean, D. H., Orrego, J. C., Hutchison, K. W., and Halvorson, H. O. (1976). *J. Virol.* **20,** 509–519.

Dedonder, R., Rapoport, G., Billault, A., Fargette, F., and Klier, A. (1980). *In* "DNA: Recombination, Interactions, and Repair" (S. Zadrazil and J. Sponar, eds.), FEBS—Vol. 63, pp. 95–104. Pergamon, Oxford.

Docherty, A. *et al., J. Bacteriol.* **145,** 129–137.

Domaradskii, I. V., Levadnaia, T. B., Sitrikov, B. S., Rassadin, A. S., and Denisova, T. S. (1976). *Dokl. Acad. Nauk SSSR* **226,** 1443–1445.

Dubnau, D., Contente, S., and Gryczan, T. J. (1980a). *In* "DNA: Recombination, Interactions, and Repair" (S. Zadrazil and J. Sponar, eds.), FEBS—Vol. 63, pp. 365–386. Pergamon, Oxford.

Dubnau, D., Gryczan, T., Contente, S., and Shivakumar, A. G. (1980b). *In* "Genetic Engineering—Principles and Methods" (J. K. Setlow and A. Hollaender, eds.), Vol. 2, pp. 115–131. Plenum Press, New York.

Dubnau, E., Ramakrishna, N., Cabane, K., and Smith, I. (1981). *J. Bacteriol.* **147,** 622–632.

Duncan, C. H., Wilson, G. A., and Young, F. E. (1977). *Gene* **1,** 153–167.

Duncan, C. H., Wilson, G. A., and Young, F. E. (1978). *Proc. Nat. Acad. Sci. USA* **75,** 3664–3668.

Ehrlich, S. D. (1977). *Proc. Nat. Acad. Sci. USA* **74,** 1680–1682.

Ehrlich, S. D. (1978). *Proc. Nat. Acad. Sci. USA* **75,** 1433–1436.

Ehrlich, S. D., Bursztyn-Pettegrew, H., Stroynowski, I., and Lederberg, J. (1976). *Proc. Nat. Acad. Sci. USA* **73,** 4145–4149.

Ehrlich, S. D., Bursztyn-Pettegrew, H., Stroynowski, I., and Lederberg, J. (1977). *In* "Recombinant Molecules: Impact on Science and Society" (R. F. Beers, Jr. and E. G. Bassett, eds.), pp. 69–80. Raven Press, New York.

Ehrlich, S. D., Jupp, S., Niaudet, B., and Goze, A. (1978). *In* "Genetic Engineering" (H. W. Boyer and S. Nicosia, eds.), pp. 25–32. Elsevier/North-Holland Publ., Amsterdam.

Erickson, R. J. (1976). *In* "Microbiology—1976" (D. Schlessinger, ed.), pp. 406–419. American Society for Microbiology, Washington, D. C.

Ermakova, L. M. *et al.* (1978). *J. Gen. Microbiol.* **107,** 169–171.

Fargette, F., Grelet, N., and Rapoport, G. (1979). *Biochimie* **61,** 93–100.

Feitelson, J. S., and Lederberg, J. (1980). *J. Gen. Microbiol.* **116,** 545–547.

Goebel, W., Kreft, J., Bernhard, K., Schrempf, H., and Weidinger, G. (1978). *In* "Genetic Engineering" (H. W. Boyer and S. Nicosia, eds.), pp. 47–58. Elsevier/North-Holland Publ., Amsterdam.

Goebel, W., Kreft, J., and Burger, K. J. (1979). *In* "Plasmids of Medical, Environmental, and Commercial Importance" (K. N. Timmis, and A. Puhler, eds.), pp. 471–480. Elsevier/North Holland Publ., Amsterdam.

Gonzalez, J. M., Jr., and Carlton, B. C. (1980). *Plasmid* **3,** 92–98.

Graham, J. B., and Istock, C. A. (1978). *Mol. Gen. Genet.* **166,** 287–290.

Graham, J. B., and Istock, C. A. (1979). *Science* **204,** 637–639.

Gray, O., and Chang, S. (1981). *J. Bacteriol.* **145,** 422–428.

Gryczan, T. J., and Dubnau, D. (1978). *Proc. Nat. Acad. Sci. USA* **75,** 1428–1432.

Gryczan, T. J., Contente, S., and Dubnau, D. (1978). *J. Bacteriol.* **134,** 318–329.

Gryczan, T., Contente, S., and Dubnau, D. (1980a). *Mol. Gen. Genet.* **177,** 459–467.

Gryczan, T., Shivakumar, A. G., and Dubnau, D. (1980b). *J. Bacteriol.* **141,** 246–253.

Gunthert, U., Stutz, J., and Klotz, G. (1975). *Mol. Gen. Genet.* **142,** 185–191.

Haldenwang, W. G. *et al.* (1980). *J. Bacteriol.* **142,** 90–98.

Harris-Warrick, R. M., Elkana, Y., Ehrlich, S. D., and Lederberg, J. (1975). *Proc. Nat. Acad. Sci. USA* **72,** 2207–2211.

Hemphill, H. E., and Whiteley, H. R. (1975). *Bacteriol. Rev.* **39,** 257–315.

Henner, D. J., and Hoch, J. A. (1980). *Microbiol. Rev.* **44,** 57–82.
Horinouchi, S. *et al.* (1977). *Mol. Gen. Genet.* **157,** 175–182.
Hoshino, T., Uozumi, T., Beppu, T., and Arima, K. (1980). *Agr. Biol. Chem.* **44,** 621–623.
Hutchison, K. W., and Halvorson, H. O. (1980). *Gene* **8,** 267–278.
Iijima, T., Kawamura, F., Saito, H., and Ikeda, Y. (1980). *Gene* **9,** 115–126.
Ikawa, S., Shibata, T., Ando, T., and Saito, H. (1979). *Mol. Gen. Genet.* **170,** 123–127.
Inciarte, M. R., Lazaro, J. M., Salas, M., and Vinuela, E. (1976). *Virology* **74,** 314–323.
Iordanescu, S. (1975). *J. Bacteriol.* **124,** 597–601.
Iordanescu, S. (1976). *Arch. Roum. Pathol. Exp. Biol.* **35,** 111–118.
Ito, J., Kawamura, F., and Yanofsky, S. (1976). *Virology* **70,** 37–51.
Kane, J. F., and Hill, R. J. (1980). *J. Bacteriol.* **141,** 993–995.
Kawamura, F., and Ito, J. (1977). *Virology* **83,** 233–245.
Kawamura, F., Saito, H., and Ikeda, Y. (1979a). *Gene* **5,** 87–91.
Kawamura, F., Saito, H., Ikeda, Y., and Ito, J. (1979b). *J. Gen. Appl. Microbiol.* **25,** 223–236.
Keggins, K. M., Duvall, E. J., and Lovett, P. S. (1978a). *J. Bacteriol.* **134,** 514–520.
Keggins, K. M., Lovett, P. S., and Duvall, E. J. (1978b). *Proc. Nat. Acad. Sci. USA* **75,** 1423–1427.
Keggins, K. M., Lovett, P. S., Marrero, R., and Hoch, S. O. (1979). *J. Bacteriol.* **139,** 1001–1006.
Klier, A. F., Kunst, F., and Rapoport, G. (1979). *Nucleic Acids Res.* **7,** 997–1010.
Kono, M., Sasatsu, M., and Hanashima, H. (1978). *Microbios Letters* **5,** 55–59.
Kreft, J., Bernhard, K., and Goebel, W. (1978). *Mol. Gen. Genet.* **162,** 59–67.
Kroyer, J. M., and Dean, D. H. (1979). *Current Microbiol.* **3,** 45–50.
Kroyer, J. M., Perkins, J. B., Rudinski, M. S., and Dean, D. H. (1980). *Mol. Gen. Genet.* **177,** 511–517.
Le Hegarat, J. -C., and Anagnostopoulos, C. (1977). *Mol. Gen. Genet.* **157,** 167–174.
Leventhal, J. M., and Chambliss, G. H. (1979). *Biochim. Biophys. Acta* **564,** 162–171.
Löfdahl, S., Sjöstrom, J. -E., and Philipson, L. (1978a). *Gene* **3,** 149–159.
Löfdahl, S., Sjöstrom, J. -E., and Philipson, L. (1978b). *Gene* **3,** 161–172.
Lovett, P. S., and Bramucci, M. G. (1975). *J. Bacteriol.* **124,** 484–490.
Lovett, P. S., and Bramucci, M. G. (1976). *In* "Microbiology, 1976" (D. Schlessinger, ed.), pp. 388–393. American Society for Microbiology, Washington, D.C.
Lovett, P. S., and Keggins, K. M. (1979). *Methods Enzymol.* **68,** 342–357.
Lovett, P. S., Duvall, E. J., and Keggins, K. M. (1976). *J. Bacteriol.* **127,** 817–828.
Mahler, I., and Halvorson, H. O. (1977). *J. Bacteriol.* **131,** 374–377.
Moran, Jr., C. P., Losick, R., and Sonenshein, A. L. (1980). *J. Bacteriol.* **142,** 331–334.
Mottes, M., Grandi, G., Sgaramella, V., Canosi, U., Morelli, G., and Trautner, T. A. (1979). *Mol. Gen. Genet.* **174,** 281–286.
Nagahari, K., and Sakaguchi, K. (1978). *Mol. Gen. Genet.* **158,** 263–270.
Naumov, L. S., Numerov, V. K., Bozkko, A. N., Lushnikov, I. V., Tyurin, V. S., and Domarodskii, J. V. (1977). *J. Microbiol. Epidemiol. Immunobiol.* **5,** 33–37.
Niaudet, B., and Ehrlich, S. D. (1979). *Plasmid* **2,** 48–58.
Nickerson, K. W. (1980). *Biotechnol. Bioeng.* **22,** 1305–1333.
Nomura, S. *et al.* (1979). *Agric. Biol. Chem.* **43,** 2637–2638.
Perkins, J. B., Zarley, C. D., and Dean, D. H. (1978). *J. Virol.* **28,** 403–407.
Potter, S. P., Bott, K. F., and Newbold, J. E. (1977). *J. Bacteriol.* **129,** 492–500.
Prozorov, A. A., and Glumova, E. F. (1980). *Plasmid* **3,** 231–232.
Rapoport, G., Klier, A., Billault, A., Fargette, F., and Dedonder, R. (1979). *Mol. Gen. Genet.* **176,** 239–245.

Reeve, J. N., Mendelson, N. H., Coyne, S. I., Hallock, L. L., and Cole, R. M. (1973). *J. Bacteriol.* **114,** 860–873.
Sadaie, Y., Burtis, K. C., and Doi, R. H. (1980). *J. Bacteriol.* **141,** 1178–1182.
Saito, H., Shibata, T., and Ando, T. (1979). *Mol. Gen. Genet.* **170,** 117–122.
Scher, B. M., Law, M. F., and Garro, A. J. (1978). *J. Virol.* **28,** 395–402.
Scherzinger, E., Grieder-Lauppe, H., Voll, N., and Wanke, M. (1980). *Nucleic Acids Res.* **8,** 1287–1305.
Segall, J., and Losick, R. (1977). *Cell* **11,** 751–761.
Sgaramella, V. *et al.* (1978). *In* "Genetic Engineering" (H. W. Boyer and S. Nicosia, eds.), pp. 33–45. Elsevier/North-Holland Publ., Amsterdam.
Shibata, T., and Ando, T. (1974). *Mol. Gen. Genet.* **131,** 275–280.
Shibata, T., and Ando, T. (1975). *Mol. Gen. Genet.* **138,** 269–279.
Shibata, T., Ikawa, S., Komatsu, Y., Ando, T., and Saito, H. (1979). *J. Bacteriol.* **139,** 308–310.
Shivakumar, A. G., and Dubnau, D. (1978a). *J. Bacteriol.* **136,** 1205–1207.
Shivakumar, A. G., and Dubnau, D. (1978b). *Plasmid* **1,** 405–416.
Shivakumar, A. G., Hahn, J., and Dubnau, D. (1979). *Plasmid* **2,** 279–289.
Shivakumar, A. G., Gryczan, T. J., Kozlov, Y. I., and Dubnau, D. (1980a). *Mol. Gen. Genet.* **179,** 241–252.
Shivakumar, A. G., Hahn, J., Grandi, G., Kozlov, Y., and Dubnau, D. (1980b). *Proc. Nat. Acad. Sci. USA* **77,** 3903–3907.
Stahly, D. P., Dingman, R. W., Irgens, B. L., Field, C. C., Feiss, M. G., and Smith, G. L. (1978) *FEMS Microbiol. Lett.* **3,** 139–141.
Tanaka, T. (1979a). *J. Bacteriol.* **139,** 775–782.
Tanaka, T. (1979b). *Mol. Gen. Genet.* **175,** 235–237.
Tanaka, T., and Koshikawa, T. (1977). *J. Bacteriol.* **131,** 699–701.
Tanaka, T., and Sakaguchi, K. (1978). *Mol. Gen. Genet.* **165,** 269–276.
Tanaka, T., Kuroda, M., and Sakaguchi, K. (1977). *J. Bacteriol.* **129,** 1487–1494.
Tanaka, T., and Kawano, N. (1980). *Gene* **10,** 131–136.
Trautner, T. A., Pawlek, B., Bron, S., and Anagnostopoulos, C. (1974). *Mol. Gen. Genet.* **131,** 181–191.
Tucker, R. G. (1969). *J. Gen. Virol.* **4,** 485–504.
Uozumi, T., Hoshino, T., Miwa, K., Horinouchi, S., Beppu, T., and Arima, K. (1977). *Mol. Gen. Genet.* **152,** 65–69.
Uozumi, T., Ozaki, A., Beppu, T., and Arima, K. (1980). *J. Bacteriol.* **142,** 315–318.
Vorobjeva, I. P., Khmel, I. A., and Alföldi, I. (1980). *FEMS Microbiol. Lett.* **7,** 261–264.
Weisblum, B., Graham, M. Y., Gryczan, T., and Dubnau, D. (1979). *J. Bacteriol.* **137,** 635–643.
Wilson, G. A., and Young, F. E. (1976). *In* "Microbiology, 1976" (D. Schlessinger, ed.), pp. 350–357. American Society for Microbiology, Washington, D.C.
Yagi, Y., McLellan, T. S., Frez, W. A., and Clewell, D. B. (1978). *Antimicrob. Agents Chemother.* **13,** 884–887.
Yoneda, Y., Graham, S., and Young, F. E. (1979a). *Gene* **7,** 51–68.
Yoneda, Y., Graham, S., and Young, F. E. (1979b). *Biochem. Biophys. Res. Commun.* **91,** 1556–1564.
Young, F. E., and Wilson, G. A. (1979). *In* "Genetic Engineering" (A. M. Chakrabarty, ed.), pp. 145–157. Chemical Rubber Press, Boca Raton, Florida.
Young, F. E., Duncan, C., and Wilson, G. A. (1977). *In* "Recombinant Molecules: Impact on Science and Society" (R. F. Beers, Jr., and E. G. Bassett, eds.), pp. 33–43. Raven Press, New York.

11

The Industrial Use of Bacilli

V. G. DEBABOV

I. Introduction

The microbiological industry has become an important element affecting technological progress during the last 30 years. Consequences of the technological revolution may not always be planned or predicted, but it appears certain that the development of an industry producing inexpensive and effective antibiotics has played a role in the demographical explosion on this planet. We can now be sure that in the near future industrial microbiology will be even more involved in the solution of

The Molecular Biology
of the Bacilli

ISBN 0-12-222701-8

global problems such as those concerned with protection of the environment, agriculture, and the production of foods for humanity and will contribute to solution of the energy problem and increase the availability of mineral resources.

The microbiological industry of today produces antibiotics, enzymes, vitamins, amino acids, nucleotides, polysaccharides, biological insecticides, single-cell protein, and a number of other products that find application in the pharmaceutical, food, and agricultural industries. The total sales of microbiological products, worldwide, amount to several billion dollars a year.

The genus *Bacillus,* which comprises the aerobic spore-forming soil saprophytes, is widely used in the microbiological industry. Selection of a microorganism for industrial application until now has been based on the screening of naturally occurring species for desired properties, followed by attempts at further improvement by mutagenesis and suitable selection.

Bacilli possess a number of valuable properties, including (1) the ability to excrete several hydrolytic enzymes into the cultivation medium, (2) the ability to produce antibiotics, (3) pathogenicity of certain species for insects, (4) the lack of pathogenicity of most species for humans and animals, and (5) acceptable cultivation conditions, including simple substrates and relatively high growth rates.

The achievements of the last decade in the basic sciences are no less impressive than those in industry. Some of the achievements in the area of molecular biology and molecular genetics are reviewed in this volume. Basic and applied studies have, until recently, been developing independently. However, at present, basic genetic and biochemical information and the ability to manipulate genetic material are more and more important in the construction of highly productive strains of microorganisms. Certain bacilli, particularly *Bacillus subtilis* 168, are among the most intensively studied microorganisms. The resulting availability of reliable genetic and biochemical information represents an advantage of this group for industrial use. Such information was critical in the selection of bacilli producing nucleosides and riboflavin for industrial application, despite the absence of overproducers of these substances among natural strains.

In this chapter we shall limit ourselves to a detailed consideration of the most economically important areas of the microbiological industry employing bacilli, including the production of enzymes, nucleosides, riboflavin, and preparations pathogenic to insects. We shall concentrate our attention on regulatory mechanisms of biosynthetic processes and possible approaches to the construction of strains, using the methods of genetics and genetic engineering.

II. Enzymes

A. Industrial Enzymes

The worldwide sales of enzyme preparations in 1978 equalled 175–180 million dollars and appeared to be increasing by 10–15% a year. Japan, producing 13,000 tons/yr, and the United States, producing 4500 tons/yr, are the main producing countries (Chemical Week, 1978).

A substantial proportion of these industrial enzymes is accounted for by two exoenzymes of bacilli: α-amylases and proteinases. Table I presents the main areas of use of these enzymes and the species of bacilli employed in their manufacture.

Other enzymes produced by bacilli are also promising for industry; these are substitutes for renin in the cheese industry (U.S. Patent, 1977), cellulolytic enzymes (U.S. Patent, 1974a), glucose isomerase used in the food industry to produce syrups (BRD Patent, 1977), and β-galactosidase used in the processing of whey in the dairy industry (Takeshi *et al.*, 1978). These enzymes of bacilli have not yet found as wide an application as α-amylases and proteinases, and for this reason we shall limit our discussion to the latter products. Information about other enzymes of bacilli can be found in a number of reviews (Fogarty *et al.*, 1974a,b; Ingle and Boyer, 1976; Erickson, 1976; Priest, 1977).

It can be seen from Table I that α-amylases and proteinases produced by different species of bacilli have somewhat different properties. It should be pointed out, in addition, that as a rule bacilli excrete not one

TABLE I Enzymes Produced by Bacilli and Their Industrial Applications

Enzyme	Bacilli	Application
Amylase		
Liquefying	*B. amyloliquefaciens*	Brewing, bread baking, paper industry
Saccharifying	*B. subtilis* var. *amylosacchariticus*, *B. natto*	Textile industry
Alkaline proteases		
Subtilopeptidases A (subtilisin Karlsberg)	*B. licheniformis*, *B. pumilis*, *B. subtilis*	Detergents, brewing, food industry
Subtilopeptidases B (subtilisin Novo, BPN, or Nagarse)	*B. amyloliquefaciens*	—
Neutroproteases		
Heat-stable	*B. amyloliquefaciens*, *B. subtilis*, *B. thermoproteolyticus*	Leather manufacturing, food industry

but several enzymes, components of an enzymatic complex. The presence of different enzymes in this complex at different levels contributes to the usefulness of a given complex and defines the area of its industrial application.

α-Amylases hydrolyze α-1,4-glycosidic bonds in starch and related polysaccharides. Bacilli produce two types of α-amylases: liquefying and saccharifying (Matsuzaki *et al.,* 1979). The first type finds a wider application in industry. Alkaline proteinases resemble in their properties such animal enzymes as chymotrypsin and trypsin (pH optimum 9.0–11.0, molecular weights in the range of 25,000–30,000). The activity of these enzymes does not require divalent metal ions, and the active site contains a serine residue. Another proposed name for these enzymes is serine proteinases.

Neutral proteinases are optimally active near a pH of about 7.0, and their activity shows an absolute dependence on the presence of divalent metal ions; therefore these proteinases are frequently called metalloproteinases (Matsubara and Feder, 1971).

Although almost all species of bacilli are able to secrete proteinases or α-amylases into the medium, natural isolates strongly differ both in the properties of the secreted enzymes and in the levels of activity achieved. Several strains can produce 5–40 times more exoenzyme than *B. subtilis* 168 (Ingle and Boyer, 1976). Usually *Bacillus amyloliquefaciens,* a species that is taxonomically distinct from *B. subtilis* 168, is used for the industrial production of α-amylases and proteinases (Welker and Campbell, 1967; Borgia and Campbell, 1978). The serine proteinase from *B. amyloliquefaciens* (Novo or BPN) differs from the *B. subtilis* (Carlsberg) proteinase by 83 amino acid changes (Smith *et al.,* 1966). At present, strains isolated from nature are subjected to screening not only for high productivity, but also for enzymes with new properties essential for practical purposes, such as heat-stable α-amylases (Madeson *et al.,* 1973; Chiang *et al.,* 1979), alkaline α-amylases having a pH optimum of 9.0–11.5 (Japanese Patent, 1977), serine proteases active at high pH up to 12 (Aunstrup *et al.,* 1972), α-amylases with altered specificity, and heat-stable neutral proteinases (U.S. Patent, 1974c).

B. Synthesis of Exoenzymes and Sporulation

The maximal rate of exoenzyme synthesis is usually observed in the late exponential or early stationary phase of growth (Schaeffer, 1969), at which time the cells begin preparation for sporulation. A possible relationship between the synthesis of exoenzymes and sporulation has been discussed in the literature for a long period of time, and perhaps it would be appropriate to formulate the following two pertinent questions: (1) Are exoenzymes really necessary for sporulation? (2) Are certain steps of sporulation necessary for the triggering of exoenzyme synthesis?

A negative answer should probably be given to the first question. There are no convincing data in the literature demonstrating the involvement of exoenzymes in the sporulation of bacilli. Mutants in the structural gene of α-amylase (*amyE*$^-$), for example grow and sporulate normally (Schaeffer, 1969). Mutants of *B. subtilis* (Hageman and Carlton, 1973), *B. megaterium* (Millet and Aubert, 1969), and *B. cereus* (Aronson *et al.*, 1971) have been isolated that are completely devoid of metalloproteinases but show normal growth and sporulation characteristics. Certain natural spore-forming isolates of *B. megaterium* and *B. cereus* synthesize only the neutral proteinase (Millet, 1969). It should be pointed out that a large group of pleiotropic mutants have been isolated with a blockage in sporulation at stage 0 (*spo0*) and which do not synthesize any alkaline protease (Hoch and Spizizen, 1969). However, the causal relationship between the absence of sporulation and the enzyme deficiency would be proven only if a *spo0* mutation were mapped in the structural gene of the alkaline protease. Leighton *et al.* (1973) described the *ts5* mutation in *B. subtilis* leading to a blockage in sporulation at stage 0 and having a thermosensitive serine proteinase. It was demonstrated later, however, that *ts5* carried two tightly linked mutations. These mutations have been separated, and Spo$^+$ variants carrying the defective serine protease have been obtained (Millet *et al.*, 1976).

Analysis of a possible role for extracellular serine proteinase in sporulation is complicated by the presence of intracellular serine proteases in cells of bacilli (Higerd and Spizizen, 1973; Daucer and Mandelstam, 1975; Lecadet *et al.*, 1977). The structural analysis of these latter enzymes has demonstrated that, although they have a certain degree of homology with the extracellular proteases, nevertheless they represent different proteins, are not precursors of exoenzymes, and are coded by distinct structural genes (Strongin *et al.*, 1978, 1979). It cannot be rigorously excluded that the intracellular serine proteases are indispensable at certain stages of sporulation. Evidence for this suggestion has been adduced from studies with inhibitors. The addition of inhibitors of serine proteases such as phenylmethylsulfonyl fluoride and *n*-aminobenzylboronic acid to *B. subtilis* in concentrations not affecting vegetative cells results in a complete suppression of sporulation (Geele *et al.*, 1975; Piggot and Coote, 1976).

An answer to the second question, concerning the importance of sporulation genes in the synthesis of exoenzymes, may be different for different enzyme proteins. The expression of sporulation genes appears to be unnecessary for the synthesis of α-amylase. For example, with starch as a carbon source, *B. subtilis* NE23 produces α-amylase at a constant rate throughout vegetative growth (Welker and Campbell, 1967). Inhibition of sporulation by thymidine starvation, for example, results in blockage of the synthesis of exogenous proteinases, RNase, and alkaline phosphatase but does not affect the formation of α-amylase (Daucer and Mandelstam, 1975). This example can be regarded as an argument for the suggestion that the synthesis of exogenous proteinases, unlike that of α-amylase, requires a certain level of cell differentiation. The genetic blockage of sporulation at stage 0 usually does not affect the synthesis of α-amylase but results in the inhibition of proteinase synthesis (Hoch and Spizizen, 1969). It may be concluded that, although exoenzymes are not directly involved in the process of differentiation, the synthesis of most of these enzymes (excluding α-amylase) requires triggering mechanisms similar to those responsible for the initiation of sporulation.

C. General Regulatory Mechanisms of Exoenzyme Synthesis

The maximal rate of synthesis of exogenous enzymes corresponds to the stage when the culture enters the stationary phase of growth. At this time, substrates are depleted and energy metabolism undergoes profound changes.

It is generally recognized that the regulation of biosynthesis of exogenous enzymes involves the phenomenon of catabolite repression. Catabolite repression is defined as the inhibition of synthesis of constitutive or inducible enzymes in the presence of glucose or other rapidly metabolizable carbon sources (Schaeffer, 1969). Phenomenologically an inverse proportional relationship has been observed between the growth rate and the rate of exoenzyme synthesis. For example, the growth rate of *B. subtilis* 168 increases with changes in the carbon source in the following sequence: lactate < glutamate < maltose < glycerol < glucose; the rate of α-amylase synthesis *decreases* in the same order (Sekiguchi and Okada, 1972). A similar correlation has been observed for the synthesis of α-amylase in *Bacillus amyloliquefaciens* (Coleman and Grant, 1966) and *B. stearothermophilus* (Welker and Campbell, 1963) and for the synthesis of proteinases in *B. licheniformis* (Laishley and Bernlohr, 1968), *B. subtilis* (Doi, 1972) and *B. amyloliquefaciens* Heineken and O'Connor, 1972). However, if the growth rate of a culture is limited by the availability of nitrogen, no depression of the synthesis of α-amylase by readily metabolizable carbon sources is observed (Meers, 1972).

The mechanism of catabolite repression in bacilli remains unclear and appears not to coincide precisely with the mechanism well known for *Escherichia coli.* In the latter organism depletion of glucose in the medium results in an increased intracellular concentration of cyclic adenosine 3′,5′-monophosphate (cAMP) which, in a complex with a specific protein (cAMP receptor protein, CRP) acts as a positive regulator of the transcription of genes sensitive to catabolite repression (Rickenberg, 1974). However, neither cAMP nor enzymes responsible for its synthesis and degradation, such as adenylate cyclase and cAMP phosphodiesterase, have been found in bacilli (Ide, 1971; Setlow, 1973; Bernlohr *et al.*, 1974; Hanson, 1976). It may well turn out that cAMP is not a universal mediator of catabolite repression and that cyclic guanosine monophosphate (cGMP) or some other molecule plays an analogous role in bacilli. Incidentally, cGMP has been found in *B. licheniformis,* and its intracellular level varies inversely with the bacterial growth rate (Hanson, 1976). Any final conclusion about the role of cGMP in the regulation of genes sensitive to catabolite repression in bacilli will be possible only after the construction and study of mutants defective in the synthesis and degradation of this nucleotide.

A number of unusual highly phosphorylated nucleotides such as ppApp, pppApp, pppAppp, ppZpUp, where A is adenosine and Z is a moiety of unknown structure, have been reported to exist in cells of *B. subtilis* under conditions of catabolite repression (Rhaese *et al.*, 1972). Adenosine hexaphosphate may accumulate in the culture medium, while the other compounds appear to be located intracellularly. The mutation in *spo0F,* leading to a blockage in sporulation, is reported to lead also to the cessation of pppAppp synthesis (Rhaese *et al.*, 1977). Ribosomes isolated from stationary-phase *B. subtilis* synthesized penta- and hexa-adenosine phosphates *in vitro,* whereas ribosomes isolated during the exponential phase of growth were devoid of such a capacity (Rhaese, and Groseurth 1974). These observations have been challenged (Nishino *et al.*, 1979; Smith *et al.*, 1980). It has been suggested by Rhaese and his co-workers that adenosine hexaphosphate may serve as a universal trigger of sporulation. However, it seems that there are not sufficient grounds for this suggestion. It would be interesting to examine the effect of highly phosphorylated nucleotides on the synthesis of exogenous enzymes, but at present there are no data of this kind.

Much more definite information has been obtained concerning the function of the so-called stringent control system in bacilli. *Escherichia coli* cells under conditions of amino acid starvation are known to synthesize guanosine tetra- and pentaphosphates, (p)ppGpp (Cashel and Gallant, 1969). The synthesis of these compounds is accomplished when codon-specific nonacylated tRNA enters the A side of the programmed ribosome (Sy and Lipmann, 1973). This activates a specific enzyme termed the stringent control factor, which synthesizes pppGpp from ATP and GTP. This highly phosphorylated guanosine is a precursor of guanosine tetraphosphate (Sy and Lipmann, 1973; Fill *et al.*, 1977; Pedersen and Kjeldgaard, 1977).

Mutations in the gene *relA*, which is the structural gene for the stringent control factor, impair the ability of cells to accumulate ppGpp under conditions of aminoacyl-tRNA deficiency (Cashel and Gallant, 1969). An increase in the intracellular ppGpp concentration in *E. coli* results in a significant rearrangement of metabolism, including inhibition of transcription of genes coding for the structure of ribosomal components and translation machinery (Dennis, 1977), activation of genes coding for enzymes of amino acid biosynthesis (Somerville and Ahmad, 1977), inhibition of phospholipid biosynthesis (Nunn and Cronan, 1976), inhibition of the formation of peptidoglycans (Ramey and Ishiguro, 1978) and of polyamines (Gould *et al.*, 1975), impairment of transport (Sokawa and Kaziro, 1969; Fast and Sköld, 1977), activation of intracellular proteolysis (John *et al.*, 1978), and changes in many other physiological processes. Stephens *et al.*, (1975) advanced a concept according to which ppGpp represents a signal molecule, or ''alarmone,'' producing deviation of the cell from the state of balanced growth. The accumulation of ppGpp enables the cell to accomplish coordinated rearrangements of metabolism necessary for adaptation to new environmental conditions.

The accumulation of (p)ppGpp under conditions of aminoacyl-tRNA deficiency has been demonstrated in *B. subtilis*. In the presence of a *rel* mutation, however, the synthesis of stable RNA continues during amino acid starvation (Swanton and Edlin, 1972). Conflicting data have been published concerning the levels and the proportion of penta- and tetraphosphates in bacilli. A marked increase in the level of (p)ppGpp under conditions of amino acid starvation has been demonstrated in a number of studies (Swanton and Edlin, 1972; Gallant and Margason, 1972; Ogilvie *et al.*, 1975; Arnold and Ogilvie, 1977; Smith *et al.*, 1978). This, however, has not been confirmed by other workers (Fortnagel and Bergman, 1972; Thomas and Copeland, 1973). Recently Shakulov and Belitskiy (1980) have revised earlier studies and have demonstrated that the concentration of (p)ppGpp in *B. subtilis* increases under conditions of amino acid starvation and that the level of (p)ppGpp accumulation depends on the species of aminoacyl-tRNA that is deficient, varying from 800 nmol (A_{600}) to 150 nmol (A_{600}) (Shakulov and Belitskiy, 1980). These differences are adequate for explaining the conflicting results reported in the literature.

Despite these fluctuations in (p)ppGpp content, the effectiveness of the stringent control of RNA synthesis does not depend on which species of aminoacyl-tRNA is deficient and limiting growth, thus indicating that even a minimal level of guanosine tetraphosphate is sufficient for effective regulation. In contrast to *E. coli,* the pentaphosphate is the predominant metabolite in *B. subtilis* cells; but, as in *E. coli,* it is a precursor of tetraphosphate (Shakulov and Belitsky, 1980).

Relaxed strains of *B. subtilis* (rel^-) are capable of normal sporulation (Rhaese *et al.*, 1976; Smith *et al.*, 1980), suggesting that the stringent control mechanism is not required for differentiation. The physiological role of ppGpp appears to involve compensation for an aminoacyl-tRNA deficit; this is achieved on the one hand by the activation of operons coding for enzymes of amino acid biosynthesis, and on the other hand by the activation of intracellular proteolysis (St. John *et al.*, 1978). It can be inferred therefore that overproduction of amino acids and proteinases in certain producing strains might be regulated by an elevated intracellular level of ppGpp. The correctness of this assumption has been demonstrated, in the case of overproduction of L-threonine, by a reference strain of *E. coli* (Gusyatiner *et al.*, 1978). Since the synthesis of exogenous proteinases is inhibited by the addition of exogenous amino acid (Neumark and Citri, 1962; Chaloupka *et al.*, 1963), it is also reasonable to suggest that the genes coding for proteinases are activated by increased ppGpp levels, as long as the amino acid deficiency is moderate, and a sufficiently high level of protein synthesis can be maintained. Similar suggestions may also be advanced to explain the induction of synthesis of amylases, lipases, and other enzymes. The formation of these enzymes is known to occur when the carbon source is limiting, a condition known to cause the accumulation of ppGpp (Chalorer-Larson and Vamazaki, 1978). The industrial use of these concepts concerning the role of ppGpp in the regulation of overproduction of exoenzymes and metabolites in bacilli appears promising.

It is an interesting characteristic of exoenzyme synthesis that the inhibition of transcription by rifamicin, by actinomycin D, or by a mutation (Kinoshita *et al.*, 1968; Both *et al.*, 1972) does not result in an immediate arrest of enzyme formation. The translation and excretion of extracellular enzymes is known to continue for 30–60 min (Kinoshita *et al.*, 1968; Both *et al.*, 1972) after transcriptional blockage. This suggests the existence of a pool of relatively stable mRNA species that are specific for the synthesis of extracellular enzymes; it appears that the tight coupling of transcription and translation generally observed in prokaryotic systems is not present in the case of extracellular enzyme synthesis. A plausible explanation for the stability of mRNA specific for extracellular enzymes involves its compartmentalization in the cell associated with the peculiar features of excretion of such enzymes from the cells of bacilli.

D. Excretion of Extracellular Proteins

The mechanisms of excretion of extracellular industrially significant proteins remain unelucidated. Most hypotheses utilize the concepts developed for the secretory proteins of eukaryotes, i.e., the signal hypothesis (Blobel and Dobberstein, 1975), according to which the synthesis of excreted proteins proceeds on ribosomes located on membranes. The primary precursors of these proteins contains a hydrophobic amino acid sequence at the N-terminus. This is the "signal sequence" that is recognized by and becomes associated with receptor membrane proteins, following which a tunnel or a pore is formed in the membrane, facilitating secretion of the growing polypeptide chain. The signal sequence is cleaved proteolytically on the external side of the membrane, and the excreted protein folds and acquires a stable hydrophilic conformation. This hypothesis is supported by a number of experimental observations. Membrane-bound ribosomes have been found in various bacilli (Schlessinger *et al.*, 1965; Coleman, 1969; van Dijk-Salkinoja *et al.*, 1970). It has been shown that precursors of many extracellular proteins are indeed associated with membranes and that these precursors are larger than the final products. Such precursors have been described for *B. licheniformis* penicillinase (Yamamoto and Lampen, 1976), α-amylase and proteinase of *B. amyloliquefaciens* (Nagata *et al.*, 1974; Sanders and May, 1975), and the levanesaccharase of *B. subtilis* (Caulfield *et al.*, 1979).

The excretion of *B. licheniformis* penicillinase has been studied in detail by Lampen (1976). This author has found that penicillinase precursors have a signal sequence at the N-terminus that contains a phospholipopeptide. Other authors, however (Simons *et al.*, 1978), have been unable to confirm the presence of a phospholipid moiety in the membrane-bound penicillinase. The latter authors have established that the precursor contains a hydrophobic amino acid sequence at the N-terminus. More recently, it has been shown that gylceride is attached to an N-terminal cysteine by a thioether linkage, probably during translation (Lai *et al.*, 1981; Smith *et al.*, 1981; Nielson *et al.*, 1981). The proteinase involved in the mobilization of exopenicillinase has been identified and purified (Aiyappa *et al.*, 1977). Exoenzymes generally are not accumulated within cells; this has led to the idea that synthesis and excretion are tightly coupled. Excretion can be inhibited by blocking lipid synthesis with the antibiotic cerulinine at concentrations that do not affect the bulk synthesis of RNA and proteins. Such an inhibition not only arrests excretion but also leads to the cessation of synthesis of extracellular enzymes, again providing evidence for a tight coupling of these processes (Caulfield *et al.*, 1979).

It has been suggested that extracellular enzymes are accumulated in the space between the external surface of the membrane and the cell wall. For example, when protein synthesis in *B. amyloliquefaciens* is completely inhibited by chloramphenicol, the cells continue to excrete proteinases and α-amylase for about 15 min, this process being independent of energy supply (Gould *et al.*, 1975). It is not known whether transfer through the cell wall in bacilli represents passive diffusion or active transport. It has been shown that the excretion of hemolysin by *E. coli* is mediated by two proteins, one of which is responsible for transport through the membrane into the periplasmic space, whereas the other participates in transport from the periplasmic space to the external medium (Goebel *et al.*, 1979; Noegel *et al.*, 1979). Evidence that excretory systems for different proteins may differ from each other has been reported. These data will be considered in the next section.

E. Genetic Regulation of Synthesis and Excretion of Extracellular Enzymes

Figure 1 presents in a schematic form the genetic map of the circular chromosome of *B. subtilis*, showing several loci known to be important for the synthesis of α-amylase and proteinases. The map order of these markers is in agreement with the map of Young and Wilson (1976).

The structural gene of α-amylase has been mapped and found to be tightly linked to the *aro-116* marker (cotransformation index 30–40%) (Yuki and Ueda, 1968). The mutation *aro-116* is located in the *aroI* locus (Yuki, 1975). Two related species, *Bacillus subtilis* 168 (Marburg) and *B. natto* synthesize different levels of α-amylase, equal to 10 and 50 units/mg of cells, respectively. These two enzymes can be distinguished by physical and chemical properties. When *B. subtilis* is transformed by DNA from *B. natto*, 10% of the transformants showing high levels of α-amylase (Amy^h) synthesize α-amylase of the recipient type, and 90% produce the donor-type enzyme (Yamaguchi *et al.*, 1974a). It has been demonstrated that the structural genes *amyE* of these two species are allelic and that the higher or lower levels of synthesis depend on the regulatory gene *amyR*. The gene *amyR1* of *B. subtilis* 168 is responsible for the slow rate of synthesis, and the gene *amyR2* of *B. natto* provides the higher rate. All four possible combinations *amyR1* $amyE_n$, *amyR1* $amyE_m$, *amyR2* $amyE^n$, and *amyR2* $amyE^m$ have been found, and the

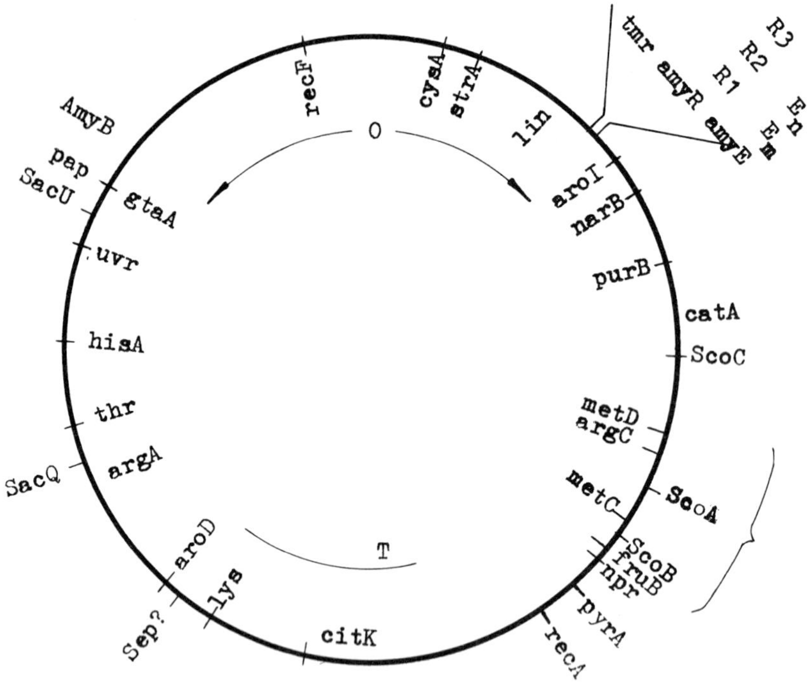

Fig. 1. Schematic representation of the *B. subtilis* 168 chromosomal map. O, Origin of replication; T, terminus of replication. Distances in this scheme do not precisely reflect either genetic or physical distances between markers. Loci important for overproduction of enzymes are indicated outside the circle; other genetic markers are shown inside the circle.

order of markers on the chromosome has been mapped (Yamaguchi *et al.*, 1974b) (Fig. 1). The gene *amyR* appears to be directly linked to the structural gene and may carry a promoter for α-amylase. In *B. subtilis* var. *amylosacchariticus* (SAC) the gene *amyR3* has been found to be apparently identical to *amyR2* (Yoneda *et al.*, 1974). Sazaki *et al.* (1976) described mutants resistant to tunicamycin (*tmr*). Two out of seven of these mutants showed a two- to- five fold increase in α-amylase synthesis. The mutation *tmr-7,* which mapped near *amyR,* did not affect the synthesis of other extracellular enzymes. The order of markers was as follows: *lin-2, tmr-7, amyR2, amyE, aro-1, narB* (Nomura *et al.*, 1978).

Uehara *et al.* (1974) were able to transform the gene(s) for overproduction of the neutral proteinase (Npr^h) from *B. subtilis natto* to *B. subtilis* Marburg. Both these strains possess identical proteinases and consequently carry identical alleles of the structural gene (*nprE*). It has been suggested that in this case, as in that of α-amylase, regulatory genes (*nprR1* and *nprR2*) are responsible for the low or high level of exoenzyme synthesis.

The structural gene for the neutral proteinase has been recently localized on the genetic map of *B. subtilis* (Uehara *et al.*, 1979) between *pyrA* and *fruB* (Fig. 1). The structural gene for serine proteinase is apparently between *aroD* and *lys* (Leighton *et al.*, 1972, 1973). No mutations have been obtained so far that could specifically increase the rate of serine proteinase synthesis.

A large group of pleiotropic mutations lead to a modification of the production level of several extracellular enzymes. The mutation *pap* (protease and amylase productivity) increases the level of α-amylase 5–16 times, increases the level of proteinases up to 3 times, and affects a number of other extracellular enzymes (Yoneda and Maruo, 1975) without affecting the level of RNase (Yoneda *et al.*, 1973). The presence of this mutation leads to changes in a number of other properties, including loss of flagella, autolytic enzyme activity and alterations in cellular morphology (Aynsawa *et al.*, 1975). Several mutations described by other authors, such as *amyB*, *sacU* and *amyC* (Kuninaka *et al.*, 1964; Sekiguchi *et al.*, 1975), cause phenotypic changes similar to those of the *pap* mutation. The *amyB*, *sacU*, and *pap* mutations appear to map in a single locus between *uvr* and *gtaA* (Steinmetz *et al.*, 1976). The mutation *sacQ* is less pleiotropic and leads to the overproduction of levansucrase and proteases but not of α-amylase (Kunst *et al.*, 1974). Another mutation leading to the elevated production of proteases but having no effect on the synthesis of α-amylase is called *hpr* and maps in the *argC* region (Higerd *et al.*, 1972). A series of mutations controlling sporulation in *B. subtilis* (*sco*) has been described (Milhaud *et al.*, 1978; Dod *et al.*, 1978; Dod and Balassa, 1978). These mutations were selected on the basis of their ability to secrete high quantities of proteinases into the medium. Six such mutations have been mapped in three loci in the *argC–metC* region (Fig. 1). The mutation *scoC* appears to define the same locus defined by the *catA* mutation (Ito and Spizizen, 1972). The latter determines the capacity of *B. subtilis* cells to sporulate in the presence of high concentrations of glucose; it impairs the catabolite repression of sporulation and leads to the increased production of alkaline proteinase (Dobrzhanskaya *et al.*, 1978).

It appears from the above data that mutations such as *pap* affect the production of many but not all extracellular enzymes. There is a general agreement that this mutation affects mainly excretory mechanisms. If this is indeed the case, the mechanisms of excretion must be different for different enzymes. This is consistent with the observation that a mutation (*RAN1*) in *B. subtilis* results in the conversion of membrane-bound alkaline phosphatase into the extracellular form (Hitotsuyanagi *et al.,* 1978). Introduction of the *pap* mutation into a strain carrying *RAN1* or into the parental strain by transformation did not in any way affect the level of production or the intracellular localization of alkaline phosphatase (Hitotsuyanagi *et al.,* 1978).

The construction of enzyme-producing industrial strains of bacilli is generally based on screening for useful properties and on further improvement of these strains by mutagenesis and selection of highly productive variants (Fogarty *et al.,* 1974a,b; Ingle and Boyer, 1976; Priest, 1977). Selection may make use of the above-described concepts concerning the mechanism of regulation of biosynthesis of extracellular enzymes as well as of a number of additional empirical findings. It has been established, for example, that *spo*$^-$ mutants frequently have an increased capacity to synthesize proteases. In these cases, enzyme production is sometimes elevated by an order of magnitude (Levinson and Aronson, 1967; Cook and Ramaley, 1978). It has been demonstrated that increased synthesis of extracellular enzymes frequently accompanies mutations conferring resistance to a number of antibiotics such as rifampicin (Shinke *et al.,* 1979), ampicillin, cycloserine (Maruo *et al.,* 1978), and streptomycin (Ermakova *et al.,* 1978).

The combination of several mutations in a given strain frequently has a synergistic effect. Work published by Hitotsuyanagi *et al.* (1979) illustrates such effects. In this study a mutation conferring resistance to cycloserine (*CS108*) was isolated in *B. subtilis* 6160. Mutations in *B. natto* and *B. subtilis* var. *amylosaccharitícus* leading to overproduction of α-amylase were introduced by transformation into this strain. Following this, repeated mutagenesis and selection led to the isolation of strain T2N26, which was able to synthesize up to 16,000 units/ml of α-amylase, whereas the parental strain synthesized only 10 units. Genetic analysis has shown that the high productivity of T2N26 was associated with the presence of six genetic markers: *amyR3, amyS, tmr-7, papSAC, CS108,* and *N26* (Fig. 1).

Insufficient information about the synthesis or excretion of extracellular enzymes and insufficient genetic data concerning many of the species of bacilli currently used in industry represent the main obstacles to the genetic construction of useful strains producing high levels of enzymes. In Section IV we shall discuss possible approaches to the purposeful construction of useful strains, including the techniques of interspecific genetic exchange.

F. Synthesis of Purine Nucleosides and Riboflavin

The ability of inosinic acid–glutamic acid mixtures to improve the taste properties of certain food products is well known. Guanylic acid has a similar taste-enhancing property and is 2.3 times more active than inosinic acid (Kuninaka *et al.*, 1964). Purine nucleotides and nucleosides have also found application in medicine.

At present, the production of these two nucleotides in Japan reaches 3000 tons/yr (2000 tons of IMP and 1000 tons of GMP) (Yamada, 1977). The most economic approach involves the synthesis of nucleosides by fermentation, with subsequent chemical phosphorylation (Yoshikawa *et al.*, 1969). The production of nucleotides by direct fermentation is also possible (Japanese Patent, 1974), but the yield is markedly lower than that of nucleosides, apparently because of lower permeability of the bacterial membrane for nucleotides and the presence of active enzymes of nucleotide degradation in the cells (Shiio, 1979). The production of inosine and guanosine in the microbiological industry makes use mainly of *B. subtilis* mutants. The regulatory pathway of purine nucleotide biosynthesis in *B. subtilis* is presented in Fig. 2.

The main characteristic of the regulation of purine nucleotide biosynthesis in *B. subtilis* is the asymmetry of the control: The first enzyme of the common pathway, PRPP amidotransferase, is subject to feedback regulation by AMP but not by GMP; in many other organisms this enzyme is controlled simultaneously by the two end products. Another point of asymmetric control is the branching of the pathway from IMP to GMP and AMP. The first enzyme of the GMP pathway, IMP dehydrogenase, has a specific activity 10–30 times greater than that of the first enzyme of the AMP pathway, adenylosuccinate synthetase. In addition, AMP does not regulate the first enzyme of its own pathway, whereas IMP dehydrogenase is subject to feedback inhibition by GMP (Ishii and Shiio, 1970). A number of enzymes between PRPP and IMP are repressible by the joint action of GMP and AMP. The strategy of construction of strains producing inosine and guanosine is based on information concerning the regulation of synthesis, as well as on a number of empirical findings made during selection work with analogs of purine bases.

All inosine-producing strains are also adenine auxotrophs because of a blockage in adenyl succinate synthetase; this relieves the feedback inhibition of PRPP amidotransferase and prevents the conversion of IMP to AMP. It should be pointed out that simple adenine auxotrophs of *B. subtilis* predominantly accumulate hypoxanthine but not inosine. Blocking inosine degradation by an addditional mutation leads to strains that produce only inosine (Aoki *et al.*, 1963). Since the repressing action of

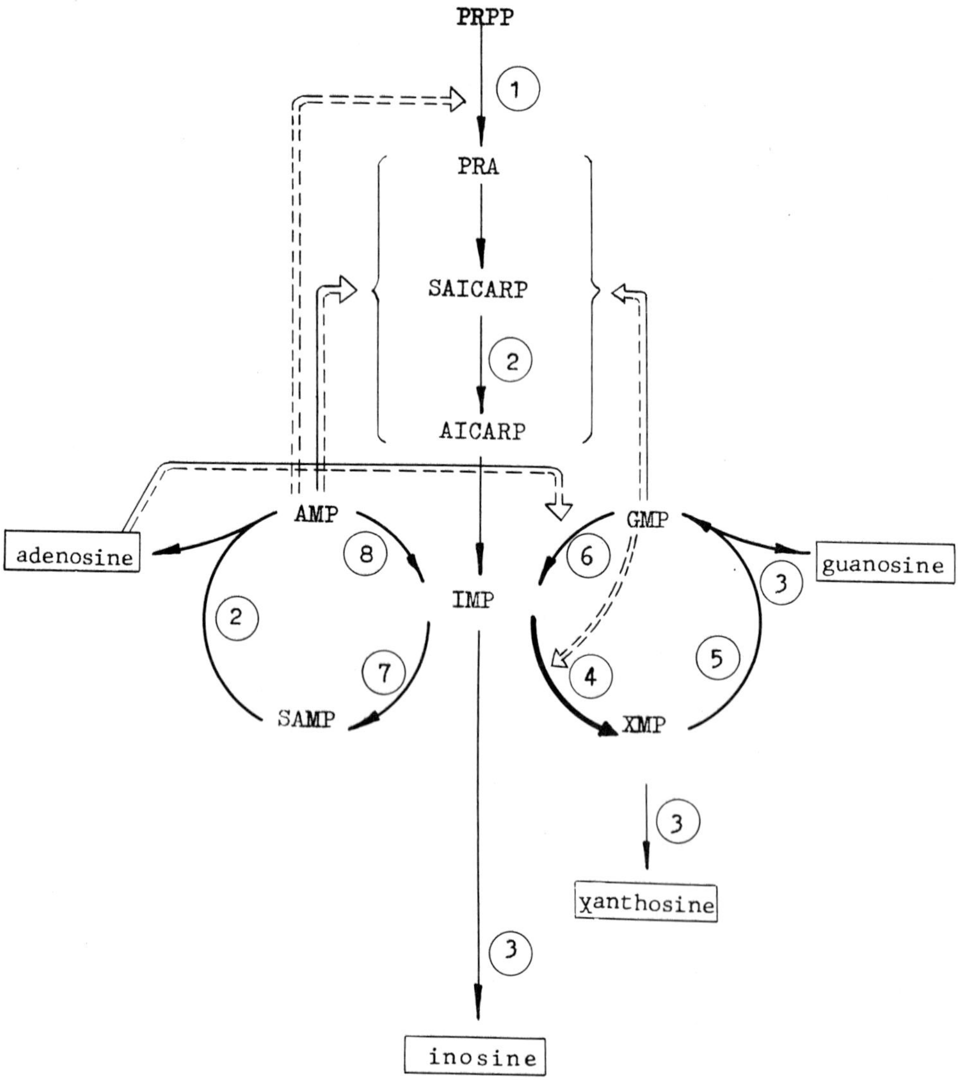

Fig. 2. Biosynthesis of purine nucleotides in *B. subtilis*. ⇢, Feedback control; ⇢ repression; ➡ preferential synthesis. (1) Amidotransferase; (2) succino-AMP lyase (catalyzes two reactions, SAMP → AMP and SAICARP → AICARP); (3) 5-nucleotidase; (4) IMP dehydrogenase; (5) GMP synthetase; (6) GMP reductase; (7) adenylosuccinate synthetase; (8) AMP deaminase; PRPP, 5′-Phosphoribosylpyrophosphate; PRA, phosphoribosylimine; SAICARP, succinoaminoimidazolecarboxyamidoriboside; AICARP, aminoimidazolcarboxyamidoriboside; IMP, inosine monophosphate; GMP, guanosine monophosphate; SAMP, succinyladenine monophosphate, XMP, xanthine monophosphate.

AMP and GMP is synergistic, the enzymes of AMP biosynthesis are markedly derepressed in the presence of only GMP. However, a blockage of GMP synthesis at the level of IMP dehydrogenase results in decreased IMP consumption and also relieves the residual repression of enzymes involved in IMP biosynthesis, leading to a 50% increase in inosine production (Table II). In practice, however, a certain pool of AMP and GMP in the cell prevents complete elimination of repression. One can further derepress the system by constructing strains resistant to the purine analogs 8-azaguanine (Ishii and Shiio, 1972) and 6-thioguanine (USSR Patent No. 2801828). This results in a further increase in productivity (Table II).

Strains producing guanosine have been obtained from strains producing inosine. An important factor is the blockage of GMP reductase, an enzyme catalyzing the conversion of GMP to IMP (Momose *et al.*, 1965). Furthermore, it has been demonstrated that GMP synthetase is inhibited by adenosine (Udaka and Moyed, 1963), and therefore mutants with alterations in this enzyme conferring resistance to inhibition by the adenosine analogs psicofuranine and decoyinine preferentially synthesize guanosine (Table II). Mutations leading to resistance to methionine sulfoxide predominantly affect 5′-nucleotidase and may result in a shift in the proportion of guanosine, inosine, and xanthine (Matsui *et al.*, 1977).

TABLE II The Production of Inosine and Guanosine by Mutants of *B. subtilis*

		Production (g/liter)		
Mutant	Phenotype	Inosine	Guanosine	Reference
RDA-3	Ade^-	10	—	Shiio (1975)
XA-23	Ade^- Xan^-	15	—	Shiio (1975)
N231-6	Ade_l^- Xan^- Ag^r	19	—	Shiio (1975)
CMPM B-1321	Ade^- His^- Tyr^-	12	—	USSR Author's Certificate (1977a)
CMPM B-1796	Ade^- His^- Tyr^- Tg^r	40	—	USSR Author's Application (1977b)
1411	Age_l^- His_l^- Red^-	11	5.5	Matsui *et al.* (1977)
MG-1	Ade^- His^- Red^- Ms^r Ps^r De^r	0	16	Matsui *et al.* (1979)

Ade^-, Adenine auxotroph; Xan^-, xanthine auxotroph; Ag^r, 8-azaguanine resistance; Ms^r, methyonine sulfoxide resistance; Ps^r, psicofuranine resistance; De^r, decoinine resistance; Red^-, devoid of GMP reductase; His^-, histidine auxotroph; Tyr^-, tyrosine auxotroph; Tg^r, 6-thioguanine resistance.

Mutants resistant to sulfaguanidine have been shown to possess an increased level of IMP dehydrogenase and GMP synthetase (Matsui *et al.*, 1978). A strain able to produce up to 16 g/liter of guanosine without contamination by other nucleosides has been obtained recently (Matsui *et al.*, 1979).

The biosynthesis of riboflavin (vitamin B_2), which finds wide application in medicine and in animal farming and is produced in hundreds of tons by microbiological processes, is closely associated with purine metabolism. Species isolated from nature and able to overproduce riboflavin are used as industrial organisms. *Ashbyii gossipii* can produce up to 5.1 g/liter (U.S. Patent, 1969), and *Eremothecium ashbyii* up to 3.5 g/liter (BRD Patent, 1975) of riboflavin. The development of *B. subtilis* riboflavin-producing industrial strains would be advantageous because bacilli have higher rates of growth than these strains and can use simpler synthetic media, which simplifies the production of crystalline vitamin B_2 (up to 0.6 g/liter) (U.S. Patent, 1975). The pathway of riboflavin biosynthesis has been best studied in the yeasts *Saccharomyces cerevisiae* (Bacher and Lingens, 1970, 1971) and *Pichia guilliermondii* (Lang *et al.*, 1974), as well as in *B. subtilis* (Bresler *et al.*, 1970, 1977, 1978; Glazunov *et al.*, 1974). The steps of riboflavin biosynthesis are generally similar in all these organisms (Fig. 3). The pathway branches off at the stage of guanosine triphosphate, and after removal of the C-8 atom, 6,7-dimethyl-8-ribityllumasine (DRL) is formed through a number of intermediary steps. At the last stage, two molecules of DRL are converted to riboflavin by riboflavin synthetase. However, many details of the formation of DRL from purines remain unelucidated.

The genetic control of riboflavin biosynthesis has been studied in *B. subtilis* by the group of Bresler (Bresler *et al.*, 1970a,b, 1973, 1977; Chernik *et al.*, 1979). The riboflavin operon consisting of the operator and a group of structural genes has been identified. Its composition is as follows: *lys-42, rib-627, rib-628, rib-629, rib-630, rib0, ribA, ribT, ribD*. The regulatory gene *ribC* is not closely linked to the *rib* operon but is located in the following sequence: *aroD, acf, ribC, strC, lys-42* (Chernik *et al.*, 1979). The product of the *ribB* gene is riboflavin synthetase which catalyzes the conversion of DRL to riboflavin. This enzyme is active in the form of trimer containing three identical (α) subunits (Higerd and Spizizen, 1973). Another form, 3α, 60β, has also been found that possesses riboflavin synthetase activity and probably participates in the early stages of flavinogenesis (Bacher and Mailänder, 1978). The β subunit alone shows no riboflavin synthetase activity. The product of the *ribD* gene participates in the conversion of ribityl to ribulose. (Bresler *et al.*, 1977). It has been assumed that the mutations *rib-627, rib-628, rib-629,*

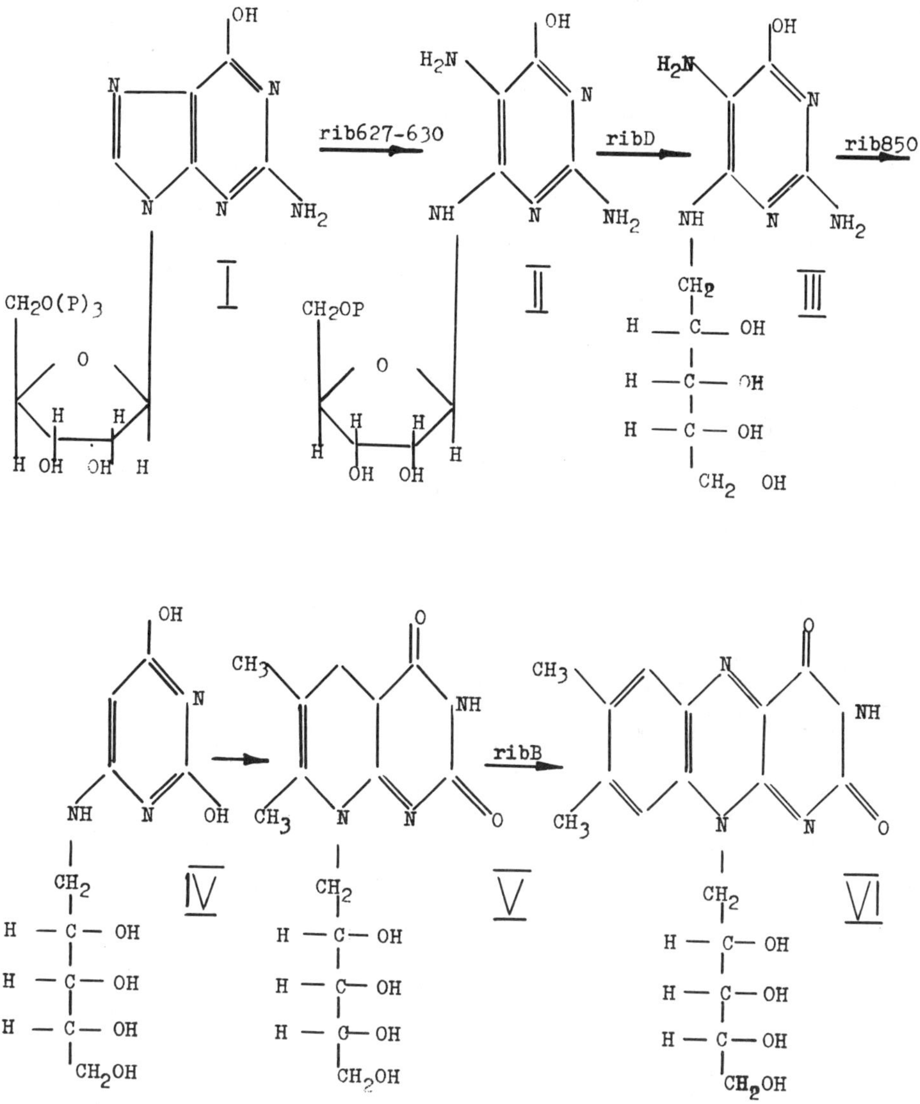

Fig. 3. The formation of riboflavin in *B. subtilis*. (I) Guanosine triphosphate; (II) 2,5-diamino-6-hydroxy-4-(D-ribosylamino)pyrimidine 5′-phosphate; (III) 2,5-diamino-6-hydroxy-4-(D-ribitylamino)pyrimidine; (IV) 5-amino-2,6-dihydroxy-4-(D-ribitylamino)pyrimidine; (V) 6,7-dimethyl-8-(ribityl)lumasine (DRL); (VI) riboflavin.

and *rib-630* create a blockage at the first stages of flavinogenesis, whereas the mutation *rib-850* affects the stages III–IV (Fig. 3). The product of the *ribC* gene serves as a universal protein regulator of flavinogenesis in *B. subtilis*. This protein exerts its action as a negative controlling element. It is able to interact with different effectors such as riboflavin, flavine mononucleotide (FMN), flavine adenine dinucleotide (FAD), lumichrome, and roseoflavin. These ligands control the capacity of the *ribC* gene product to interact with several operators, including *ribO* and the operators of genes coding for riboflavin kinase and FMN adenyltransferase, involved in the synthesis of FMN and FAD, respectively (Bresler *et al.*, 1973, 1977; Glazunov *et al.*, 1974). The existence of other operators, in addition to *ribO,* for genes involved in the early stages of vitamin B_2 synthesis has been postulated (Chernik *et al.*, 1979).

The selection of high-producing strains makes use of variants resistant to purine analogs (U.S. Patent, 1975) or to riboflavin analogs such as 8-amino(nor)riboflavin, 8-ribitylamino(nor)riboflavin, and roseoflavin (Stepanov *et al.*, 1975, 1977). More than 80% of *B. subtilis* mutants selected for resistance to roseoflavin (100 μg/ml) are able to overproduce vitamin B_2 and carry mutations in the *ribO* region (Stepanov *et al.*, 1977). When the mutations *ribO* and *ribC* are present in the same strain, the capacity for riboflavin production is markedly increased.

III. Insect Pathogens (Entomopathogens)

Bacilli are nowadays widely used for the industrial production of bacterial insecticides (Faust, 1978). Different variants of *Bacillus thuringiensis, B. popillae, B. cereus, B. sphaericus,* and other related species are pathogenic to insects.

Several variants of *B. thuringiensis* provide the basis for the production of industrial insecticides known under different trade names: Dipel (Abbott Laboratories, United States), Biotrol-BTB (Nutrile Products, Inc., United States), Turicide (International Minerals and Chemical Corporation, Sandoz Ltd, Switzerland), enterobacterine and dendrobacilline (USSR).

The main active ingredient of the insecticide activity produced by *B. thuringiensis* is a crystalline protein material formed during sporulation and variously called the crystal, the parosporal crystal, the parosporal body, or δ-endotoxin (Heimpel, 1967). The mass of crystal attains more than 30% of the dry weight of the mature sporangium (Prasad and Shethna, 1976). In addition to the crystal, at least three substances toxic to insects are synthesized by different strains of crystal-forming bacteria. Based on Heimpel's nomenclature (Heimpel, 1967) these are α-exotoxin, or phospholipase C, which hydrolyzes essential phospholipids of insect tissues, β-exotoxin, a small water-soluble temperature-resistant nucleotide which is a structural analog of ATP, and γ-exotoxin, probably a phospholipase.

β-Exotoxin is poisonous to humans and animals. Many variants of *B. thuringiensis* are unable to produce β-exotoxin, and these strains are used in industry (Bassand and Carpy, 1977). Commercial preparations represent a mixture of viable spores and crystals.

The main or perhaps the only component present in the crystal is protein; no lipid or nucleic acid components have been found (Bulla *et al.*, 1977). Various authors have detected carbohydrates in the crystals, ranging from 0.5 to 12% by weight (Holmes and Monro, 1965; Bateson and Stainsby, 1970; Bulla *et al.*, 1977). These were predominantly glucose or glucose plus mannose. Amino sugars have not been detected. No covalent linkage between carbohydrates and protein has been demonstrated and no correlations between sugar content and toxicity have been observed. The data from electron microscopy and x-ray analysis have demonstrated that the crystals consist of 9 × 13.5 nm subunits (Holmes and Monro, 1965), corresponding to a calculated molecular weight of about 230,000.

The crystal represents a protoxin. After ingestion by insect larvae the protoxin is cleaved by gastric juice and forms the active toxin that destroys the intestinal epithelium and apparently creates an ionic imbalance in the insect, leading to death (Cooksey, 1971). An additional harmful effect is due to the transfer of viable spores to the hemolymph where they germinate and multiply, leading to septicemia.

The specificity of these bacterial insecticides for different insect species is due on the one hand to the chemical and structural characteristics of the δ-endotoxin, and on the other hand is dependent on the efficiency of protoxin activation in the intestinal tract of the insect. The conversion of δ-endotoxin to the active form requires high pH values (about 9–10) and the presence of proteinases. The order Lepidoptera, containing more than 130 insect species, is particularly sensitive to the δ-endotoxin of *B. thuringiensis* (Fast, 1974).

Different variants of *B. thuringiensis* appear to contain chemically similar but nonidentical proteins (Table III). Physical and chemical studies on the crystals are associated with marked problems because of the relatively drastic conditions required for solubilization and the tendency of the proteins to aggregate. Conflicting data can be found in the literature concerning the number of protein components and their molecular weights. The number of components reported varies from one to six according to different authors, and reported molecular weights vary from 1000 to 700,000 (Sayles *et al.*, 1970; Akune *et al.*, 1971; Herbert *et al.*, 1971; Glatron *et al.*, 1972; Bulla *et al.*, 1977). Probably the main factor responsible for such discordance in the data is the presence of exogenous proteinases adsorbed on the crystal surface or incorporated into defects of the crystalline lattice; these proteases can cleave the proteins present in crystals during isolation and solubilization (Chestukhina *et al.*, 1978). Recently developed methods permit inactivation of the proteinases and allow characterization of the protein crystals produced by various serotypes of *B. thuringiensis* (Chestukhina *et al.*, 1977, 1979; Zalunin *et al.*, 1979; Kostina *et al.*, 1979).

It can be seen from the data presented in Table III that each serotype listed contains only one protein, the molecular weight of which varies slightly from one strain to another. Limited proteolysis of the crystalline proteins using trypsin, chymotrypsin, subtilisin, and the enzymes present in the gastric juice of *Galleria mellonella* larvae results in the formation of relatively stable fragments with molecular weights of about 70,000. Such fragments, stable under the conditions of limited proteolysis, are referred to as RP particles or as toxin, since they are toxic to insects. The size of the RP particle appears to correlate with the size of the protein molecule (Table III) (Chestukhina *et al.*, 1979). Differences in the RP particles can be invoked to explain differences in the action spectrum of δ-toxins, produced by different strains, upon different species of insects (Angus, 1967). A strain of *B. thuringiensis* var. *israelensis* has been recently isolated. While its parosporal crystal was toxic for the larvae of such dangerous insects as the mosquitoes *Aedes aegypti* and *Anopheles stephens,* the crystals were devoid of any toxicity for the larvae of Lepidoptera (De Barjac, 1978).

Bacillus cereus strains are also often pathogenic for insects. This common soil saprophyte possesses a very wide action spectrum against insects, including the orders Coleoptera, Hymenoptera, and Leipdoptera (Heimpel and Angus, 1963). *Bacillus cereus* produces phospholipase C(α-exotoxin) which permits the bacteria to pass through the barrier of the intestinal epithelial cells. Bacteria kill the insect by penetrating further into the hemolymph and multiplying there. A disadvantage of *B. cereus,* however, is its relatively weak pathogenicity.

Bacillus sphaericus represents a promising agent potentially useful for the control of mosquito populations. The active product is an endotoxin specific for a given insect species (Davidson *et al.*, 1975; Myers *et al.*, 1979). The toxin appears to be located at the cell wall (Myers and Yon-

TABLE III Molecular Weight of Crystal Protein from Different Variants of *B. thuringiensis*[a]

		Molecular weight	
Serotype	Variant	Protoxin	RP particle
1	*thuringiensia* (Berliner)	145,000	75,000
	insectus	145,000	—
2	*finitimus*	130,000	—
3a	*alesti*	135,000	70,000
3a, 3b	*kurshaki*	130,000	—
4a, 3b	*dendrolimus*	135,000	—
	sotto	—	—
4a, 4c	*kenyal*	135,000	—
5a, 5b	*galleriae*	135,000	—
	galleriae	130,000	65,000
5a, 5c	*canadiensis*	—	—
6	*entomocidus*	—	—
	subfoxicus	—	—
7	*aizawai*	—	—
8	*morrisoni*	135,000	—
9	*tolworth*	130,000	—
10	*darmstadtiensis*	135,000	—
11	*toumanoffi*	—	—
12	*thompsoni*	130,000	—
13	*israelensis*	—	—

[a] Chestukhina *et al.* (1977); Zalunin *et al.* (1979).

sten, 1978). In contrast to *B. thuringiensis,* in which δ-endotoxin is synthesized during sporulation, between stages II and VI (Ribier and Lacadet, 1973; Bectel and Bulla, 1976), the synthesis of the toxin in *B. sphaericus* takes place during vegetative growth. Vegetative cells and asporogenic mutants are capable of killing insect larvae (Bulla, 1978).

Bacillus popilliae is pathogenic to various scarabaeid beetles. However, there are no data concerning the possible toxicity of the crystals produced by this species for insects. A preparation of *B. popilliae,* called Milki Spore, is produced commercially in limited amounts and is effective in control of the Japanese beetle. A problem with this species of bacilli is its inability to grow on simple media. The commercial preparation is manufactured by infecting the larvae of the Japanese beetle, growing these larvae, and then homogenizing them in the presence of talcum (Faust, 1978).

An important advantage of various bacillus products in the control of insects is their high selectivity and the absence of any harmful effects on plants, warm-blooded animals, or humans, as well as on pollinating insects. Particularly broad, diverse information is available on preparations manufactured from *B. thuringiensis* (Faust, 1978). Preparations produced from other bacilli should be subjected to thorough testing, particularly in light of the results obtained recently showing that certain strains of the soil saprophyte *B. cereus* may be responsible for a number of intestinal and burn infections in humans (Melling and Capel, 1978; Turnbull *et al.*, 1979a,b). It should be noted that the δ-endotoxin of *B. thuringiensis* does not represent a completely neutral substance for warm-blooded animals, because it stimulates the immune response and has been reported to delay the growth of several tumors (Prasad and Shethna, 1976).

Further development of the practical use of these biological preparations toxic to insects requires a decrease in their production cost and an increase in their effectiveness. It is desirable to construct *Bacillus* strains having (1) increased pathogenicity for insects, (2) a modified action spectrum, (3) increased resistance to ultraviolet (UV) light and other environmental factors, (4) ability to grow on simple media in submerged culture, (5) resistance to phages, (6) no harmful side effects (such as the production of β-exotoxin in *B. thuringiensis* and the pathogenicity of certain strains of *B. cereus* for humans.

Some of these problems may be solved by wider screening programs for new strains in nature. Also, the combination within one organism of several useful traits can provide more useful strains. Such genetic construction requires the availability of some means of genetic exchange. Unfortunately, genetic transformation is unknown in the entomopathogenic group of bacilli. Transduction, however, has been demonstrated in *B. thuringiensis* (De Barjac, 1970; Aleshkin *et al.*, 1979). Transduction frequencies for different markers range from 10^{-6} to 10^{-7} per phage particle. The low frequency of transfer for different markers (such as Str^r, Met^+, Ade^+) is evidence for generalized transduction (Pogosbekova and Azizbekyan, 1979). The isolation of auxotrophic mutants in bacteria pathogenic to insects is also associated with certain difficulties (Belykh *et al.*, 1970), explaining the relative scarcity of genetically marked strains of *B. thuringiensis* at our disposal. Transduction is limited to only a few strains, probably because of the limited host range of the phages, as well as other factors that are not quite clear. For example, the phage CR-54 was able to accomplish generalized transduction with *B. thuringiensis* var. *finitimus* or *alesti* as recipient and with *B. thuringiensis* var. *finitimus, alesti, sotto,* or *galleriae* as donor. The use of other variants as donors did not lead to positive results (Thorne, 1978). The phage CR-51p can transduce the chromosomal markers Met^+ and Ade^+ in *B. thu-*

ringiensis var. *thompsoni* and is also able to transfer these markers to *B. cereus* (Yelton and Thorne, 1970). It is clear from these data that the available techniques for the exchange of genetic information in *B. thuringiensis* are very limited.

The technique of protoplast fusion developed for *B. subtilis* and *B. megaterium* (Schaefer *et al.*, 1976; Foder and Alföldi, 1976) provides another promising approach in the exchange of genetic information. Grigoryeva and Azizbekyan (in press) adapted this method for *B. thuringiensis*. The yield of spheroplasts in their experiments was equal to 98–99%, and the extent of protoplast conversion into the bacillary form was 0.5–1.0%. The strains used carried mutations in the genes *strR*, *his*$^-$, and *dra* (coding for deoxyriboaldolase) and *drm* (coding for phosphodeoxyribomutase) (Grigoryeva *et al.*, 1978). The frequency of formation of primary colonies was equal to 5×10^{-4}, but these primary colonies were nonuniform and frequently contained unstable recombinants or a mixture of parental forms. The proportion of stable recombinants among primary colonies was 20%. Thus the technique of protoplast fusion may yield a relatively high frequency of recombinants and be a useful technique for strain construction. However, the use of this method for genetic mapping is difficult because of the formation of unstable heterogenotes.

Plasmids have been discovered in *B. thuringiensis* strains belonging to several different serotypes. Frequently several circular DNA species having different sizes coexist in a single strain (Table IV). The composition of these DNA molecules appears to be unique, since defined fragments have been demonstrated for many of them by restriction analysis (Miteva, 1978). The mode of replication and the biological role of these plasmids remain unclear. Plasmids have not been detected in certain acrystalline and asporogenic mutants. This suggests the simultaneous loss of all types of circular DNA (Debabov *et al.*, 1977; Stahley *et al.*, 1978). The correlation of plasmid loss with loss of protein crystal production prompted the suggestion that the crystals are plasmid determined (Debabov *et al.*, 1977; Stahly *et al.*, 1978). Recently, direct evidence in support of this homothesis has been presented (Schnepf and Whiteley, 1981). These authors have constructed a recombinant plasmid using the *Bam*HI site of the vector pBR322 and a partial *Sau3A1* digest of *B. thuringiensis* var. *kurstaki* HD-1 plasmid DNA. The resulting recombinant strain of *Escherichia coli* produces a polypeptide (molecular weight 130,000, see Table III) which reacts with anticrystal protein antibody. Extracts of the *E. coli* strain are toxic for larvae of the tobacco hornworm, *Manduca sexta*. Interestingly, the recombinant plasmid carries sequences which are homologous to the 30- and 47-Mdal plasmids of *B. thuringiensis* var. *kurstaki* HD-1 (see Table IV).

TABLE IV Plasmids of *B. thuringiensis*

Variant	Strain	Serotype	Plasmid molecular weights ($\times 10^6$)	References
galleria	13P	5a, 5b	5.5, 5.9	Galushko (1979)
galleria	351	5a, 5b	5.1, 5.9, 10, 10.9	Debabov *et al.* (1977)
tolworth	—	9	4.0, 5.9, 7.7, 9.2	Galushko (1979)
kurstaki	—	3a, 3b	1.32, 4.9, 5.45, 9.64, 30.1, 47.13	Stahly *et al.* (1978)
alesti	—	3a	2.6, 3.8, 4.75, 5.32, 6.29, 7.29, 8.10, 8.67, 9.96, 18.12, 36.72, 44.50	Stahly *et al.* (1978)
thuringiensis (Berlinger)	—	1	5.3, 6.2, 7.1	Miteva (1978)
sotto	—	4a, 4b	0.62, 0.8, 23.6	Faust *et al.* (1979)
finitimus	—	2	0.79, 0.98, 50	Faust (1978)
kurstaki	—	3a, 3b	0.74, 0.8, 0.87, 1.1, 3.6, 3.9, 4.2, 17.7, 29.9, 45, 50	Faust *et al.* (1979)

In view of the progress achieved in the development of vector systems for the genus *Bacillus* (Chapter 10, this volume), one may hope that recombinant DNA techniques will be successfully applied to the construction of new strains pathogenic to insects. Possibly useful for this attempt is the recent observation that protoplast transformation by plasmid DNA can occur in *B. thuringiensis* (Alikhanian *et al.*, 1981).

Other Areas of Application

Bacteria belonging to the genus *Bacillus* are able to excrete amino acids into the culture medium. Among 38 strains studied, 31 produced arginine, 30 lysine, 26 leucine, and 26 aspartic acid (Romero and Ramos-Cornmenzana, 1978). Sometimes strains obtained from nature produce marked quantities of amino acids. For example, a soil isolate identified as *B. subtilis* was able to produce up to 4.54 g/liter of L-valine (Chattopadhay and Banerjee, 1978). The capacity to produce amino acids, quite frequent among bacilli, may indicate that a relatively simple system regulates amino acid biosynthesis in bacilli compared to other microorganisms. For example, the key synthetic step in formation of the aspartic acid family in *E. coli* is carried out by three aspartokinases, each of which is regulated by a different end product (lysine, methionine or threonine). In contrast, only two aspartokinases have been found in *B. cereus*, which are regulated by L-lysine and diaminopimelic acid. One aspartokinase functions only during sporulation, and the other (lysine-regulated) aspartokinase functions during vegetative growth (Hoganson, 1978).

The first step in the pathway of aromatic amono acid synthesis in *B. subtilis* is catalyzed by only one enzyme—3-deoxy-D-arabinoheptuloso-7-phosphate synthetase (DAHP synthetase), the activity of which is controlled by biosynthetic intermediates (Jensen and Nester, 1965). In contrast, in *E. coli* three isoenzymes of DAPH synthetase are repressible by three aromatic amino acids (Camacaris and Piffard, 1971). Another peculiar feature of this enzyme in *B. subtilis* is the formation of a complex with two other enzymes in the pathway, one of which, shikimate kinase, appears to be the key enzyme in the regulation of the common part of the aromatic pathway (Huang *et al.,* 1975).

Marked differences between *B. subtilis* and other microorganisms appear also in the properties and regulation of chorismate mutase which has two forms with different activities in *B. subtilis*. This enzyme is involved in the distribution of flow of the common precursor in the synthesis of different aromatic amino acids (Kane *et al.,* 1971). The enzymes of the tryptophan pathway in *B. subtilis* do not aggregate with each other, in contrast to the situation in *E. coli*. These enzymes are controlled by repression of the whole operon and by feedback inhibition of the first enzyme of the pathway, anthranilate synthetase, by tryptophan (Hoch *et al.,* 1969).

Information about the capacity of bacilli to transform indole and anthranilic acid into tryptophan (British Patent, 1972; U.S. Patent, 1974b) and to synthesize L-tryptophan (U.S. Patent. 1972b), and L-tyrosine (U.S. Patent, 1973) on synthetic media without adding precursors has been documented in patents. A strain of *Bacillus* able to produce up to 30 g/liter of L-arginine, strains transforming arginine into citrulline (U.S. Patent, 1978; Jyoji *et al.,* 1977), and strains producing from 6 to 9 g/liter of L-tryptophan without any precursor (Shiio *et al.,* 1973; USSR Author's Certificate, 1975b) may also find industrial application. This information provides a background for further improvement of strains producing amino acids.

Almost all antibiotics produced by bacilli belong to the group of oligopeptide antibiotics (Sadoff, 1972), with the exception of butirosins which belong to the aminoglucoside family (Claridge *et al.,* 1974). Some of these antibiotics have found practical medical application or are used in agriculture (bacitracin, polymixin M, gramicidin S). The biological role and the genetic control of antibiotic biosynthesis have been insufficiently studied and are poorly understood.

Several possible further uses of bacilli deserve mention. It is possible that they will be used to synthesize single-cell protein (British Patent, 1979; Bough *et al.*, 1972; Schmidt *et al.*, 1977), as well as viscous acidic polysaccharides which find application in the food industry and in oil mining (French Patent, 1978; BRD Patent, 1978). The ability of bacilli to induce biodegradation of industrial wastes, pesticides, herbicides, and other minerals located in the soil is under intensive investigation. Bacilli have been shown to be capable of degrading polystyrene (Sielicki, 1978), caprolactam (USSR Author's Certificate, 1976a), hexamethylenediamine (USSR Author's Certificate, 1976b), herbicides derived from urea (Wallnöfer and Bader, 1970), nitroaniline (USSR Author's Certificate, 1975a), *p*-aminoazobenzene (Hiroyuki *et al.*, 1977), and *n*-alkanes (Kachholz and Rehm, 1978).

In this chapter we have not attempted to analyze all possible aspects of the industrial use of bacilli. We instead demonstrated the broad potential open for this genus of microorganisms in the fermentation industry. We shall now consider certain opportunities presented by genetic methodology for the improvement of industrial strains.

IV. Perspectives for the Genetic Improvement of Industrial Strains

In the foregoing we attempted to review the existing areas of application of bacilli in the fermentation industry. Various fields of application require strain improvement. For those producing nucleosides, amino acids, antibiotics, and other primary and secondary metabolites, this involves first of all increased yields of the final products, improvement in the conversion of substrate, a decrease in the fermentation time, and increased phage resistance. For entomopathogens, improvement requires increased product stability under natural conditions, improvement in epidemiological properties, and expansion or alteration of the pathogenicity spectrum for various species of insects. For several species of entomopathogenic bacilli these improvements will probably involve modification of nutritional requirements aimed at the simplification of industrial cultivation. For bacilli producing useful enzymes the requirements include not only an improvement in yield but also the construction of strains producing a complex of enzymes with properties needed by industry. For example, it would be useful to obtain enzymes capable of functioning within a broader range of temperature and pH, under different ionic conditions, etc. In the case of bacilli producing fodder and food proteins, the main requirement involves optimization of amino acid composition.

These needs have long existed in the fermentation industry, but until recently the methods of strain improvement were limited to a few traditional approaches: (1) screening of naturally occurring strains possessing the necessary properties (for example, strains producing heat-stable exoenzymes or antibiotics or strains pathogenic to given species of insects); (2) improvement of the natural isolate by mutagenesis and the stepwise selection of more productive variants; (3) selection based on resistance to structural analogs of the desired metabolites; (4) genetic crosses of productive variants aimed at obtaining a synergistic effect of different mutations.

All these approaches have been discussed in this chapter; they have been used to obtain the strains currently employed in industry. It is a characteristic feature of this strategy that the scientist does not have to understand the mechanisms regulating biosynthetic processes in the cell. The experimental work in this area is consequently based on intuition and has more in common with art than with science.

Certain bacilli have been thoroughly studied from the genetic and biochemical point of view. These are widely used in the fermentation industry, and there is every reason to suggest that the transition from traditional selection to purposeful and planned construction of strains will be made in this group of bacilli. Such a transition requires, first of all, the ability to manipulate the genome and to transfer allelic genes from one strain to another or even to introduce completely foreign genetic information. It also requires knowledge of the mechanisms of biosynthetic pathways, their regulation, processes of transport, and processes of energy metabolism.

The potential for genetic exchange in bacilli is very broad and includes transformation (Spizizen, 1958; Spizizen *et al.*, 1966), generalized transduction, specialized transduction (Shapiro *et al.*, 1974), transfection (Takahashi, 1963), and protoplast fusion (Foder and Alfoldi, 1976; Schaefer *et al.*, 1976) (see Chapters 1 and 9). It may well be that bacilli have a markedly wider capability for genetic exchange than we visualize at present. Thus, it was found that, during the joint growth in the soil of genetically marked strains of *B. subtilis,* the exchange of large blocks of chromosomal genes took place and that the linkage of genetic markers was markedly greater than under standard laboratory conditions (Graham and Istock, 1978). The exchange of genetic information that occurs in germinating spores is resistant to DNase and requires close physical contact between cells. In this sense it resembles conjugation rather than transformation (Orrego *et al.*, 1978).

Extensive application of genetic exchange in the construction of useful strains is limited, because the strains used in industry belong to different species of bacilli and, in addition, the taxonomy of bacilli is rather confused. For example, one can mention that *Bacillus natto* and *B. atterimus* appear to belong to the same species which includes *B. subtilis* 168. When DNA from *B. subtilis* and the two strains mentioned above is used for transformation, the frequency of transformation of *B. subtilis* is identical or even higher with *B. natto* and *B. atterimus* DNA (Marmur *et al.*, 1963; Prozorov *et al.*, 1977; Cherepenko and Tikhonova, 1978). By the same criterion *B. subtilis* N23 and *B. subtilis* H appear to belong to different species (Chilton and McCarthy, 1969).

Bacillus licheniformis (Gwinn and Thorne, 1964) and *B. amyloliquefaciens* (Coukoulis and Campbell, 1971) possess natural competence for transformation, just as *B. subtilis* 168 does. This has permitted several studies of interspecific genetic exchange. If the species belong to different taxonomic groups, the frequencies of interspecific or heterologous transformation are lower by several orders of magnitude than the frequency of intraspecific transformation. These frequencies show a strong marker dependence. Generally the genes coding for RNA, ribosomal proteins, and RNA polymerase are transmitted with higher frequencies. These genes provide for vitally important cellular functions and therefore appear to be represented by conserved, slowly diverging sequences. Auxotrophic markers are exchanged interspecifically with a very low frequency or not at all (Doi and Igarashi, 1965; Dubnau *et al.*, 1965).

Theoretically the following barriers to heterologous transformation can be visualized: (1) incompetence of cells; (2) systems effecting the restriction of foreign DNA; and (3) absence of DNA homology, which prevents synapsis and recombination. All these barriers can be overcome in one way or another. It has been demonstrated that a restriction–modification system does not represent an essential obstacle to transformation in *B. subtilis* (Trautner *et al.*, 1974), because transforming DNA enters the cell in the single-stranded form (Piechowska and Fox, 1971; Davidoff-Abelson and Dubnau, 1973), which is not attacked by most restriction endonucleases (see Chapter 5).

The barrier associated with a lack of competence can be circumvented if spheroplasts are used as recipients instead of whole cells. Chang and Cohen (1979) developed the technique of high-frequency spheroplast transformation using plasmid DNA. However, with this technique, restriction systems may limit transformation frequency, since the processing of DNA may be different from that in competent cells. The main barrier to interspecific transformation is the nonhomology of DNA sequences (Harris-Warrick and Lederberg, 1978a; Wilson and Young, 1972). However, this barrier can be overcome as well. Wilson and Young (1972) demonstrated that fragments of heterologous integrated DNA flanked by homologous sequences, known as intergenotes, were able to transform recipient strains with frequencies close to those observed for homologous DNA. Harris-Warrick and Lederberg (1978b) have observed analogous effects of flanking homology. Theoretically it is possible to use these techniques to introduce foreign sequences. The conservative portion of the genome common to different bacilli can be considered a basic initial region of homology, or recombinant DNA techniques can be used to join homologous and nonhomologous segments. The exchange of regulatory alleles for α-amylase (see Section II,E and Maruo *et al.*, 1978) can be regarded as an example of heterologous transformation. We may conclude that heterologous transformation opens new perspectives for use of the genetic pool of all bacilli, as well as of other organisms, for the construction of useful strains (Erickson and Young, 1978).

Prognoses are even more optimistic for the use of recombinant DNA techniques for the construction of practically important strains. The potential of bacilli as a cloning system has been reviewed in this volume (Chapter 10), as well as by Young and Wilson (1978) and by Dubnau *et al.* (1980). Gene cloning will result in rapid progress in the understanding of the structural organization and regulatory systems of bacilli. The isolation and cloning of *Bacillus* genes is facilitated by the expression of many of these genes in *E. coli*, although the reverse frequently does not appear to be true (Ehrlich and Sgaramella, 1978). Such important DNA regions as the riboflavine region (Rabinovich *et al.*, 1978) and tryptophan operon (Gryczan *et al.*, 1980) have already been cloned. *Bacillus subtilis* possesses a powerful pentose cycle supplying high levels of intermediate metabolites (this is expressed in the construction of strains excreting high levels of purine nucleotides). One may assume that, since these metabolites will not be limiting, the amplification or some regulatory alteration of genes in the tryptophan or riboflavine operons, for instance, may result in the construction of strains producing high levels of these metabolites. So far, we are aware of only one example in which gene amplification has led to a commercially useful producing strain. A strain derived from *E. coli* K12 is able to produce up to 30 g/liter of L-threonine during 24–30 hr of fermentation (USSR Author's Application, 1978).

A particularly promising approach is the use of recombinant DNA techniques together with systems of biosynthesis and excretion of extracellular enzymes for the microbiological synthesis of specific proteins. The advantages of this system are self-evident: they include high productivity (the amount of α-amylase, for example, can attain 80% of the total cell protein (Priest, 1977) and the excretion of the product into the cultivation medium, facilitating further isolation and purification. Such an approach appears to be rather promising for the production of proteins used as fodder or food. The production of "monoprotein" opens perspectives for the regulation of amino acid composition. This would simplify the standardization of the final product and the technology of preparation of the protein isolate. The industrial production of polypeptide hormones and other human or animal proteins in the future may also take advantage of *Bacillus* systems rather than *E. coli*. The success of this trend will be largely determined by the development of knowledge about the mechanisms of biosynthesis and excretion of exoenzymes.

Certain industrial problems may require the introduction of new genes into the chromosomes of bacilli. This may facilitate the modification of nutritional requirements or the synthesis of new extracellular enzymes. In addition to the approaches suggested above, new properties may be brought into a strain by integration into the chromosome of partially heterologous circular DNA molecules having a region of homology, via a mechanism similar to that operating in the integration of λ DNA (Duncan *et al.*, 1978).

If we regard the microbial cell as a machine designed for the transformation of a substrate into a product, the potential genetic redundancy of this machine is striking. When cultivated in a defined medium under defined conditions, the cell of any microorganism carries genetic information that is not used under these conditions. This is particularly noticeable in bacilli during vegetative growth. In this case, all sporulation genes, which comprise 40–50 loci (Hranueli *et al.*, 1974), and many other genes activated during differentiation are "unnecessary." Thus, it may be feasible to create a "minicell" obtained by the deletion of all or most unnecessary genes from the genome. Such an organism might possess an increased growth rate and a better substrate conversion and would be absolutely nonviable in the natural environment. Of course, this task is one for the future. It is equivalent to the construction of a new living organism.

IV. Conclusions

The availability of reliable genetic and biochemical information on certain species of bacilli, the well-documented absence of any harmful biological effects, a long history of use in the fermentation industry, rapid progress in the adaptation of recombinant DNA technique to these organisms, and rapid progress in developing techniques for genetic exchange make bacilli a unique microbial group for scientists dealing with the construction of industrial strains of microorganisms. It may well be that the transition from the screening of useful strains in nature and empirical selection to the more rational construction of industrial producing strains will be made soon in the genus *Bacillus*.

References

Aiyappa, P. S., Traficante, L. J., and Lampen, J. O. (1977). *J. Bacteriol.* **129,** 191–197.

Akune, S., Watanabe, T., Mukoi, J., Tsutsui, R., and Abe, K. (1971). *Jpn. J. Med. Sci. Biol.* **24,** 57–59.

Aleshkin, G. L., Burtseva, L. I., and Skovronskaya, A. G. (1979). *Genetika (USSR:cp* **15,** 1186–1190.

Angus, T. A. (1967). *J. Invertebr. Pathol.* **9,** 256–260.

Aoki, F., Momose, H., Kondo, Y., Mutamatsu, N., and Tsuchiga, Y. (1963). *J. Gen. Appl. Microbiol.* **9,** 387–392.

Arnold, H. H., and Ogilvie, A. (1977). *Biochem. Biophys. Res. Commun.* **74,** 343–349.

Aronson, A. Y., Angelo, N., and Holt, S. C. (1971). *J. Bacteriol.* **106,** 1016–1025.

Aunstrup, K., Outtrup, H., Andersen, O., and Danbmann, C. (1972). *Proc. Int. Fermentat. Technol., 4th* (G. Terui, ed.), pp. 299–305. Kyoto, Japan.

Aynsawa, C., Yoneda, Y., Yamane, K., and Maruo, B. (1975). *J. Bacteriol.* **124,** 459–469.

Bacher, A., and Lingens, F. (1970). *J. Biol. Chem.* **245,** 4647–4656.

Bacher, A., and Lingens, F. (1971). *J. Biol. Chem.* **246,** 7018–7022.

Bacher, A., and Mailänder, B. (1978). *J. Bacteriol.* **134,** 476–482.

Bassand, D., and Carpy, S. (1977). *Experientia* **33,** 1545–1551.

Bateson, J. B., and Stainsby, G. (1970). *J. Fd. Technol.* **5,** 403–415.

Bectel, D. B., and Bulla, L. A. (1976). *J. Bacteriol.* **127,** 1472–1481.

Belykh, R. A., Yakubova, R. M., and Azizbekyan, R. R. (1979). *Genetika (USSR)* **15,** 837–844.

Bernlohr, R. W., Maddox, M. K., and Goldberg, N. D. (1974). *J. Biol. Chem.* **249,** 4329–4331.

Blobel, G., and Dobberstein, B. (1975). *J. Cell. Biol.* **67,** 835–851.

Borgia, P. T., and Campbell, L. L. (1978). *J. Bacteriol.* **134,** 389–393.

Both, G. W., McInnes, J. L., Hanlon, J. E., May, B. K., and Elliot, W. H. (1972). *J. Mol. Biol.* **67,** 199–207.

Bough, W. A., Brown, W. L., Porsche, J. D., Doy, D. M. (1972). *Appl. Microbiol.* **24,** 226–235.

BRD Patent (1975). No. 2453827.

BRD Patent (1977). No. 2223864.

BRD Patent (1978). No. 2810279.

Bresler, S. Ye., Cherepenko, Ye. I., Chernik, T. P., Kalinin, V. L., and Perumov, D. A. (1970a). *Genetika (USSR)* **6,** 116–124.

Bresler, S. Ye., Cherepenko, Ye. I., and Perumov, D. A. (1970b). *Genetika (USSR)* **6,** 126–139.

Bresler, S. Ye., Glazunov, Y. A., Chernik, T. P., Shevchenko, T. N., and Perumov, D. A. (1973). *Genetika (USSR)* **9,** 84–91.

Bresler, S. Ye., Perumov, D. A., Glazunov, Y. A., Shevchenko, T. N., and Chernik, T. P. (1977a). *Genetika (USSR)* **13,** 880–887.

Bresler, S. Ye., Glazunov, Ye. A., Perumov, D. A., and Chernik, T. P. (1977b). *Genetika (USSR)* **13,** 2007–2016.

Bresler, S. Ye., Gorinchuk, T. F., Chernik, T. P., and Perumov, D. A. (1978). *Genetika (USSR)* **14,** 2082–2090.

Bulla, L. A. Jr. (1978). *In* "Spores VII" (G. Chambless and J. C. Vary, eds.), pp. 16. American Society for Microbiology, Washington, D. C.
Bulla, L. A. Jr., Kramer, K. J., and Davidson, L. J. (1977). *J. Bacteriol.* **130,** 375–383.
Camacaris, J., and Piffard, J. (1971). *J. Bacteriol.* **107,** 406–414.
Cashel, M., and Gallant, J. (1969). *Nature (London)* **221,** 838–841.
Caulfield, M. P., Berkeley, R. C., Pepper, E. A., and Melling, J. (1979). *J. Bacteriol.* **138,** 345–351.
Chaloupka, J., Kreckova, P., and Rihova, L. (1963). *Biochim. Biophys. Res. Commun.* **12,** 380–382.
Chalorer-Larson, J., and Yamazaki, H. (1978). *Can. J. Biochem.* **54,** 264–272.
Chang, S., and Cohen, S. N. (1979). *Mole. Gen. Genet.* **168,** 111–115.
Chattopadhyay, S. P., and Banerjee, A. K. (1978). *Z. Allg. Mikrobiol.* **18,** 243–254.
Chemical Week (1978). **123,** 41–42.
Cherepenko, Ye., and Tikhonova, T. (1978). *Genetika (USSR)* **12,** 109–114.
Chernik, T. P., Bresler, S. Ye., Mankovsky, Y. V., and Perumov, D. A. (1979). *Genetika (USSR)* **15,** 1569–1577.
Chestukhina, G. G. *et al.* (1977). *Biochimiya (USSR)* **42,** 1660–1667.
Chestukhina, G. G. *et al.* (1978). *Biochimiya (USSR)* **43,** 857–864.
Chestukhina, G. G., Kotova, T. S., Zalunin, I. A., and Stepanov, V. M. (1979). *Biochimiya (USSR)* **44,** 796–802.
Chiang, J. P., Alter, J. E., and Sternberg, M. (1979). *Stärke* **31,** 86–92.
Chilton, M. D., and McCarthy, B. J. (1969). *Genetics* **62,** 697–710.
Claridge, C. A., Bush, J. A., Defuria, M. D., and Price, K. E. (1974). *Dev. Ind. Microbiol.* **15,** 101–113.
Coleman, G. (1969). *Biochem. J.* **122,** 533–539.
Coleman, G., and Grant, M. A. (1966). *Nature (London)* **211,** 306–307.
Cook, R., and Ramaley, R. F. (1978). *In* "Spores VII" (G. Chambliss and J. C. Vary, eds.), pp. 171–175. American Society for Microbiology, Washington, D.C.
Cooksey, K. E. (1971). *In* "Microbial Control of Insects and Mites" (H. D. Burgess and H. W. Hussey, eds.), pp. 247–274. Academic Press, New York.
Coukoulis, H., and Campbell, L. L. (1971). *J. Bacteriol.* **105,** 319–322.
Davidoff-Abelson, R., and Dubnau, D. (1973). *J. Bacteriol.* **116,** 154–162.
Daucer, B. N., and Mandelstam, J. (1975). *J. Bacteriol.* **121,** 406–411.
Davidson, E. W., Singer, S., and Brigges, J. D. (1975). *J. Invert. Pathol.* **25,** 179–184.
Debabov, V. G., Azizbekyan, R. R., Khlebalina, O. I., Dyachenko, V. V., Galushka, F. P., and Belykh, R. A. (1977). *Genetika (USSR)* **13,** 496–500.
De Barjac, H. (1970). *C. R. Acad. Sci. Paris* **18,** 227–231.
De Barjac, H. (1978). *C. R. Acad. Sci. Paris* **286,** 797–800.
Dennis, P. P. (1977). *J. Bacteriol.* **129,** 580–588.
Dobrzhanskaya, Ye. O., Erokhina, L. I., and Bolshakova, T. N. (1978). *Genetika (USSR)* **14,** 1175–1184.
Doi, R. H. (1972). *Curr. Top. Cell Regul.* **6,** 1–20.
Dod, B., and Balassa, G. (1978). *Mol. Gen. Genet.* **163,** 57–63.
Dod, B., Balassa, G., Raulet, E., and Jeannoda, V. (1978). *Mol. Gen. Genet.* **163,** 45–56.
Doi, R. H., and Igarashi, R. I. (1965). *J. Bacteriol.* **90,** 387–390.
Dubnau, D., Smith, J., Morell, P., and Marmur, J. (1965). *Proc. Nat. Acad. Sci. USA* **54,** 491–498.
Dubnau, D., Gryczan, T., Contente, S., and Shivakumar, A. G. (1980). *In* "Genetic Engineering: Principles and Methods" (J. Setlow and A. Hollaender, eds.), Vol. 2, pp. 115–132. Plenum Press, New York.

Duncan, C. H., Wilson, G. A., and Young, F. E. (1978). *Proc. Nat. Acad. Sci. USA* **75,** 3664–3668.
Ehrlich, S. D., and Sgaramella, V. (1978). *Trends Biol. Sci.* (*TIBS*) **3,** 259–261.
Erickson, R. J. (1976). *In* "Microbiology-76" (D. Schlessinger ed.), pp. 406–419. American Society for Microbiology, Washington D.C.
Erickson, R. J., and Young, F. E. (1978). *Dev. Ind. Microbiol.* **19,** 245–252.
Ermakova, L. M. *et al.* (1978). *Genetika* (*USSR*) **14,** 1900–1907.
Fast, P. G. (1974). *Dev. Ind. Microbiol.* **15,** 195–198.
Fast, R., and Sköld, O. (1977). *J. Biol. Chem.* **252,** 7620–7624.
Faust, R. M. (1978). *Recombinant DNA Tech. Bull.* **1,** 1–14.
Faust, R., Spizizen, J., Gage, V., and Travers, R. (1979). *J. Invert. Pathol.* **33,** 233–238.
Fill, N. P., Willumsen, B. M., Friesen, J. D., and von Meyenburg, K. (1977). *Mol. Gen. Genet.* **150,** 87–101.
Foder, K., and Alfoldi, L. (1976). *Proc. Nat. Acad. Sci. USA* **73,** 2147–2150.
Fogarty, W. M., Griffin, P. J., and Joyce, A. M. (1974a). *Process Biochem.* **9,** Part I, 11–24.
Fogarty, W. M., Griffin, P. J., and Joyce, A. M. (1974b). *Process Biochem.* **9,** Part II, 27–35.
Fortnagel, P., and Bergman, R. (1972). *Biochem. Biophys. Res. Commun.* **56,** 264–272.
French Patent (1978). No. 2383231.
Gallant, J., and Margason, C. (1972). *J. Biol. Chem.* **247,** 2289–2294.
Galushka, F. P. (1979). Candidate's Dissertation. Institute of Genetics and Selection of Industrial Microorganisms, Moscow.
Geele, G., Garrett, E., and Hageman, J. H. (1975). *In* "Spores VI" (P. Gerhardt, R. N. Costilow, and H. L. Sadoff, eds.), pp. 391–396. American Society for Microbiology, Washington D.C.
Glatron, M. F., Lecadet, M. M., and Dedonder, R. (1972). *Eur. J. Biochem.* **30,** 330–338.
Glazunov, Ye. A., Bresler, S. Ye., and Perumov, D. A. (1974). *Genetika* (*USSR*) **10,** 83–92.
Goebel, W. A., Noegel, A., Rdest, U., Springer, W., Kreff, J., and Burger Hopje-Seyler's, K. J. (1979). *Z. Physiol. Chem.* **360,** 1029–1033.
Gould, A. R., May, B. K., and Elliot, W. H. (1975). *J. Bacteriol.* **122,** 34–40.
Graham, J. B., and Istock, C. A. (1978). *Mol. Gen. Genet.* **166,** 287–290.
Grigoryeva, T. M., Smirnov, Yu. V., Sukhodolets, V. V. (1978). *Genetika* (*USSR*) **14,** 1311–1318.
Great Britain Patent (1979) No. 1546621.
Gryczan, T. J., Contente, S., and Dubnau, D. (1980). *Mol. Gen. Genet.* **177,** 459–467.
Gusyatiner, M. M., Ahdanova, N. I., Livshits, V. A., Zaigraeva, G. G., and Shakulov, R. S. (1978). *Genetika (USSR)* **14,** 957–968.
Gwinn, D. D., and Thorne, C. B. (1964). *J. Bacteriol.* **87,** 519–526.
Hageman, J. H., and Carlton, B. C. (1973). *J. Bacteriol.* **114,** 612–617.
Hanson, R. S. (1976). *In* "Spores VI" (P. Gerhardt, H. L. Sadoff, and R. N. Gostilow, eds.), pp. 318–326. American Society for Microbiology. Washington, D.C.
Harris-Warrick, R. M., and Lederberg, J. (1978a). *J. Bacteriol.* **133,** 1237–1245.
Harris-Warrick, R. M., and Lederberg, J. (1978b). *J. Bacteriol.* **133,** 1246–1253.
Heimpel, A. M. (1967). *Rev. Ent.* **12,** 287–322.
Heineken, F. G., and O'Connor, R. J. (1972). *J. Gen. Microbiol.* **73,** 35–70.
Heimpel, A. M., and Angus, T. A. (1963). *In* "Insect Pathology" (E. A. Steinhaus, ed.), pp. 21–73. Academic Press, New York.
Herbert, B. N., Gaeld, A. J., and Chain, E. B. (1971). *Eur. J. Biochem.* **24,** 366–375.

Higerd, T. B., and Spizizen, J. (1973). *J. Bacteriol.* **114,** 1184–1192.
Higerd, T. B., Hoch, J. A., and Spizizen, J. (1972). *J. Bacteriol.* **112,** 1026–1028.
Hiroyuki, H., Mitsuo, T., Eiichi, J., Mikio, T., and Toshihiko, O. (1977). *Eur. J. Appl. Microbiol.* **4,** 217–224.
Hitotsuyanagi, K., Yamane, K., and Maruo, B. (1978). *Agr. Biol. Chem.* **42,** 585–592.
Hitotsuyanagi, K., Yamane, K., and Maruo, B. (1979). *Agr. Biol. Chem.* **43,** 2343–2349.
Hoch, J. A., and Spizizen, J. (1969). *In* "Spores IV" (L. L. Campbell, ed.), pp. 112–123. American Society for Microbiology, Washington, D.C.
Hoch, S. O., Anaguostopoulos, C., and Crowford, I. P. (1969). *Biochem. Biophys. Res. Commun.* **35,** 838–843.
Hoganson, D. A., Smith, C. D., and Stahly, D. P. (1978). *In* "Spores VII" (G. Chambless and J. C. Vary, eds.), pp. 304–307. American Society for Microbiology, Washington D.C.
Holmes, K. C., and Monro, R. E. (1965). *J. Mol. Biol.* **14,** 572–581.
Hranueli, D., Piggot, P., and Mandelstam, J. (1974). *J. Bacteriol.* **119,** 684–690.
Huang, J., Montoya, A., and Nester, E. (1975). *J. Biol. Chem.* **250,** 7675–7683.
Ide, M. (1971). *Arch. Biochem. Biophys.* **144,** 262–268.
Ingle, M. B., and Boyer, E. W. (1976). *In* "Microbiology, 76" (D. Schlessinger, ed.), pp. 420–426. American Society for Microbiology, Washington D.C.
Ishii, K., and Shiio, I. (1970). *J. Biochem.* **68,** 171–176.
Ishii, K., and Shiio, I. (1972). *Agr. Biol. Chem.* **36,** 1511–1519.
Ito, J., and Spizizen, J. (1972). *In* "Spores V" (H. O. Halvorson, R. Hanson and L. L. Campbell, eds.), pp. 107–112. American Society for Microbiology, Washington D.C.
Japanese Patent (1977). No. 4022666.
Japanese Patent (1974). No. 49-27356.
Jensen, R. A., and Nester, E. W. (1965). *J. Mol. Biol.* **12,** 468–472.
John, A. C., Conklin, K., Rosenthal, E., and Goldberg, A. L. (1978). *J. Biol. Chem.* **253,** 3945–3951.
Jyoji, K., Masahiko, K., Tsutomu, T., and Ichiro, C. (1977). *Appl. Environ. Microbiol.* **34,** 689–694.
Kachholz, T., and Rehm, H. J. (1978). *Eur. J. Appl. Microbiol. Biotechnol.* **6,** 39–54.
Kane, I. F., Steumark, S. L., Calhoun, D. M., and Iensen, R. A. (1971). *J. Biol. Chem.* **246,** 4308–4318.
Kinoshita, S., Okada, H., and Terui, G. (1968). *J. Ferment. Technol.* **46,** 427–436.
Kostina, L. I., Mikhailova, A. L., Zalunin, I. A., Katrukha, S. P., and Chestukhina, G. G. (1979). *All-Union Symp. Chem. Proteolytic Enzymes, 2nd* pp. 92–93. Uglich, USSR.
Kuninaka, A., Kibi, M., and Sakaguchi, K. (1964). *Food Technol.* **18,** 287–293.
Kunst, F., Pascal, M., Lepesant-Kejzlarova, J., Billault, A., and Dedonder, D. (1974). *Biochimie* **56,** . 1481–1489.
Lai, J. S., Sarvas, M., Brammar, W. J. Neugebauer, K., and Wu, H. C. (1981). *Proc. Nat. Acad. Sci. USA.* **78,** 3506–3510.
Laishley, E. J., and Bernlohr, R. W. (1968). *J. Bacteriol.* **96,** 322–329.
Lampen, J. O. (1976). *In* "Microbiology, 1976" (D. Schlessinger, ed.), pp. 540–542. American Society for Microbiology, Washington D.C.
Lecadet, M., Lescourret, M., and Klier, A. (1977). *Eur. J. Biochem.* **79,** 329–338.
Leighton, T. J., Freese, P. K., Doi, H., Warren, R. A., and Kollin, R. A. (1972). *In* "Spores V" (H. O. Halvorson, R. Hanson and L. L. Campbell, eds.), pp. 238–246. American Society for Microbiology, Washington D.C.
Leighton, T. J., Doi, K. H., Warren, R. A. J., and Kella, R. A. (1973). *J. Mol. Biol.* **76,** 103–122.

Levinson, S. S., and Aronson, I. (1967). *J. Bacteriol.* **93,** 1023–1030.
Madeson, G. B., Norman, B. E., and Slott, S. (1973). *Stärke* **25,** 304–308.
Marmur, J., Seaman, E., and Levine, J. (1963). *J. Bacteriol.* **85,** 461–467.
Maruo, B., Yamane, K., Yoneda, Y., Hitotsuyanagi, K. (1978). *Proc. Jap. Acad.* **54B,** 435–439.
Matsubara, H., and Feder, J. (1971). *In* "The Enzymes" (P. D. Boyer, ed.), pp. 721–792. Academic Press, New York.
Matsui, H., Sato, K., Enei, H., and Hirose, Y. (1977). *Appl. Environ. Microbiol.* **34,** 337–341.
Matsui, H., Sato, K., Enei, H., and Hirose, Y. (1978). *Agr. Biol. Chem.* **42,** 637–641.
Matsui, H., Sato, K., Enei, H., and Hirose, Y. (1979). *Agr. Biol. Chem.* **43,** 393–394.
Matsuzaki, H., Yamane, K., Yamaguchi, K., Nagata, Y., and Maruo, B. (1979). *Biochim. Biophys. Acta* **356,** 235–247.
Meers, J. L. (1972). *J. Microbiol. Serol.* **38,** 585–590.
Melling, J., and Capel, B. J. (1978). *FEMS-Microbiol. Lett.* **4,** 133–135.
Milhaud, P., Balassa, G., and Zucca, J. (1978). *Mol. Gen. Genet.* **163,** 35–44.
Millet, J. (1969). *Bull. Soc. Chim. Biol.* **51,** 61–70.
Millet, J., and Aubert, J. P. (1969). *Ann. Inst. Pasteur (Paris)* **117,** 461–473.
Millet, J., Larribe, M., and Aubert, J. P. (1976). *Biochim.* **58,** 109–117.
Miteva, V. J. (1978). *Rep. Bulg. Acad. Sci.* **31,** 1059–1062.
Momose, H., Nishikawa, H., and Katsuya, N. (1965). *J. Gen. Appl. Microbiol.* **11,** 211–219.
Myers, P., and Yousten, A. A. (1978). *In* "Spores VII" (G. Chambless and J. C. Vary, eds.), pp. 31–33. American Society for Microbiology, Washington, D.C.
Myers, P., Yousten, A., and Davidson, E. (1979). *Can. J. Microbiol.* **25,** 1227–1238.
Nagata, J., Yamaguchi, K., and Maruo, B. (1974). *J. Bacteriol.* **119,** 425–430.
Nielson, J. B. K., Caulfield, M. P. and Lampen, J. O. (1981). *Proc. Nat. Acad. Sci. USA.* **78,** 3511–3515.
Neumark, R., and Citri, N. (1962). *Biochim. Biophys. Acta* **59,** 749–751.
Nishino, T., Gallant, J., Shalit, P., Palmer, L., and Wehr, T. (1979). *J. Bacteriol.* **140,** 671–679.
Noegel, A., Rdest, U., Springer, W., and Goebel, W. (1979). *Mol. Gen. Genet.* **175,** 343–350.
Nomura, S., Yamane, K., Sasaki, T., Yamasaki, M., Tamura, G., and Maruo, B. (1978). *J. Bacteriol.* **136,** 818–821.
Nunn, W. D., and Cronan, J. E. Jr. (1976). *Biochemistry* **15,** 2546–2550.
Ogilvie, A., Wiebaner, K., and Kersten, W. (1975). *Biochem. J.* **152,** 517–522.
Orrego, C., Arnaud, M., and Halvorson, H. O. (1978). *J. Bacteriol.* **134,** 973–981.
Pedersen, E. S., and Kjeldgaard, N. O. (1977). *Eur. J. Biochem.* **76,** 91–97.
Piechowska, M., and Fox, M. (1971). *J. Bacteriol.* **108,** 680–689.
Piggot, P. J., and Coote, J. G. (1976). *Bacteriol. Rev.* **40,** 908–962.
Pogosbekova, M. P., and Azizbekyan, R. R. (1979). *Genetika (USSR)* **15,** 1980–1986.
Prasad, S. S. V., and Shethna, Y. J. (1976). *J. Sci. Ind. Res.* **35,** 626–632.
Priest, F. G. (1977). *Bacteriol. Rev.* **41,** 711–753.
Prozorov, A. A., Gening, L. B., and Debabov, V. G. (1977). *Genetika (USSR)* **13,** 1976–1980.
Rabinovich, P. M., Beburov, M. Yu., Linevich, Z. K., and Stepanov, A. I. (1978). *Genetika (USSR)* **14,** 1696–1705.
Ramey, W. D., and Ishiguro, E. E. (1978). *J. Bacteriol.* **135,** 71–77.
Rhaese, H. J., and Groscurth, R. (1974). *FEBS Lett.* **44,** 87–93.
Rhaese, H. J., Dichtelmüller, F. M., and Groscurt, R. (1972). *In* "Spore V" (H. Halvorson,

R. Hanson, and L. L. Campbell, eds.). American Society for Microbiology, Washington D.C.
Rhaese, H. J., Grade, R., and Dichtelmüller, H. (1976). *Eur. J. Biochem.* **64,** 205–213.
Rhaese, H. J., Hoch, J. A. and Groscurth, R. (1977). *Proc. Nat. Acad. Sci. USA* **74,** 1125–1130.
Ribier, J., and Lacadet, M. M. (1973). *Ann. Microbiol.* **124A,** 311–344.
Rickenberg, H. V. (1974). *Ann. Rev. Microbiol.* **28,** 353–369.
Romero, P., and Ramos-Cormenzana, A. (1978). *Ars. Pharm. Rev. Fac. Farm.* **19,** 361–368.
Sadoff, H. L. (1972). *Prog. Ind. Microbiol.* **11,** 1–27.
Sanders, R. L., and May, B. K. (1975). *J. Bacteriol.* **123,** 806–814.
Sayles, V. B., Aronson, J. N., and Rosenthal, A. (1970). *Biochem. Biophys. Res. Commun.* **41,** 1126–1133.
Sazaki, T., Yamasaki, M., Yoneda, Y., Yamane, K., Takatsuki, A., and Tamura, G. (1976). *Biochem. Biophys. Res. Commun.* **70,** 125–131.
Schaeffer, P. (1969). *Bacteriol. Rev.* **33,** 48–71.
Schaeffer, P., Cami, B., and Hotchkiss, R. (1976). *Proc. Nat. Acad. Sci. USA* **73,** 2151–2155.
Schlessinger, D., Marchesi, V. T., and Kwan, B. C. K. (1965). *J. Bacteriol.* **90,** 456–466.
Schmidt, B. E., Busta, F. E., and Mo Kay, L. L. (1977). *J. Food Process. Preserv.* **1,** 13–29.
Schnepf, H. E., and Whiteley, H. R. (1981). *Proc. Natl. Acad. Sci. USA.* **78,** 2893–2897.
Sekiguchi, J., and Okada, H. (1972). *J. Ferment. Technol.* **50,** 801–809.
Sekiguchi, J., Takada, N., and Okada, H. (1975). *J. Bacteriol.* **116,** 531–534.
Setlow, P. (1973). *Biochem. Biophys. Res. Commun.* **52,** 365–372.
Shakulov, R. S., and Belitsky, B. R. (1980). *Biochimiya (USSR)* **45,** 155–162.
Shapiro, J. A., Dean, D. H., and Halvorson, H. O. (1974). *Virology* **62,** 393–403.
Shinke, R., Aoki, K., Nishira, H., and Yuki, S. (1979). *J. Ferment. Technol.* **57,** 53–55.
Shiio, J. (1975). *Int. Cong. JAMS, 1st, Tokyo.*
Shiio, I. (1979). *Int. Cong. Microbiol. Food Ind., Paris.*
Shiio, I., Ishii, K., and Yokozeki, K. (1973). *Agr. Biol. Chem.* **37,** 1991–2000.
Sielicki, M., Focht, D., Martin, D., and Can, J. P. (1978). *J. Microbiol.* **24,** 798–803.
Simons, K., Sarvas, M., Garoff, H., and Helenius, A. (1978). *J. Mol. Biol.* **126,** 673–690.
Smith, E. L., Markland, F. S., Kasper, C. B., Delange, R. J., London, M., and Evans, W. H. (1966). *J. Biol. Chem.* **241,** 5974–5976.
Smith, I., Paress, P., and Pestka, S. (1978). *Proc. Nat. Acad. Sci. USA* **75,** 5993–5997.
Smith, I., Paress, P., Cabane, K., and Dubnau, E. (1980). *Mol. Gen. Genet.* **178,** 271–279.
Smith, W. P., Tai, P. C., and Davis, B. D. (1981). *Proc. Nat. Acad. Sci. USA.* **78,** 3501–3505.
Sokawa, Y., and Kaziro, Y. (1969). *Biochem. Biophys. Res. Commun.* **34,** 99–103.
Somerville, C. R., and Ahmed, A. (1977). *J. Mol. Biol.* **111,** 77–81.
Spizizen, J. (1958). *Proc. Nat. Acad. Sci. USA* **44,** 1072–1078.
Spizizen, J., Reilly, B. E., and Evans, E. H. (1966). *Ann. Rev. Microbiol.* **20,** 371–400.
St. John, A. C., Conklin, K., Rosenthal, E., and Goldberg, A. L. (1978). *J. Biol. Chem.* **253,** 3945–3951.
Stahly, D. P., Dingman, D. W., Bulla, L. A., and Aronson, A. I. (1978). *Biochem. Biophys. Res. Commun.* **84,** 581–588.
Stahly, D. P., Dingman, D. W., Irgens, R. L., Field, C. C., Feiss, M. G., and Smith, G. L. (1978). *FEMS—Microbiol. Lett.* **3,** 139–146.
Steinmetz, M., Kunst, F., and Dedonder, R. (1976). *Mol. Gen. Genet.* **148,** 281–285.

Stepanov, A. I., Tulchinskaya, L. S., Berezovsky, V. N., and Kukanova, A. Ya. (1975). *Genetika (USSR)* **11,** 116–123.
Stepanov, A. I., Kukanova, A. Ya., Glazunov, Ye. A., and Zhdanov, V. G. (1977). *Genetika (USSR)* **13,** 490–495.
Stephens, J. C., Artz, S. W., and Ames, B. N. (1975). *Proc. Nat. Acad. Sci. USA* **72,** 4389–4393.
Strongin, A. Ya., Izotova, L. S., Abramov, Z. T., Gorodetsky, D. I., and Stepanov, V. M. (1978). *Mol. Gen. Genet.* **159,** 337–345.
Strongin, A. Ya. *et al.* (1979). *J. Bacteriol.* **137,** 1017–1019.
Swanton, M., and Edlin, G. (1972). *Biochem. Biophys. Res. Commun.* **46,** 583–588.
Sy, J., and Lipmann, F. (1973). *Proc. Nat. Acad. Sci. USA* **70,** 306–309.
Takahashi, J. (1963). *J. Gen. Microbiol.* **31,** 211–217.
Takeshi, K., Yoshimi, H., Funio, O., Shoishi, S., and Fuji, U. (1978). *J. Ferment. Technol.* **56,** 309–314.
Thomas, J. E., and Copeland, J. C. (1973). *J. Bacteriol.* **116,** 938–943.
Thorne, C. B. (1978). *Appl. Environ. Microbiol.* **35,** 1109–1118.
Trautner, T. A., Paulek, B., Bron, S., and Anagnostopoulos, C. (1974). *Mol. Gen. Genet.* **131,** 181–191.
Turnbull, P. C. B., Kramer, J. M., Jörgensen, K., Gilbert, R. J., and Melling, J. (1979a). *Am. J. Clin. Nutr.* **32,** 219–228.
Turnbull, P. C. B., Jörgensen, K., Kramer, J. M., Gilbert, R. J., and Jennifer, P. (1979b). *J. Clin. Pathol.* **32,** 289–293.
Udaka, S., and Moyed, H. S. (1963). *J. Biol. Chem.* **238,** 2797–2803.
Uehara, H., Yoneda, Y., Yamane, K., and Maruo, B. (1974). *J. Bacteriol.* **119,** 82–91.
Uehara, H., Yamane, K., and Maruo, B. (1979). *J. Bacteriol.* **139,** 583–590.
U.S. Patent (1969). No. 3475274.
Great Britain Patent (1972a). No. 1298499.
U.S. Patent (1972a). No. 3700558.
U.S. Patent (1972b). No. 3660235.
U.S. Patent (1973). No. 3709785.
U.S. Patent (1974a). No. 3844890.
U.S. Patent (1974b). No. 3801457.
U.S. Patent (1974c). No. 3796635.
U.S. Patent (1975). No. 3900368.
U.S. Patent (1977). No. 4048339.
U.S. Patent (1978). No. 4086137.
USSR Author's Certificate (1975a). No. 488857.
USSR Author's Certificate (1975b). No. 480758.
USSR Author's Certificate (1976a). No. 529210.
USSR Author's Certificate (1976b). No. 531843.
USSR Author's Certificate (1977a). No. 594769.
USSR Author's Application (1977b). No. 2801828.
USSR Author's Application (1978). No. 096037.
USSR Patent. No. 2801828.
Van Dijk-Salkinoja, M. S., Stoof, T. J., and Planta, R. J. (1970). *Eur. J. Biochem.* **12,** 474–482.
Wallnöfer, P. R., Bader, J. (1970). *Appl. Microbiol.* **19,** 714–717.
Welker, N. F., and Campbell, L. L. (1963). *J. Bacteriol.* **86,** 687–691.
Welker, N. E., and Campbell, L. L. (1967). *J. Bacteriol.* **94,** 1124–1130.

Wilson, G. A., and Young, F. E. (1972). *J. Bacteriol.* **111,** 705–716.
Yamada, K. (1977). *Biotechnol. Bioeng.* **19,** 1563–1621.
Yamaguchi, K., Nagata, Y., and Maruo, B. (1974a). *J. Bacteriol.* **119,** 410–415.
Yamaguchi, K., Nagata, Y., and Maruo, B. (1974b). *J. Bacteriol.* **119,** 416–424.
Yamamoto, S., and Lampen, J. O. (1976). *J. Biol. Chem.* **251,** 4095–4101.
Yelton, D. B., and Thorne, C. B. (1970). *J. Bacteriol.* **102,** 573–579.
Yoneda, Y, and Maruo, B. (1975). *J. Bacteriol.* **124,** 48–54.
Yoneda, Y., Yamane, K., and Maruo, B. (1973). *Biochem. Biophys. Res. Commun.* **50,** 765–770.
Yoneda, Y., Yamane, K., Yamaguchi, K., Nagata, Y., and Maruo, B. (1974). *J. Bacteriol.* **120,** 1144–1150.
Yoshikawa, M., Kato, T., and Takenishi, T. (1969). *Bull. Chem. Soc. Jpn.* **42,** 3505–3511.
Young, F. E., and Wilson, G. A. (1976). ''Handbook of Biochemistry and Molecular Biology'' (G. D. Fasman, ed.), Vol. 2, 3rd ed., pp. 686–703. Chemical Rubber Co., Cleveland, Ohio.
Young, F. E., and Wilson, G. A. (1978). *In* ''Genetic Engineering'' (A. P. Chakrabarty, ed.), pp. 145–154. CRC Press, Cleveland, Ohio.
Yuki, S. (1975). *Jpn. J. Genet.* **50,** 155–157.
Yuki, S., and Ueda, Y. (1968). *Jpn. J. Genet.* **43,** 121–128.
Zalunin, I. A., Chestukhina, G. G., and Stepanov, V. M. (1979). *Biochimiya* (*USSR*) **44,** 693–698.

Index

D

E

F

G

H

I

S

T

U

V

Molecular Biology

An International Series of Monographs and Textbooks

Editors

HAROLD A. SCHERAGA. Protein Structure. 1961

STUART A. RICE AND MITSURU NAGASAWA. Polyelectrolyte Solutions: A Theoretical Introduction, *with a contribution by Herbert Morawetz*. 1961

SIDNEY UDENFRIEND. Fluorescence Assay in Biology and Medicine. Volume I–1962. Volume II–1969

J. HERBERT TAYLOR (Editor). Molecular Genetics. Part I–1963. Part II–1967. Part III–Chromosome Structure–1979

ARTHUR VEIS. The Macromolecular Chemistry of Gelatin. 1964

M. JOLY. A Physico-chemical Approach to the Denaturation of Proteins. 1965

SYDNEY J. LEACH (Editor). Physical Principles and Techniques of Protein Chemistry. Part A–1969. Part B–1970. Part C–1973

KENDRIC C. SMITH AND PHILIP C. HANAWALT. Molecular Photobiology: Inactivation and Recovery. 1969

RONALD BENTLEY. Molecular Asymmetry in Biology. Volume I–1969. Volume II–1970

JACINTO STEINHARDT AND JACQUELINE A. REYNOLDS. Multiple Equilibria in Protein. 1969

DOUGLAS POLAND AND HAROLD A. SCHERAGA. Theory of Helix-Coil Transitions in Biopolymers. 1970

JOHN R. CANN. Interacting Macromolecules: The Theory and Practice of Their Electrophoresis, Ultracentrifugation, and Chromatography. 1970

WALTER W. WAINIO. The Mammalian Mitochondrial Respiratory Chain. 1970

LAWRENCE I. ROTHFIELD (Editor). Structure and Function of Biological Membranes. 1971

ALAN G. WALTON AND JOHN BLACKWELL. Biopolymers. 1973

WALTER LOVENBERG (Editor). Iron-Sulfur Proteins. Volume I, Biological Properties–1973. Volume II, Molecular Properties–1973. Volume III, Structure and Metabolic Mechanisms–1977

A. J. HOPFINGER. Conformational Properties of Macromolecules. 1973

R. D. B. FRASER AND T. P. MACRAE. Conformation in Fibrous Proteins. 1973

OSAMU HAYAISHI (Editor). Molecular Mechanisms of Oxygen Activation. 1974

FUMIO OOSAWA AND SHO ASAKURA. Thermodynamics of the Polymerization of Protein. 1975

LAWRENCE J. BERLINER (Editor). Spin Labeling: Theory and Applications. Volume I, 1976. Volume II, 1978

T. BLUNDELL AND L. JOHNSON. Protein Crystallography. 1976

HERBERT WEISSBACH AND SIDNEY PESTKA (Editors). Molecular Mechanisms of Protein Biosynthesis. 1977

TERRANCE LEIGHTON AND WILLIAM F. LOOMIS, JR. (Editors). The Molecular Genetics of Development: An Introduction to Recent Research on Experimental Systems. 1980

ROBERT B. FREEDMAN AND HILARY C. HAWKINS (Editors). The Enzymology of Post-Translational Modification of Proteins, Volume 1. 1980

WAI YIU CHEUNG (Editor). Calcium and Cell Function, Volume I: Calmodulin. 1980

OLEG JARDETZKY and G. C. K. ROBERTS. NMR in Molecular Biology. 1981

DAVID A. DUBNAU (Editor). The Molecular Biology of the *Bacilli,* Volume I: *Bacillus subtilis.* 1982

In preparation

CHARIS GHELIS and JEANNINE YON. Protein Folding. 1982

WAI YIU CHEUNG (Editor). Calcium and Cell Function, Volume II. 1982

GORDON G. HAMMES. Enzyme Catalysis and Regulation. 1982

P. R. CAREY. Biochemical Applications of Raman and Resonance Raman Spectroscopies. 1982

GUNTER KAHL and JOSEF S. SCHELL (Editors). Molecular Biology of Plant Tumors. 1982